**Fast Solvers for
Flow Problems**

Edited by
Wolfgang Hackbusch and
Gabriel Wittum

Notes on Numerical Fluid Mechanics (NNFM)   Volume 49

Series Editors: Ernst Heinrich Hirschel, München (General Editor)
Kozo Fujii, Tokyo
Bram van Leer, Ann Arbor
Keith William Morton, Oxford
Maurizio Pandolfi, Torino
Arthur Rizzi, Stockholm
Bernard Roux, Marseille

Volume 26   Numerical Solution of Compressible Euler Flows (A. Dervieux / B. van Leer / J. Periaux / A. Rizzi, Eds.)
Volume 27   Numerical Simulation of Oscillatory Convection in Low-Pr Fluids (B. Roux, Ed.)
Volume 28   Vortical Solution of the Conical Euler Equations (K. G. Powell)
Volume 29   Proceedings of the Eighth GAMM-Conference on Numerical Methods in Fluid Mechanics (P. Wesseling, Ed.)
Volume 30   Numerical Treatment of the Navier-Stokes Equations (W. Hackbusch / R. Rannacher, Eds.)
Volume 31   Parallel Algorithms for Partial Differential Equations (W. Hackbusch, Ed.)
Volume 32   Adaptive Finite Element Solution Algorithm for the Euler Equations (R. A. Shapiro)
Volume 33   Numerical Techniques for Boundary Element Methods (W. Hackbusch, Ed.)
Volume 34   Numerical Solutions of the Euler Equations for Steady Flow Problems (A. Eberle / A. Rizzi / E. H. Hirschel)
Volume 35   Proceedings of the Ninth GAMM-Conference on Numerical Methods in Fluid Mechanics (J. B. Vos / A. Rizzi / I. L. Ryhming, Eds.)
Volume 36   Numerical Simulation of 3-D Incompressible Unsteady Viscous Laminar Flows (M. Deville / T.-H. Lê / Y. Morchoisne, Eds.)
Volume 37   Supercomputers and Their Performance in Computational Fluid Mechanics (K. Fujii, Ed.)
Volume 38   Flow Simulation on High-Performance Computers I (E. H. Hirschel, Ed.)
Volume 39   3-D Computation of Incompressible Internal Flows (G. Sottas / I. L. Ryhming, Eds.)
Volume 40   Physics of Separated Flow – Numerical, Experimental, and Theoretical Aspects (K. Gersten, Ed.)
Volume 41   Incomplete Decompositions (ILU) – Algorithms, Theory and Applications (W. Hackbusch / G. Wittum, Eds.)
Volume 42   EUROVAL – A European Initiative on Validation of CFD Codes (W. Haase / F. Brandsma / E. Elsholz / M. Leschziner / D. Schwamborn, Eds.)
Volume 43   Nonlinear Hyperbolic Problems: Theoretical, Applied, and Computational Aspects Proceedings of the Fourth International Conference on Hyperbolic Problems, Taormina, Italy, April 3 to 8, 1992 (A. Donato / F. Oliveri, Eds.)

Volumes 1 to 25 are out of print.
The addresses of the Editors and further titles of the series are listed at the end of the book.

Fast Solvers for Flow Problems

Proceedings of the Tenth GAMM-Seminar
Kiel, January 14–16, 1994

Edited by
Wolfgang Hackbusch and
Gabriel Wittum

CIP-Codierung angefordert

All rights reserved
© Friedr. Vieweg & Sohn Verlagsgesellschaft mbH, Braunschweig/Wiesbaden, 1995

Vieweg ist a subsidiary company of Bertelsmann Professional Information.

No part of this publication may be reproduced, stored in a retrieval system or transmitted, mechanical, photocopying or otherwise, without prior permission of the copyright holder.

Produced by Langelüddecke, Braunschweig
Printed on acid-free paper
Printed in Germany

ISSN 0179-9614
ISBN 3-531-07649-6

10th GAMM-SEMINAR KIEL: A SCIENTIFIC COMMENT

In January 1994 the tenth workshop in a series of seminars on the numerical treatment of partial differential equations was held. A decade had passed since the first one, and the topics of the seminars provide a good insight into the progress during this period. The series was started by a proposal of the GAMM-Committee for "Efficient Numerical Methods for Partial Differential Equations" (GAMM-Fachausschuss "Effiziente Verfahren für partielle Differentialgleichungen"). They were organized by Wolfgang Hackbusch together with different international specialists in this field.

The first seminar took place only a short time after the multigrid method had become well-known to a broader audience in mathematics and engineering. Therefore at that time there was less experience with large scale computations in partial differential equations. So the first seminars were still concerned with the question for which method of discretization the multigrid method can be successfully used. Since the multigrid idea makes use of the relation between the discretization on different levels, it is clear that the discretization method must have good approximation properties. It is less obvious that the approximation properties must be accompanied by the corresponding inverse estimates in order to get good smoothing properties.

At that time this was well understood for finite differences and for conforming finite elements when applied to scalar differential equations. But ten years later it has almost been forgotten that the situation was less promising for problems in structural mechanics. For other problems there were suitable algorithms but without proofs. So, in particular, the robustness of mixed methods which are important in continuum and fluid mechanics was often an important (though sometimes hidden) question during the first seminars.

The seminars were not restricted to multigrid algorithms. In the next years the topic changed, and the sum of them provides a broad image of the state of the art through the variation of the topics. There were finite elements and, on the other hand, boundary elements and panel methods. Since the latter lead to non-sparse matrices, a different idea brought a reduction of the complexity, namely the separation of the (complicated) influence of the neighbours from that of far-away, which can be handled by using asymptotics.

There were not only multigrid methods with the (now) classical coarse-grid-corrections and the standard smoothers. The robust multigrid methods were designed for the cases in which the functions on the 2h-grid do not coincide with the smooth functions in the original space. Moreover, incomplete decompositions provide advanced smoothing methods.

In the beginning of the multigrid area the analysis was often performed for rectangular and uniform grids. The seminars show that the tendency now leads in the opposite direction. Adaptive mesh-refinement may produce grids which are far from having uniform meshsizes. At first glance, the effort for the solution seems to increase faster than linearly in these cases. Now the connection with Schwarz' alternating method has led to variants without this deficiency. The alternating method also plays an important role in the analysis of methods for parallel computers which were the topic of the sixth conference. For a mathematician, parallel computing does not only mean that known methods are implemented on a parallel computer in an intelligent way. New methods have been designed which are suitable for these computers. The most important of them are the domain decomposition methods which are interesting both from the theoretical and the practical viewpoint.

The tenth seminar, in particular, shows that now problems in fluid mechanics having a very large number of unknowns can be treated. To this end different advanced techniques are combined (and among them parallel methods or special block methods or special grids are only an incomplete enumeration of the methods used).

We hope that we will have further exciting seminars in Kiel which will also in only 2 1/2 days provide such excellent overviews of the state of the art.

Bochum, August 1994   Dietrich Braess

Piet Hemker (Amsterdam), Owe Axelsson (Nijmegen)

Ralf Kornhuber (Berlin), Gabriel Wittum (Stuttgart),
Randy Bank (San Diego), Arnold Reusken (Eindhoven) (Left to right)

GREETINGS

for the proceedings of the 10th GAMM-Seminar Kiel

Dear ladies and gentlemen,

in behalf of the town council of Kiel I'd like to send my regards to the participants of the 10th GAMM-Seminar Kiel. As Lord mayor I'm delighted to see that this seminar series has become a firm tradition and that in this way it has been made known our nice city to mathematicians all over the world. I hope the participants had time apart from talks on their field to get to known the city of the meeting better, and that they are already waiting for the next GAMM-Seminar at the Christian-Albrechts-University to deepen this acquaintance.

Nowadays mathematics provide solutions for problems we didn't need to set in former times. Kiels GAMM-Seminars can help to find a way through the jungle of complex problems in a world getting more complicated every day. Therefore I wish much success in the future.

Kiel, August 1994   Otto Kelling, Lord mayor of Kiel

Left to right:
Jens Burmeister (Kiel),
Wolfgang Hackbusch (Kiel)
Otto Kelling (Lord mayor of Kiel)

FOREWORD

The GAMM Committee for "Efficient Numerical Methods for Partial Differential Equations" organises seminars on subjects concerning the algorithmic treatment of partial differential equations. The topics are discretisation methods like the finite element and the boundary element method for various types of applications in structural and fluid mechanics. Particular attention is devoted to the advanced solution methods.

The series of such seminars was continued in 1994, January 14–16, with the 10th Kiel-Seminar on the special topic

Fast Solvers for Flow Problems

at the Christian-Albrechts-University of Kiel. The seminar was attended by 100 scientist from 9 countries. 23 lectures were given, including two invited survey lectures.

The central topic of most of the contributions are the Navier-Stokes equations. The authors discuss robust methods, parallel implementations, defect correction techniques, adaptive methods including error estimators, domain decompositions and FEM-BEM coupling. The invited lectures concern the hierarchical multigrid method for unstructured grids and structured adaptive finite-volume multigrid for compressible flows.

Kiel, August 1994
W. Hackbusch
G. Wittum

Wolfgang Hackbusch (Kiel), Otto Kelling (Lord mayor of Kiel)

CONTENTS

R.E. BANK, J. XU (Invited lecture I):
A hierarchical basis multigrid method for unstructured grids 1

H.T.M. VAN DER MAAREL, P.W. HEMKER (Invited lecture II):
Structured adaptive finite-volume multigrid for compressible flows 14

O. AXELSSON, I.E. KAPORIN: On the solution of nonlinear equations
for nondifferentiable mappings 38

R. BECKER, R. RANNACHER: Finite element solution of the incompressible
Navier-Stokes equations on anisotropically refined meshes 52

J. BEY, G. WITTUM: Downwind numbering: A robust multigrid method for
convection-diffusion problems on unstructured grids 63

G. ENGL: A fast solver for gas flow networks 74

P.A. FARRELL, A.F. HEGARTY, J.J.H. MILLER, E. O'RIORDAN,
G.I. SHISHKIN: On the design of piecewise uniform meshes for solving
advection-dominated transport equations to a prescribed accuracy 86

M. GEIBEN: Numerical simulation of three-dimensional non-stationary
compressible flow in complex geometries 96

U. GROH, ST. MEINEL, A. MEYER: On the numerical simulation of coupled
transient problems on MIMD parallel systems 109

M. HAHNE, E.P. STEPHAN, W. THIES: Fast solvers for coupled FEM-BEM
equations I 121

M. HILGENSTOCK, J. GROENNER, E. VON LAVANTE:
Numerical simulations of compressible Navier-Stokes equations
with nonequilibrium chemistry 131

GH. JUNCU: Preconditioned cg-like methods and defect correction for
solving steady incompressible Navier-Stokes equations 142

A. KAPURKIN, G. LUBE: A domain decomposition method for singularly
perturbed elliptic problems 151

E. KATZER: A parallel subspace decomposition method for elliptic and
hyperbolic systems 163

S. MÜLLER, A. PROHL, R. RANNACHER, S. TUREK: Implicit time-
discretization of the nonstationary incompressible Navier-Stokes equations 175

C.W. OOSTERLEE, H. RITZDORF, A. SCHÜLLER, B. STECKEL:
Experiences with a parallel multiblock multigrid solution technique
for the Euler equations 192

M. RAW: A coupled algebraic multigrid method for the 3d Navier-Stokes equations 204

H. REICHERT, G. WITTUM: Robust multigrid methods for the incompressible Navier-Stokes equations 216

A. REUSKEN: A new robust multigrid method for 2d convection-diffusion problems 229

M. RUMPF: Adapting meshes by deformation. Numerical examples and applications 241

F. SCHIEWECK: Multigrid convergence rates of a sequential and a parallel Navier-Stokes solver 251

O. STEINBACH: Parallel iterative solvers for symmetric boundary element domain decompostion methods 263

R. VERFÜRTH: The equivalence of a posteriori error estimators 273

List of participants 284

How to contact the authors by email 289

A HIERARCHICAL BASIS MULTIGRID METHOD FOR UNSTRUCTURED GRIDS

RANDOLPH E. BANK* AND JINCHAO XU[†]

Abstract

This paper is concerned with the application of the hierarchical basis method to completely unstructured grids in two dimensional space. A two level method is presented and analyzed. Generalization to the multilevel case is also discussed with numerical examples demonstrating the efficiency of the algorithm.

Key words. Finite element, hierarchical basis, multigrid.

AMS(MOS) subject classifications. 65F10, 65N20

1. Introduction

Hierarchical basis methods have proved to be one of the more robust classes of methods for solving broad classes of elliptic partial differential equations, especially the large systems arising in conjunction with adaptive local mesh refinement techniques [5, 2]; they have been shown to be strongly connected to subspace correction methods and to classical multigrid methods [16, 17, 4, 11]. As with typical multigrid methods, classical hierarchical basis methods are usually defined in terms of an underlying refinement structure of a sequence of nested meshes. In many cases this is no disadvantage, but it limits the applicability of the methods to truly unstructured meshes, which may be highly nonuniform but *not* derived from some grid refinement process. A major goal of our study is to generalize the construction of hierarchical bases to such meshes, allowing HBMG and other hierarchical basis methods to be applied. Some work on multigrid methods on non-nested meshes is reported in Bramble, Pasciak and Xu [9], Xu [16], and Zhang [20].

The generalized HBMG under our study is closely related to certain incomplete LU decompositions and this connection is discussed in Bank and Xu [7]. See [12] for some discussion of recent developments in the area of incomplete LU decompositions, and Kornhuber [13] for some discussion of a related method for variational inequalities. In this paper, we investigate the theoretical aspects of the algorithm, although the presentation will be mainly restricted to the two level case. There are many interesting recent developments in the theory of classical hierarchical basis and multigrid methods; see for example Griebel and Oswald [10], Oswald [14, 15], Bornemann and Yserentant [8], Yserentant [19] and Xu [17].

We shall use the letter C or c, with or without subscript, to denote generic constants that may not be the same at different occurrences. To avoid writing these constants repeatedly, we shall use the notation $\lesssim, \gtrsim$ and $\eqsim$. When we write $x_1 \lesssim y_1, x_2 \gtrsim y_2$, and $x_3 \eqsim y_3$, then, there exist constants C_1, c_2, c_3 and C_3 such that $x_1 \leq C_1 y_1$, $x_2 \geq c_2 y_2$ and $c_3 x_3 \leq y_3 \leq C_3 x_3$.

---

* Department of Mathematics, University of California at San Diego, La Jolla, CA 92093. The work of this author was supported by the Office of Naval Research under contract N00014-89J-1440.

[†] Department of Mathematics, Penn State University, University Park, PA 16802. The work of this author was partially supported by National Science Foundation and Schlumberger Foundation.

The rest of the paper is organized as follows. In §2, we give a brief description of the model boundary value problem and finite element spaces. in §3, we present the generalized hierarchical basis, and discuss some algorithmic aspects of the method. In §4, we present a theoretical analysis of the method. Generalization of the method to multilevel case and numerical experiments are reported in §5.

2. Model problems and finite element spaces

We consider the following model boundary-value problem:

$$-\Delta U = F \quad \text{in } \Omega,$$
$$U = 0 \quad \text{on } \partial\Omega$$

where $\Omega \subset \mathbf{R}^2$ is a bounded domain with a Lipschitz boundary.

Let $H^1(\Omega)$ be the standard Sobolev space consisting of square integrable functions with square integrable derivatives of first order and $H_0^1(\Omega)$ be a subspace of $H^1(\Omega)$ consisting of functions that vanish on $\partial\Omega$ (in an appropriate sense). It is well-known that $U \in H_0^1(\Omega)$ is the solution of

$$a(U, \chi) = (F, \chi) \quad \forall \chi \in H_0^1(\Omega)$$

where $(F, \chi) = \int_\Omega F\chi dx$ and $a(U, \chi) = \int_\Omega \nabla U \cdot \nabla \chi dx$.

We approximate the above problem by finite element discretizations on a quasi-uniform triangulation $\mathcal{T}$ of Ω. Corresponding to the triangulation $\mathcal{T}$, let $V \subset H_0^1(\Omega)$ be the finite element space consisting of piecewise linear functions. The finite element discretization of U on the space V is finding a solution $u \in V$ such that

$$a(u, v) = (F, v) \quad \forall v \in V.$$

The space V has a natural nodal basis $\{\phi_i\}_{i=1}^n$ ($n = \dim V$) that satisfies

$$\phi_i(x_l) = \delta_{il} \quad \forall i, l = 1, \ldots, n,$$

where $\{x_l : l = 1, \ldots, n\} \equiv \mathcal{N}$ is the set of all interior nodal points of the triangulation $\mathcal{T}$. By means of these nodal basis functions, the finite element solution can be written as $u = \sum \alpha_i \phi_i$ and $\alpha = (\alpha_i)_{i=1}^n$ satisfies

$$\hat{A}\alpha = \beta$$

where $\hat{A} = (a(\nabla\phi_i, \nabla\phi_l))_{n \times n}$ and $\beta = ((F, \phi_i))_{i=1}^n$. $\hat{A}$ is often called the stiffness matrix, and is symmetric, positive definite.

It is well-known that $\hat{A}$ is ill-conditioned and $cond(\hat{A})$, the ratio of the extreme eigenvalues of $\hat{A}$, is $O(h^{-2})$, where h is the mesh size of $\mathcal{T}$. The purpose of this paper is to construct some preconditioner for the matrix $\hat{A}$ by means of some generalized hierarchical basis.

3. Generalized hierarchical basis

In this section, we shall study generalized hierarchical basis method by unrefining general unstructured grids. The first step of our method is to create a certain hierarchical structure based on the given triangulation $\mathcal{T}$. This is accomplished by unrefining the triangulation $\mathcal{T}$ to obtain a sequence of coarse grids and corresponding sequence of finite element subspaces. The sequence of the multiple grid points is nested in the sense that the set of nodes in one grid contains the nodes of its subsequent coarser grid. Nevertheless the sequence of finite element subspaces are in general not nested. The efficiency of our algorithm has a lot to do with how this unrefinement is done. Although we have developed a systematical method and code for this purpose, the detail of this unrefinement process is not of the primary interest in this paper.

For simplicity, our discussion will be confined in a two level case, namely the triangulation $\mathcal{T}$ is unrefined once and a coarse triangulation $\mathcal{T}_c$ of Ω is obtained. Assume that $\mathcal{N}_c$ is the set of interior nodal points of $\mathcal{T}_c$. Note that $\mathcal{N}_c \subset \mathcal{N}$. Assume that the nodes in $\mathcal{N}$ are ordered in such a way that $\mathcal{N} \setminus \mathcal{N}_c = \{x_i : i = 1 : n_1\}$, where $n_1 = n - n_2$ and $n_2 = n_c$.

Given $x \in \mathcal{N} \setminus \mathcal{N}_c$, the vertices of the triangle or the edge in $\mathcal{T}_c$ that contains x as an interior point will be called the *parent nodes* of the vertex x. Thus x has two parents if it is on an edge and has three parents if it is inside a triangle. In the case of nested multilevel spaces, each x is the midpoint of its two parents. Given $x \in \mathcal{N}_c$, the *child nodes* of x will be those nodal points in $\mathcal{N}$ that are in the interior of the support of the coarse grid nodal basis function associated with x.

Let $I_f : C(\bar{\Omega}) \mapsto V$ and $I_c : C(\bar{\Omega}) \mapsto V_c$ be the nodal value interpolants, ϕ_i ($i = 1 : n, , x_i \in \mathcal{N}$) the nodal basis functions in V, and ϕ_j^c ($j = 1 : n_2, x_j \in \mathcal{N}_c$) the nodal basis functions in V_c.

The following simple result is fundamental to our new method.

LEMMA 3.1. *Define*

$$V_1 = (I - I_f I_c)V \quad \text{and} \quad V_2 = I_f I_c V.$$

Then V_1 and V_2 are two subspaces of V such that

$$V_1 = \text{span}\{\phi_i : x_i \in \mathcal{N} \setminus \mathcal{N}_c\} \quad \text{and} \quad V_2 = \text{span}\{I_f \phi_i^c : x_i \in \mathcal{N}_c\}$$

and $V = V_1 \oplus V_2$.

Proof. By counting the dimension, it suffices to prove that $\{I_f \phi_i^c : x_i \in \mathcal{N}_c\}$ is linear independent. To see that, assume that

$$\sum_{x_i \in \mathcal{N}_c} \alpha_i I_f \phi_i^c(x) = 0 \quad \forall \, x \in \Omega.$$

This means that

$$v \equiv \sum_{x_i \in \mathcal{N}_c} \alpha_i \phi_i^c(x) = 0 \quad \forall \, x \in \mathcal{N}.$$

But $v \in V_c$ and $\mathcal{N}_c \subset \mathcal{N}$. Hence $\alpha_i = 0$ for all i. □

The proof of the above lemma makes use of the fact that $\mathcal{N}_c \subset \mathcal{N}$ which is crucial in the whole theory. As a result of the above lemma and the relation that $V = V_1 \oplus V_2$, the union of bases of V_1 and V_2 given in the above lemma forms a basis of V, which will be called the generalized hierarchical basis and denoted by $\{\psi_i\}$. This basis of course coincides with the classic hierarchical basis in the nested case ($V_c \subset V$). For simplicity, we assume that the index is arranged such that $\psi_i = \phi_i \quad 1 \le i \le n_1$.

Let $\mathcal{A}$ denote the stiffness matrix under the aforementioned hierarchical basis. Then

$$\mathcal{A} = \mathcal{S}\hat{\mathcal{A}}\mathcal{S}^t$$

where $\mathcal{S} = (s_{ij})$ is the transformation matrix from hierarchical basis to nodal basis functions. Namely

$$\psi_j = \sum_{i=1}^{n} s_{ij}\phi_i.$$

Obviously the matrix $\mathcal{S}$ is of the following form:

$$\mathcal{S} = \begin{bmatrix} \mathcal{I}_{n_1} & 0 \\ \mathcal{R} & \mathcal{I}_{n_2} \end{bmatrix}$$

where $\mathcal{I}_k$ stands for $k \times k$ identity matrix.

The matrix $\mathcal{R} = (r_{ij})$ is sparse. Given $x_i = (\xi_1, \eta_1) \in \mathcal{N}_c$, the number of nonzeros on the row $i-m$ of R is exactly the number of child nodes of x_i. Assume that $x_j = (\bar{\xi}, \bar{\eta})$ is one of x_i's child node which is on the triangle τ with vertices (ξ_l, η_l) ($l = 1, 2, 3$). then

$$r_{i-m,j} = s_{ij} = \frac{(\xi_2 - \bar{\xi})(\eta_3 - \bar{\eta}) - (\xi_3 - \bar{\xi})(\eta_2 - \bar{\eta})}{(\xi_2 - \xi_1)(\eta_3 - \eta_1) - (\xi_3 - \xi_1)(\eta_2 - \eta_1)}.$$

The above quantity is simply the barycentric coordinate of x_j with respect to x_i in τ, which equals $1/2$ in the nested case.

The stiffness matrix $\mathcal{A}$ can be written as

$$\mathcal{A} = \begin{bmatrix} \mathcal{A}_1 & * \\ * & \mathcal{A}_2 \end{bmatrix}$$

where $\mathcal{A}_1 = (a(\psi_i, \psi_j)) \in \mathbf{R}^{n_1 \times n_1}$ and $\mathcal{A}_2 = (a(I_f \phi_i^c, I_f \phi_j^c)) \in \mathbf{R}^{n_2 \times n_2}$. The matrix $\mathcal{A}_1$ is well-conditioned as in the nested case and the matrix $\mathcal{A}_2$ corresponds to the stiffness matrix on the coarse grid. A preconditioner for $\mathcal{A}_2$, say $\mathcal{R}_2$, may be obtained by further unrefining the grid, and this is the topic of the generalized hierarchical basis method in the multilevel case [7, 6]. As we shall not discuss the multilevel case here, $\mathcal{R}_2$ will not be specified in our discussion.

The main theoretical result of this paper is as follows.

THEOREM 3.2. *Assume that $\mathcal{R}_2$ is a preconditioner of $\mathcal{A}_2$ with $\lambda_0 = \lambda_{\min}(\mathcal{R}_2 \mathcal{A}_2)$ and $\lambda_1 = \lambda_{\max}(\mathcal{R}_2 \mathcal{A}_2)$, and*

$$\mathcal{B} = \begin{bmatrix} \mathcal{I}_{n_1} & 0 \\ 0 & \mathcal{R}_2 \end{bmatrix}. \qquad (1)$$

Then, there exist some constants c_1, c_2 and c_3 such that

$$cond(\mathcal{B}\mathcal{A}) \leq (c_1 + c_2/\lambda_0)(c_3 + \lambda_1).$$

In particular,

$$cond(\mathcal{A}) \eqsim cond\left(\begin{bmatrix} \mathcal{I}_{n_1} & 0 \\ 0 & \mathcal{A}_2 \end{bmatrix}\right).$$

The proof of the above result will be given in the next section.

4. Estimate of the condition number

In this section, we shall analyze the conditioning of the hierarchical basis stiffness matrix $\mathcal{A}$ and give a proof of Theorem 3.2. Theoretical analysis of the classic hierarchical basis method can be done through many different approaches, cf. Bank and Dupont [3, 1], Yserentant [18, 19], Bank, Dupont and Yserentant [4], Bornemann and Yserentant [8], Oswald [14, 15] and Xu [17]. It turns out that the analysis in our current situation fits the general framework in [17].

Let us first recall that the matrix representation of an operator T from one vector space with a basis $(v_1, \cdots, v_n)$ to another vector space with a basis $(w_1, \cdots, w_m)$ is the matrix $\tilde{T} = (t_{ij}) \in \mathbf{R}^{m \times n}$ satisfying

$$Tv_j = \sum_{i=1}^{m} t_{ij} w_i \quad 1 \leq j \leq n.$$

For the finite element spaces V, V_1 and V_2 introduced from previous section, the bases we shall use are $(\psi_i : 1 \leq i \leq n)$ (generalized hierarchical basis), $(\psi_i : 1 \leq i \leq n_1)$ and $(\psi_i : n_1 + 1 \leq i \leq n)$, respectively.

Associated with the spaces V, V_1 and V_2, define the operators $A : V \mapsto V$, $A_i : V_i \mapsto V_i$ and $Q_i, P_i : V \mapsto V_i$ $(i = 1, 2)$ by

$$(Au, v) = a(u, v) \quad \forall\, u, v \in V, \quad (A_i u_i, v_i) = a(u_i, v_i) \quad \forall\, u_i, v_i \in V_i$$

and

$$(Q_i u, v_i) = (u, v_i), \quad a(P_i u, v_i) = a(u, v_i) \quad \forall\, u \in V, v_i \in V_i.$$

Using the basis (ψ_i) for V, we have the stiffness matrix $\mathcal{A}$ defined earlier and the following mass matrix:

$$\mathcal{M} = ((\psi_i, \psi_j))_{n \times n}.$$

The relation between the stiffness matrix $\mathcal{A}$ and the matrix representation of A is given by

$$\tilde{A} = \mathcal{M}^{-1}\mathcal{A}.$$

With respect to the bases of V_1 and V_2, we define the following two mass matrices

$$\mathcal{M}_1 = ((\psi_i, \psi_j)) \in \mathbf{R}^{n_1 \times n_1} \quad \text{and} \quad \mathcal{M}_2 = ((I_f \phi_i^c, I_f \phi_j^c)) \in \mathbf{R}^{n_2 \times n_2}.$$

We use the concept of parallel subspace correction from [17] to construct a preconditioner for the operator A as follows:

$$B = R_1 Q_1 + R_2 Q_2 \tag{2}$$

where $R_i : V_i \mapsto V_i$ are certain preconditioners for A_i defined below. The preconditioner R_1 is defined by

$$R_1 v = \sum_{i=1}^{m} (v, \psi_i) \psi_i \quad v \in V_1. \tag{3}$$

R_2 is a preconditioner of A_2 such that $\widetilde{R}_2 = \mathcal{R}_2 \mathcal{M}_2$. Thus $\widetilde{R_2 A_2} = \mathcal{R}_2 \mathcal{A}_2$. Consequently

$$\lambda_{\min}(R_2 A_2) = \lambda_{\min}(\mathcal{R}_2 \mathcal{A}_2) \quad \text{and} \quad \lambda_{\max}(R_2 A_2) = \lambda_{\max}(\mathcal{R}_2 \mathcal{A}_2).$$

By definition, the matrix representation of the inclusion $I_1 : V_1 \mapsto V$ and $I_2 : V_2 \mapsto V$ are

$$\begin{bmatrix} \mathcal{I}_{n_1} \\ 0_{n_2 \times n_1} \end{bmatrix} \quad \text{and} \quad \begin{bmatrix} 0_{n_1 \times n_2} \\ \mathcal{I}_{n_2} \end{bmatrix}$$

respectively. It is also easy to see that the matrix representations of $Q_1 : V \mapsto V_1$ and $Q_2 : V \mapsto V_2$ are

$$\widetilde{Q}_1 = \mathcal{M}_1^{-1}[\mathcal{I}_{n_1}, 0_{n_1 \times n_2}]\mathcal{M} \quad \text{and} \quad \widetilde{Q}_2 = \mathcal{M}_2^{-1}[0_{n_2 \times n_1}, \mathcal{I}_{n_2}]\mathcal{M},$$

respectively.

LEMMA 4.1. *For B and $\mathcal{B}$ given by (2) and (1) respectively, we have*

$$\widetilde{BA} = \mathcal{B}\mathcal{A}.$$

Consequently

$$cond(BA) = cond(\mathcal{B}\mathcal{A}).$$

Proof. By the basic properties of matrix representation and the matrix representations for I_i, R_i and Q_i ($i = 1, 2$), it follows that

$$\widetilde{B} = \widetilde{I}_1 \widetilde{R}_1 \widetilde{Q}_1 + \widetilde{I}_2 \widetilde{R}_2 \widetilde{Q}_2 = \mathcal{B}\mathcal{M}.$$

Thus $\widetilde{BA} = \widetilde{B}\widetilde{A} = \mathcal{B}\mathcal{M}\mathcal{M}^{-1}\mathcal{A} = \mathcal{B}\mathcal{A}$. This proves the first estimate. The second estimate follows from the fact that the spectrum of an operator and its matrix representation are the same. □

As a result of the above lemma, the estimate of the $cond(\mathcal{BA})$ is reduced to the estimate of $cond(BA)$.

LEMMA 4.2. *For the operator R_1 given by (3), we have*

$$(R_1 v, v) \eqsim h^2(v, v) \quad \forall\, v \in V_1.$$

Proof. By definition, it is easy to see that

$$\tilde{R}_1 = \mathcal{M}_1.$$

Thus the estimate we need to prove is equivalent to

$$\nu^t \mathcal{M}_1 \nu \eqsim h^2 |\nu|^2 \quad \forall\, \nu \in \mathbf{R}^{n_1}.$$

The proof of the above estimate is routine as in the analysis of the conditioning of the standard finite element mass matrix and the detail is left to the interested readers. □

LEMMA 4.3. *The interpolants $I_f : V_c \mapsto V_f$ and $I_c : V_f \mapsto V_c$ admit the following estimate*

$$\|I_c v\|_1 \lesssim \|v\|_1, \quad \|(I - I_c)v\| \lesssim h_c \|v\|_1 \quad \forall\, v \in V$$

and

$$\|I_f v\|_1 \lesssim \|v\|_1, \quad \|(I - I_f)v\| \lesssim h_f \|v\|_1 \quad \forall\, v \in V_c.$$

Proof. Given $\delta \in (0, 1/2)$. For any $v \in V$, we have (cf. [16])

$$\|(I - I_c)v\| \lesssim h_c^{1+\delta} \|v\|_{1+\delta} \quad \text{and} \quad \|(I - I_c)v\|_1 \lesssim h_c^{\delta} \|v\|_{1+\delta}.$$

The first two estimates then follow from the following inverse inequality (cf. [9, 16])

$$\|v\|_{1+\delta} \lesssim h^{-\delta} \|v\|_1.$$

The proof of the last two inequalities are obviously similar. □

LEMMA 4.4.

$$(R_1^{-1} v_1, v_1) + (R_2^{-1} v_2, v_2) \leq (c_1 + c_2/\lambda_0) a(v, v) \quad \forall\, v \in V$$

with $v_1 = (I - I_f I_c)v$ and $v_2 = I_f I_c v$.

Proof. By Lemma 4.2

$$(R_1^{-1} v_1, v_1) \lesssim h^{-2} \|(I - I_f I_c)v\|^2.$$

But, by Lemma 4.3, we have

$$\|(I - I_f I_c)v\| \lesssim \|(I - I_c)v\| + \|(I - I_f) I_c v\| \lesssim h(\|v\|_1 + \|I_c v\|_1) \lesssim h\|v\|_1.$$

Therefore, for some positive constant c_1

$$(R_1^{-1}v_1, v_1) \leq c_1 a(v,v) \quad \forall\, v \in V.$$

On the other hand

$$(R_2^{-1}v_2, v_2) \leq (A_2(R_2 A_2)^{-1} v_2, v_2) \leq 1/\lambda_0 a(v_2, v_2).$$

Combining this with the estimate in Lemma 4.3, we get, for some positive constant c_2

$$(R_2^{-1} v_2, v_2) \leq c_2/\lambda_0 a(v, v) \quad \forall\, v \in V.$$

The desired estimate then follows. □

LEMMA 4.5.

$$\text{cond}(BA) \leq (c_1 + c_2/\lambda_0)(c_3 + \lambda_1).$$

Proof. By virtue of Lemmas 4.4, the estimate can be reduced from the theory in Xu [17] (see Theorem 4.3 there). For completeness, nevertheless, we shall now include a detailed proof.

First of all,

$$a(BAv, v) = (R_1 A_1 P_1 v, A_1 P_1 v) + (R_2 A_2 P_2 v, A_2 P_2 v).$$

Thus, by Lemma (4.2), for some positive constant c_3

$$a(BAv, v) \leq c_3 a(P_1 v, P_1 v) + \lambda_1 a(P_2 v, P_2 v) \leq (c_3 + \lambda_1) a(v,v).$$

This implies that $\lambda_{\max}(BA) \leq c_3 + \lambda_1$.

Given $v \in V$, let v_1 and v_2 be as in Lemma 4.4. It follows that

$$\begin{aligned}
a(v,v) &= a(v_1, v) + a(v_2, v) \\
&= a(v_1, P_1 v) + a(v_2, P_2 v) \\
&= (v_1, A_1 P_1 v) + a(v_2, A_2 P_2 v) \\
&= (R_1^{-1} v_1, v_1)^{1/2} (R_1 A_1 P_1 v, v)^{1/2} + (R_2^{-1} v_2, v_2)^{1/2} (R_2 A_2 P_2 v, v)^{1/2} \\
&= \left((R_1^{-1} v_1, v_1) + (R_2^{-1} v_2, v_2)\right)^{1/2} ((R_1 A_1 P_1 v, v) + (R_2 A_2 P_2 v, v))^{1/2} \\
&\leq ((c_1 + c_2/\lambda_0) a(v, v) a(BAv, v))^{1/2}.
\end{aligned}$$

This implies that $\lambda_{\min}(BA) \geq (c_1 + c_2/\lambda_0)^{-1}$. The desired estimate then follows. □

Theorem 3.2 is then a consequence of Lemmas 4.1 and 4.5.

5. Multilevel case and numerical examples

Although the discussion so far has been on the two level case, the algorithm can be extended to multilevel case. For the multilevel case, we have to further unrefine the coarse grid $\mathcal{N}_c$ to obtain a sequence of nested nodes:

$$\mathcal{N}_1 \subset \mathcal{N}_2 \subset \cdots \mathcal{N}_J$$

with $\mathcal{N}_J = \mathcal{N}$. Correspondingly a sequence of (nonnested) triangulations and finite element subspaces can be constructed. Algorithmically, the key issue in this general case is to appropriately construct a hierarchical basis and effectively compute the action of the transformation matrix of this hierarchical basis in terms of the nodal basis. One way to think of this generalization is to inductively apply the two level method discussed above to construct a preconditioner $\mathcal{R}_2$ for $\mathcal{A}_2$ and so on. However, the estimate of the condition number proves to be much more technical than the two level case or the nested multilevel case [6]. We shall report some numerical experiments here for the hierarchical basis multigrid method that we have developed; details of the algorithm and its theoretical analysis will be reported in some forthcoming papers. The finite element code *PLTMG* [2] was used in creating both examples.

In our first example, we solved the Dirichlet problem $-\Delta u = 1$ in Ω, with $u = 0$ on $\partial\Omega$, where Ω is the region pictured on the left of Figure 1. We triangulated this region with an unstructured mesh with $n = 3229$ vertices, pictured on the right in Figure 1. We applied a symbolic elimination algorithm similar to *ILU* to this mesh, creating a nonnested sequence of nine meshes. We stopped the elimination process with $n_c = 50$, and the coarse mesh shown on the left of Figure 2. As we mentioned above, the details of this coarsening algorithm are complicated, and will not be discussed here. However, note that our algorithm does allow for coarsening the approximation of the boundary, which can be seen in Figure 2. We then solved finite element equations using an algebraic nine level HBMG iteration based on this decomposition. The convergence history is shown on the right in Figure 2. The quantity plotted is

$$\sigma_k = \log\left\{\frac{\|r_k\|}{\|r_0\|}\right\},$$

where r_k is the residual at iteration k and $\|\cdot\|$ is the ℓ^2 norm. From the data points, we estimate by least squares that the convergence rate is approximately .39, which is fairly typical of this particular iterative method applied to a similar problem on a sequence of refined meshes.

For our second example, we illustrate the behavior of the algebraic HBMG iteration on a problem with strongly varying coefficients. We consider the Dirichlet problem $-\nabla(D\nabla U) = 1$ in Ω, with $u = 0$ on $\partial\Omega$, where now Ω is the region pictured on the right of Figure 3. This region consists of six subregions, and the coefficient D is chosen to be constant on each subregion. $D = 1$, $D = 10^{-4}$, $D = 1$, left to right, for the three smaller subregions at the top. $D = 10^{-3}$, $D = 10^{-6}$, left to right, for the two middle subregions, while $D = 1$ for the bottom subregion.

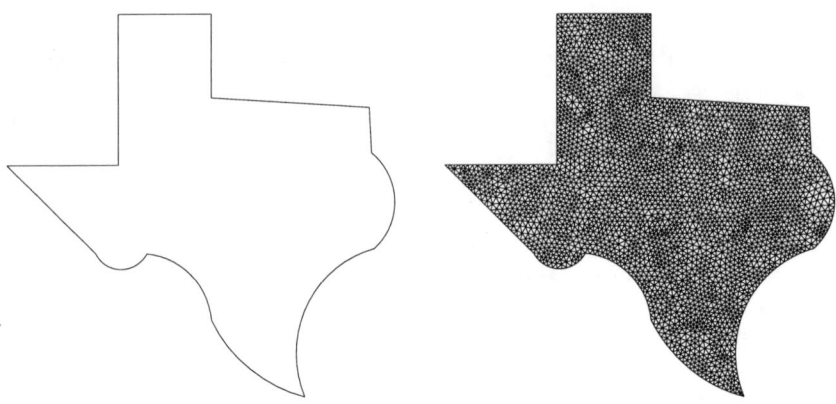

FIG. 1. *The region for the first example, and the initial triangulation with $n = 3229$ vertices*

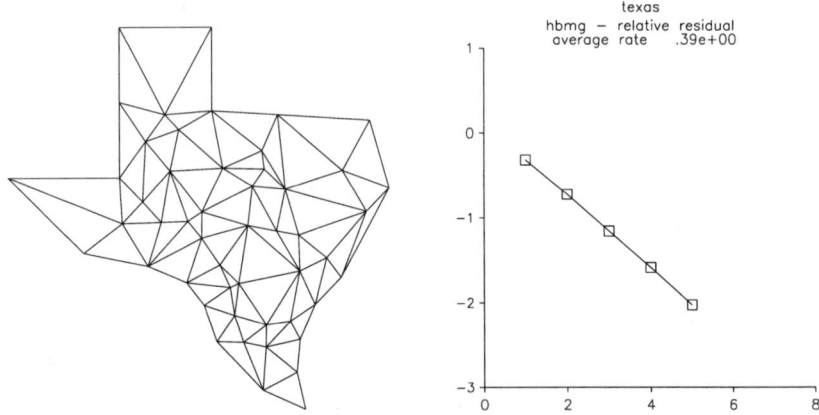

FIG. 2. *The coarse mesh with $n_c = 50$ vertices, and the convergence history of the resulting 9 level HBMG iteration*

This region was triangulated with an unstructured mesh with $n = 3124$ vertices shown in Figure 3. This mesh was coarsened using the same algorithm as in the first example, again creating nine levels and a coarsest mesh with $n_c = 75$ vertices shown in Figure 4. Note that our coarsening process approximates, but retains, the internal interfaces.

The convergence history for the nine level algebraic HBMG iteration is shown in figure 4. We note that the average convergence rate of .51 is not quite as good as the first example, but it is still well within the range of typical $HBMG$ performance for nested sequences of meshes. The difference is attributed to two sources, the hueristic used to coarsen the mesh and create the hierarchical basis, and the discontinuous coefficient. To quantify the effect of the discontinuous coefficient, we solved the problem again with $D = 1$ in all 6 regions, and obtained an average convergence rate of .46.

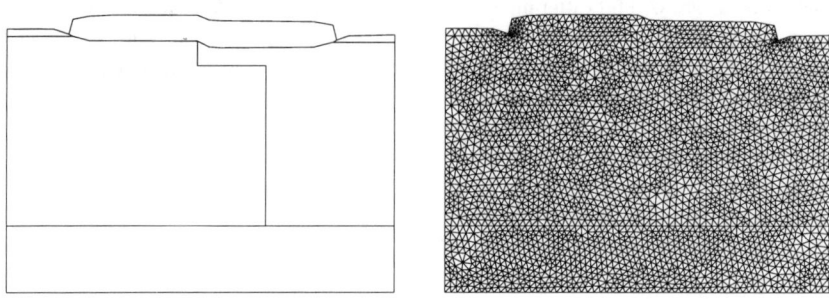

FIG. 3. *The region for the second example, and the initial triangulation with $n = 3124$ vertices*

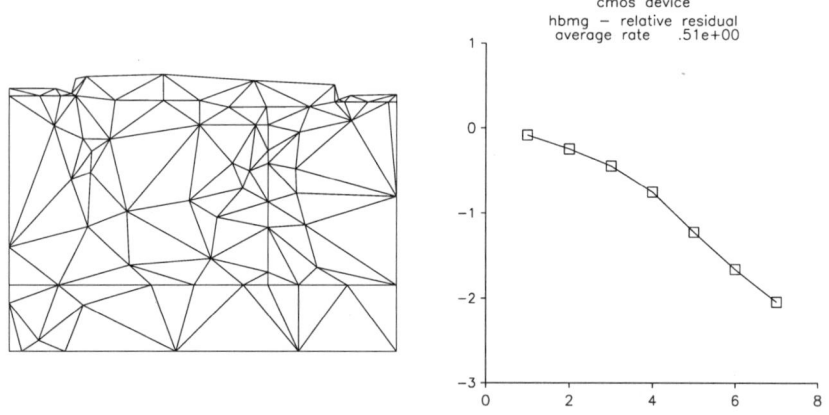

FIG. 4. *The coarse mesh with $n_c = 75$ vertices, and the convergence history of the resulting 9 level HBMG iteration*

REFERENCES

[1] R. E. BANK, *Hierarchical preconditioners for elliptic partial differential equations*, in Proceedings of the SERC Summer Workshop in Numerical Analysis, 1992.
[2] ———, *PLTMG: A Software Package for Solving Elliptic Partial Differential Equations, Users' Guide 7.0*, Frontiers in Applied Mathematics, SIAM, Philadelphia, 1994.
[3] R. E. BANK AND T. F. DUPONT, *Analysis of a two level scheme for solving finite element equations*, Tech. Report CNA-159, Center for Numerical Analysis, University of Texas at Austin, 1980.
[4] R. E. BANK, T. F. DUPONT, AND H. YSERENTANT, *The hierarchical basis multigrid method*, Numer. Math., 52 (1988), pp. 427–458.
[5] R. E. BANK, A. H. SHERMAN, AND A. WEISER, *Refinement algorithms and data structures for regular local mesh refinement*, in Scientific Computing (Applications of Mathematics and Computing to the Physical Sciences) (ed. R. S. Stepleman), North Holland, 1983, pp. 3–17.
[6] R. E. BANK AND J. XU, *Hierarchical basis multigrid methods for unstructured grids: Algorithm and theory*, (in preparation).
[7] ———, *The hierarchical basis multigrid method and incomplete LU decomposition*, in Proceedings of Seventh International Conference on Domain Decomposition. (ed. D. Keyes and J. Xu), AMS, (to appear).
[8] F. BORNEMANN AND H. YSERENTANT, *A basic norm equivalence for the theory of multigrid methods*, Numer. Math., 64 (1993), pp. 455–476.
[9] J. BRAMBLE, J. PASCIAK, AND J. XU, *The analysis of multigrid algorithms with non-imbedded spaces or non-inherited quadratic forms*, Math. Comp., 56 (1991), pp. 1–43.
[10] M. GRIEBEL AND P. OSWALD, *Remarks on the abstract theory of additive and multiplicative Schwarz algorithms*, Tech. Report TUM-19314, Institut für Informatik, Technische Universität München, 1993.
[11] W. HACKBUSCH, *Multigrid Methods and Applications*, Springer-Verlag, Berlin, 1985.
[12] W. HACKBUSCH AND G. WITTUM, EDS, *Incomplete Decompositions - Algorithms, Theory and Applications*, Notes on Numerical Fluid mechanics, vol. 41, Vieweg, Braunschweig, 1993.
[13] R. KORNHUBER, *Monotone multigrid methods for variational inequalities I*, Numer. Math., (to appear).
[14] P. OSWALD, *On discrete norm estimates related to multigrid preconditioners in the finite element method*, in Proceedings of the International Conference on Constructive Theory of Functions, Varna, 1991.
[15] ———, *Stable subspace splittings for sobolev spaces and thier aqpplications*, Tech. Report Math/93/7, Friedrich-Schiller-Universität Jena, 1993.
[16] J. XU, *Theory of Multilevel Methods*, PhD thesis, Cornell University. Report AM-48, Penn State, 1989.
[17] ———, *Iterative methods by space decomposition and subspace correction*, SIAM Review, 34 (1992), pp. 581–613.
[18] H. YSERENTANT, *On the multi-level splitting of finite element spaces*, Numer. Math., 49 (1986), pp. 379–412.
[19] ———, *Old and new convergence proofs for multigrid methods*, Acta Numerica, (1992).
[20] S. ZHANG, *Multilevel Iterative Techniques*, PhD thesis, Pennsylvania State University, Department of Mathematics Report 88020, 1988.

Structured Adaptive Finite-Volume Multigrid for Compressible Flows

H.T.M. van der Maarel*
P.W. Hemker†

Summary

For a system of hyperbolic conservation laws, such as the Euler equations of compressible flow, in this paper we give an outline of the theory necessary to derive first and second-order accurate discretisations on a structured, adaptive finite-volume mesh. The mesh is constructed so that the equations can be defined on a rather arbitrary domain, and the usual nonlinear multigrid techniques can be used for the solution of the discrete system. During the solution process the mesh can be adapted to the solution and to the accuracy of the discrete equations. This requires a sufficiently accurate estimate of the local truncation error.

After formally introducing the geometric structure and notations, we discuss the discretisation and we study the various contributions to the local discretisation error. Emphasis is put on the discretisation involving the interfaces between the coarse and the fine parts of the grid. Our analysis leads to a small set of requirements, to be satisfied in order to attain a discretisation which is first or second-order accurate (in a sense that will be specified) with respect to the mesh size of the partitioning. Then interpolations are presented which satisfy these conditions.

1 Introduction

In this paper we describe the discretisation of a system of steady conservation laws in two space dimensions, using a finite volume discretisation with a structured, locally refined partitioning of the domain of definition.

We consider a system of d conservation laws, defined on an open domain $\Omega \subset \mathbb{R}^2$, with $q : \overline{\Omega} \to \mathbb{R}^d$, $s : \Omega \to \mathbb{R}^d$, $f, g : \mathbb{R}^d \to \mathbb{R}^d$ given by

$$\frac{\partial f(q(x,y))}{\partial x} + \frac{\partial g(q(x,y))}{\partial y} = s(x,y), \quad (x,y) \in \Omega. \tag{1.1a}$$

supplemented with appropriate boundary conditions.

For the discretisation we introduce a partitioning of Ω. The partitioning forms, possibly after a smooth coordinate transformation, a set of regular quadrilaterals, called

---
*MARIN, P.O.Box 28, 6700 AA Wageningen, The Netherlands; e-mail: maarel@marin.nl
†CWI, P.O.Box 94079, 1090 GB Amsterdam, The Netherlands; e-mail: pieth@cwi.nl

the *grid*. In transformed coordinates, the locally refined grid is composed of a sequence of nested grids, where each grid is a regular partitioning of a subdomain of Ω. In non-transformed coordinates, the union of all quadrilaterals is an approximation of the domain of definition.

Each quadrilateral of the grid is used as a control volume on which the system of conservation laws, integrated over the control volume, is approximately satisfied. This leads to a discretisation of a weak formulation of problem (1.1).

The error in the approximation of the weak formulation consists of contributions from the various steps in the discretisation. The first step is the approximation of the domain by the partitioning. This approximation has consequences for the accuracy of the discrete equations defined for quadrilaterals along the boundary of the domain and for the 'coarse-grid' approximation. The approximation of the weak formulation for each quadrilateral involves the approximation of the mean flux per unit time and 'area' across each side of each quadrilateral. The next step is the approximation of the mean flux across the sides of the quadrilaterals by the flux at the mean state at each side of a face Finally the state at the cell face is approximated from the discrete data (the available numerical solution, a cell-wise constant function). The flux may be evaluated in an upwind fashion, where each approximated flux depends in an upwind biased sense on the discrete data.

The numerical solution itself approximates the mean of the exact solution of (1.1) over each cell. Hence, with each cell is associated an approximation of the mean value on the cell of the solution of the continuous problem. This is the so-called *cell-centered* approach.

The decision where to refine a given composite grid (or to remove a refinement), may be based on the local discretisation error. Therefore we study the a-posteriori estimation of the local discretisation error. However, refinement should not be based solely on the local discretisation error. Apart from a sufficiently accurate discretisation scheme, an accurate solution requires that the grid provides sufficient *resolution*. Resolution of the grid is measured by the derivatives of the exact solution, as approximated by the numerical solution. The grid should therefore also be refined on the basis of solution gradients. For the equations associated with cells near discontinuities in the solution, an estimate of the local discretisation error is superfluous, since the grid will be refined due to the approximations of the large gradient in the solution. For the smooth part of the solution an estimate of the local discretisation error may be obtained with sufficient accuracy and can be used in a refinement strategy.

In the neighbourhood of fine-coarse grid interfaces, by nature, the discretisation scheme used is different from the one used elsewhere. Estimating the local truncation error in such a situation by means of extrapolation techniques, (as in τ-extrapolation, [3, 5]), requires a careful treatment of the various contributions to the local truncation error. Here we carefully consider these contributions, with emphasis on their accurate estimation.

2 The geometric structure

We study a system of conservation laws, defined on Ω, with boundary conditions on $\partial\Omega \subset \overline{\Omega}$, where $\overline{\Omega}$ denotes the closure of Ω. We assume that a rectangular domain $\tilde{\overline{\Omega}} \subset \mathbb{R}^2$ exists and a sufficiently smooth surjective mapping $M : \tilde{\overline{\Omega}} \to \overline{\Omega}$, which is also injective in the interior $\tilde{\Omega}$. The mapping is a transformation of the Cartesian coordinates in $\tilde{\overline{\Omega}}$, the *computational* space, into Cartesian coordinates in $\overline{\Omega}$, the *physical* space.

As the system of conservation laws is discretised by a finite volume method, a regular rectangular partitioning of $\widetilde{\Omega}$ is introduced, consisting of disjoint rectangles. This defines a partitioning of Ω, through the mapping M. A rectangle in the computational space $\widetilde{\Omega}$, as well as its image in Ω, is called a *cell*. The partitioning is called the *grid*.

We consider grids on different *levels of refinement*. A level of refinement l, with $l = 0, 1, 2, \cdots$, is a regular partitioning, denoted by $\widetilde{\Omega}^l$, of a subdomain in $\widetilde{\Omega}$, and a surjective mapping $M^l : \overline{\widetilde{\Omega}} \to \overline{\widehat{\Omega}^l}$, injective in the interior $\widetilde{\Omega}^l$. We use the notation $\Omega^l = M^l(\widetilde{\Omega}^l)$ for an image under the mapping M^l. For the image $\widehat{\Omega}^l$ of $\widetilde{\Omega}$ under M^l, and the hull $\widehat{\Omega}$ of all images of $\widetilde{\Omega}$ under the sequence of mappings $\{M^l\}_{l\in\mathbb{Z}}$, we use the notation

$$\widehat{\Omega}^l = M^l(\widetilde{\Omega}), \quad \text{and} \quad \widehat{\Omega} = \bigcup_l \widehat{\Omega}^l. \tag{2.1}$$

A mapping M^l is associated with level of refinement l and is an approximation of M. Generally we take M^l so that it maps a cell vertex from the partitioning $\widetilde{\Omega}^l$ to the same point in $\mathbb{R}^2$ as the original mapping M does. Then in the sequence of mappings $\{M^l\}_{l\in\mathbb{Z}}$, M^l approaches M as $l \to \infty$. Hence, if $\overline{\widetilde{\Omega}^l} = \overline{\widetilde{\Omega}}$, then $\overline{\Omega^l}$ is an approximation of $\overline{\Omega}$.

Since the partitioning of Ω on a level l is regular quadrilateral, each cell on level l can be denoted by $\Omega^l_{i,j} \subset \widehat{\Omega}$. The set I of indices present in the approximation is

$$I = \left\{ (i, j, l) \in \mathbb{Z}^3 \mid \exists \Omega^l_{i,j} \subset \widehat{\Omega} \right\}.$$

The grid on level l is

$$\Omega^l = \left\{ \Omega^l_{i,j} \mid (i, j, l) \in I \right\}.$$

By the regularity of the partitioning each cell on Ω^l has either *none* or only *one* neighbouring cell at each side, residing at the same level. A cell $\Omega^l_{i,j}$ is the northern neighbour of $\Omega^l_{i,j-1}$ and the eastern neighbour of $\Omega^l_{i-1,j}$, provided $(i, j - 1, l), (i - 1, j, l) \in I$. The boundary $\partial \Omega^l_{i,j} \subset \overline{\Omega^l_{i,j}}$ consists of the four *cell faces* of the cell, identified through their relative locations by $\partial \Omega^l_{i,j,k}$, $k \in D$, and $\partial \Omega^l_{i,j} = \bigcup_{k\in D} \partial \Omega^l_{i,j,k}$, where $D = \{N, E, S, W\}$.

Refinements of a cell $\Omega^l_{i,j}$ are the cells obtained by subdivision of the corresponding cell $\widetilde{\Omega}^l_{i,j}$ in the computational domain into 2×2 smaller cells of equal size. By applying the mapping M^{l+1} to these refinements in the computational domain, we construct the refinements in the partitioning of the physical domain. Except for cells on the coarsest grid, each cell is one of the four *descendants* of a cell on the coarser grid. The coarse-grid cells on Ω^l and the fine-grid cells on Ω^{l+1} are coexistent (i.e. when cells appear on Ω^{l+1}, the corresponding coarse-grid cells remain part of Ω^l). A cell on the coarser grids is called *parent* and its descendants are called its *kids*. In this way all cells in the geometric structure belong to a quad-tree structure.

For the smallest integer coordinates on the coarsest grid, Ω^0, without loss of generality we take $i = 0$ and $j = 0$. The integer coordinates of a cell on Ω^{l+1} are so that the kids of $\Omega^l_{i,j}$ are denoted by $\Omega^{l+1}_{2i,2j}$, $\Omega^{l+1}_{2i+1,2j}$, $\Omega^{l+1}_{2i,2j+1}$ and $\Omega^{l+1}_{2i+1,2j+1}$. A cell vertex in the physical domain is $P^l_{i,j} = M^l(\xi^l_{i,j}, \eta^l_{i,j})$, where, without loss of generality, $\widetilde{P}^l_{i,j} = (\xi^l_{i,j}, \eta^l_{i,j}) = (2^{-l}i, 2^{-l}j)$.

In this paper, functions, function spaces and subdomains defined for the computational domain are identified by a tilde on the same symbol used for the physical domain.

2.1 The sequence of grids

The geometric structure described so far, is used for multigrid computations on a locally refined grid. The grid Ω^{l_b} on some basic level $l_b \geq 0$ covers all $M^{l_b}(\widetilde{\Omega})$. The grids Ω^l,

$l > l_b$, are adaptively constructed during the computation, when it is decided that cells should be refined or refinements should be deleted, depending on the computed solution. At some stage in the computation, a sequence of grids $\{\Omega^l\}_{l=0,\ldots,L}$ has been generated, where L is the highest level present. Thus, the cells on grid Ω^l, $l > l_b$ typically do *not* cover all of the domain Ω. Therefore, the grid $\Omega^l = \Omega_f^l \cup \Omega_c^l$ consists of a part Ω_f^l of which the cells have been refined (for which kids exist on level $l+1$) and a part Ω_c^l, with cells that have not been refined (without kids). The set of all non-refined cells is called the *composite grid* Ω_c, defined by

$$\Omega_c = \bigcup_{l=l_b}^{L} \Omega_c^l.$$

Further, we define sets of indices associated with the different grids and parts of grids by $I^l = \{(i,j) \in \mathbb{Z}^2 \mid (i,j,l) \in I\}$. $I_f^l = \{(i,j) \in I^l \mid \Omega_{i,j}^l \subset \Omega_f^l\}$, $I_f = \{(i,j,l) \in I \mid (i,j) \in I_f^l\}$. For practical purposes we also introduce K, defined by

$$K(i,j) = \{(2i,2j), (2i+1,2j), (2i,2j+1), (2i+1,2j+1)\}. \tag{2.2}$$

The boundary $\partial \Omega^l$ on level l is $\partial \Omega^l = $ boundary of $\bigcup_{(i,j) \in I^l} \overline{\Omega_{i,j}^l}$. Following [14], the part of the boundary of the subdomain Ω^l, which does neither coincide with a physical boundary nor with its discrete counterpart, is called a *green* boundary.

A grid is *uniform* if, in the computational domain, it covers all of $\widetilde{\Omega}$ and if it is not refined anywhere; it is called *locally uniform* with respect to a discrete operator in a cell, if no green boundary is involved in the definition of the operator for that cell. A grid that contains locally non-uniform cells, is called a *locally refined* grid. Also a composite grid that consists of cells from more than one level is called locally refined.

3 Finite volume discretisation on a locally refined grid

3.1 Grids and grid functions

We see that the grid Ω^l on level l consists of cells $\Omega_{i,j}^l$, $(i,j) \in I^l$. A cell $\Omega_{i,j}^l$ in the physical domain is the result of $M^l(\widetilde{\Omega}_{i,j}^l)$, where M^l is an approximation of M as described above. We assume that M^l is continuous and piecewise affine on each cell face $\partial \widetilde{\Omega}_{i,j,k}^l$. Then the grid Ω^l in the physical space is a collection of disjoint *quadrilaterals*. In that situation, M^l can be described as a vector of two continuous functions, both piecewise bilinear in each $\widetilde{\Omega}_{i,j}^l$. Furthermore we assume

$$(P_{i,j}^l)^T = M^l(\widetilde{P}_{i,j}^l) = M(\widetilde{P}_{i,j}^l), \tag{3.1}$$

i.e. M^l is exact at the vertices $\widetilde{P}_{i,j}^l$.

The boundary $\partial \Omega_{i,j}^l$ consists of the four faces $\partial \Omega_{i,j,k}^l$, $k = N, E, S, W$. A cell face $\partial \Omega_{i,j,k}^l$ has a length denoted by $s_{i,j,k}^l$. The area of a cell $\Omega_{i,j}^l$ is denoted by $A_{i,j}^l$.

If a function $u : \Omega^l \to \mathbb{R}^d$ is defined then also a function $\widetilde{u} : \widetilde{\Omega}^l \to \mathbb{R}^d$ is defined through

$$\widetilde{u}(\xi, \eta) = u(x^l(\xi, \eta), y^l(\xi, \eta)),$$

where x^l and y^l are given by

$$\begin{pmatrix} x^l(\xi,\eta) \\ y^l(\xi,\eta) \end{pmatrix} = M^l(\xi,\eta). \tag{3.2}$$

We denote the unknown vector function in our problem by q. The approximating function q^l defined for the grid on level l. is a cell-wise constant function. The value of q^l in a cell $\Omega^l_{i,j}$ is denoted by $q^l_{i,j}$, hence

$$q^l_{i,j} = q^l(x',y'), \quad (x',y') \in \Omega^l_{i,j} \cap \Omega^*.$$

The function value $q^l_{i,j}$ is called the *state* in cell $\Omega^l_{i,j}$. The space of all admissible state vectors is denoted by $X_a \subset \mathbb{R}^d$

3.2 Restrictions

In order to define the relations between the approximations on the different grids, it is appropriate to define a number of *restrictions*. The first is $\overline{R}^l$, the $L^2(\Omega)$-projection to the piecewise constant functions on Ω^l. A projection closely related to this operator is $\overline{R}^{l,l+1}$, which restricts the piecewise constant functions on level l to the refined part Ω^l_f of the grid. Next, we define the restriction $\overline{R}^l_{l+1}$, giving the piecewise constant function, which in each $\Omega^l_{i,j} \subset \Omega^l_f$ delivers the integral mean of the operand over its kids. I.e., if the collection of kids of $\Omega^l_{i,j}$ is

$$\Sigma^l_{i,j} = \bigcup_{m \in K(i,j)} \Omega^{l+1}_m,$$

then the restriction $\overline{R}^l_{l+1}$ is defined by

$$\{\overline{R}^l_{l+1} u\}^l_{i,j} = \frac{\int_{\Sigma^l_{i,j}} u\, d\Omega}{\int_{\Sigma^l_{i,j}} d\Omega}. \tag{3.3}$$

Note that $\overline{R}^{l,l+1} = \overline{R}^l_{l+1}$, if the grid is obtained by a piecewise bilinear mapping M^l and $M^{l+1} = M^l$.

Another set of three restrictions (denoted without the overbar) gives the relations between vector functions in the r.h.s. space. The first restriction is the projection R^l, defined as the operator giving the cell-wise constant function consisting of the values if the integral of the operand on each cell on level l. The second restriction is the projection $R^{l,l+1}$, defined similarly as $\overline{R}^{l,l+1}$, i.e. R^l restricted to the area covered by the fine grid Ω^l_f. Finally we define a restriction R^l_{l+1}, which is related to $\overline{R}^l_{l+1}$ in (3.3) through the operators A^l and $A^{l,l+1}$, which are defined as

$$A^l u(x',y') = A^l_{i,j} u(x',y'), \quad \forall (x',y') \in \Omega^l_{i,j},$$

and

$$A^{l,l+1} u(x',y') = A^l u(x',y'), \quad \forall (x',y') \in \Omega^l_{i,j} \text{ and } (i,j,l) \in I_f.$$

With these definitions, we define the last restriction by

$$R^l_{l+1} = A^{l,l+1} \overline{R}^l_{l+1} (A^{l+1})^{-1}. \tag{3.4}$$

3.3 The system of discrete equations

In this section we describe the system of equations obtained by the discretisation. We distinguish between equations obtained for cells on the composite grid and equations obtained for refined cells.

Equations for a cell on a composite grid

A discretisation of the set of conservation laws (1.1a) on a composite grid is obtained by considering the weak formulation of the problem: find q from the solution space, satisfying the boundary conditions and so that for all $\Omega^* \subset \Omega$

$$\int_{\Omega^*} \frac{\partial f(q(x,y))}{\partial x} + \frac{\partial g(q(x,y))}{\partial y} \, d\Omega = \int_{\Omega^*} s \, d\Omega. \qquad (3.5)$$

We assume that q and s are defined on $\hat{\Omega}$. In case $\hat{\Omega} \supset \Omega$, we assume that q and s, defined on Ω, can be extended to $\hat{\Omega}$ in a sufficiently regular way. Then (3.5) is approximated by

$$N(q) = r, \qquad (3.6)$$

where N and r are functions defined on any $\Omega^* \subset \hat{\Omega}$. For $\Omega_{i,j}^l \subset \hat{\Omega}$ we define the restriction $R^l N$ by

$$\left\{ R^l N(q) \right\}_{i,j}^l \equiv \oint_{\partial \Omega_{i,j}^l} f(q) n_x + g(q) n_y \, ds,$$

where n_x and n_y are the components of the outward unit normal n on the boundary $\partial \Omega_{i,j}^l$, in x and y direction respectively. The discretisation of the equations is obtained by requiring an approximation of (3.6) to hold for each cell on the composite grid. We first assume that for the discretisation the source term s is exactly integrated. In our notation this implies

$$r^l = R^l s. \qquad (3.7)$$

The mean value of the flux across the kth cell face $\partial \Omega_{i,j,k}^l \subset \partial \Omega_{i,j}^l$ of cell $\Omega_{i,j}^l$ is

$$f_{i,j,k}^l(q) = \frac{1}{s_{i,j,k}^l} \int_{\partial \Omega_{i,j,k}^l} f(q) n_x + g(q) n_y \, ds.$$

Hence, a solution of (3.6) exactly satisfies

$$\left\{ R^l N(q) \right\}_{i,j}^l = \sum_{k \in D} f_{i,j,k}^l(q) s_{i,j,k}^l = r_{i,j}^l, \quad \forall (i,j,l) \in I. \qquad (3.8)$$

Equations (3.8) are approximated by approximating the mean fluxes $f_{i,j,k}^l$ across each cell face $\partial \Omega_{i,j,k}^l$ by a *numerical flux*, denoted by $F_{i,j,k}^l$. In general this numerical flux depends on the functions q^m, $m = l_b, \ldots, l$. On a level l the approximation of (3.8) reads for all $\Omega_{i,j}^l \in \Omega_c^l$

$$\sum_{k \in D} F_{i,j,k}^l(q^l; q^{l-1}, \ldots, q^{l_b}) s_{i,j,k}^l = r_{i,j}^l, \qquad (3.9)$$

or in operator form

$$N^l(q^l; q^{l-1}, \ldots, q^{l_b}) = r^l,$$

where N^l is defined by

$$\{N^l(q^l; q^{l-1}, \ldots, q^{l_b})\}_{i,j}^l = \sum_{k \in D} F_{i,j,k}^l(q^l; q^{l-1}, \ldots, q^{l_b}) \, s_{i,j,k}^l.$$

Here, $q^{l-1}, \ldots, q^{l_b}$ act as parameters to N^l. These formulas define the discretisation on level l.

The numerical flux function

The numerical flux $F_{i,j,k}^l$ depends on the sequence $\{q^m\}_{m=l_b,...,l}$. Usually we assume that the numerical flux can be written as

$$F_{i,j,k}^l(q^l; q^{l-1}, \ldots, q^{l_b}) = F((q^L)_{i,j,k}^l, (q^R)_{i,j,k}^l, n_{i,j,k}^l).$$

The arguments $(q^L)_{i,j,k}^l$ and $(q^R)_{i,j,k}^l$ denote estimates of the mean of q along $\partial\Omega_{i,j,k}^l$, dependent on $\{q^m\}_{m=l_b,...,l}$, with a bias to the left and right side of $\partial\Omega_{i,j,k}^l$, respectively. The entry $n_{i,j,k}^l \in E$ denotes the unit normal on $\partial\Omega_{i,j,k}^l$, pointing outward from $\Omega_{i,j}^l$, where $E \subset \mathbb{R}^2$ is the unit circle in $\mathbb{R}^2$. The function $F(q^L, q^R, n)$ is an approximation of the flux $f(q)n_x + g(q)n_y$, with q^L and q^R in the neighbourhood of q.

There are various ways to define the states q^L and q^R and the numerical flux F. In fact, the choice of F and the states q^L and q^R determine the discretisation method and its accuracy. The left and right states are usually obtained by piecewise polynomial *reconstruction*, using discrete data, i.e. using $\{q^m\}_{m=l_b,...,l}$ (cf. Section 3.4).

For a hyperbolic set of conservation laws [10], we are interested in an upwind discretisation, which may be obtained by taking for F an (approximate) Riemann solver. The best-known approximate Riemann solvers are introduced in [13, 22, 12].

Equations for a refined cell

Discrete equations (3.9) are approximations of the conservation equation (3.5) for each cell that has not been refined. The left and right state for the computation of a numerical flux depend on the states in neighbouring cells, possibly on different levels. By definition, for a locally non-uniform grid cell, the left or right state for at least one cell face depends on coarse-grid states.

The set of equations obtained by applying the discretisation as described at the beginning of this section are under-determined for a locally refined grid. If a neighbouring cell has been refined (has kids), that neighbour is not part of the composite grid, and no equation like (3.9) has been defined. Additional equations, however, are obtained by

$$q_{i,j}^l = \{\overline{R}_{l+1}^l q^{l+1}\}_{i,j}^l, \quad \forall (i,j,l) \in I_f. \tag{3.10}$$

We use the equations (3.9) together with (3.10) to define discretisations on a locally refined (i.e. composite) grid.

3.4 Left and right states

Here we describe the computation of $(q^L)_{i,j,k}^l$ and $(q^R)_{i,j,k}^l$, the left and right state used in the numerical flux function. We consider first and second-order accurate discretisations, both for a locally uniform and a locally non-uniform situation. We use the concept of reconstruction of piecewise C^∞-functions from the cell-wise constant data, $q_{i,j}^l$, associated with each cell. This idea was introduced in [21] and [24] for one-dimensional convection and extended and applied in [1] and [2] for unstructured grids in two spatial dimensions. Contrary to this work, we do not reconstruct a single, unique (vector) function in each cell, but we take care that –in a locally non-uniform grid situation– the computation of the left and right state is so that the resulting scheme is consistent of the required order (at least in some *weak* sense, see Section 4). This is done by making a different reconstruction for each side of each cell face.

Locally uniform composite grids

The computation of the first-order as well as of the second-order consistent discretisation depends on the mean states. On a locally uniform grid, first-order consistency is obtained by applying an $\mathcal{O}(h_l)$ accurate reconstruction. Consider for example the eastern cell face $\partial\Omega^l_{i,j,E}$ of a cell $\Omega^l_{i,j}$ on a locally uniform composite grid, where $(i+1,j) \in I^l$. For this situation we take for the states, as usual in first-order Godunov-type schemes, [4, 8, 18],

$$(q^L)^l_{i,j,E} = q^l_{i,j}, \tag{3.11a}$$
$$(q^R)^l_{i,j,E} = q^l_{i+1,j}. \tag{3.11b}$$

For second-order consistency on a locally uniform grid, the states are based on $\mathcal{O}(h_l^2)$ accurate reconstructions of the state functions. This reconstruction can be done with a limiter to suppress spurious wiggling of the solution (as proposed e.g. in [15, 19] and applied in [9]), or without a limiter, like the κ-schemes [23] (as e.g. in [7, 16] and [9]). Again, for the eastern cell face of a cell $\Omega^l_{i,j}$ on a locally uniform grid, where $(i-1,j), (i+1,j), (i+2,j) \in I^l$, the limiter and κ-schemes are given by

$$(q^L)^l_{i,j,E} = C(q^l_{i-1,j}, q^l_{i,j}, q^l_{i+1,j}), \tag{3.12a}$$
$$(q^R)^l_{i,j,E} = C(q^l_{i+2,j}, q^l_{i+1,j}, q^l_{i,j}), \tag{3.12b}$$

where, $C : X_a \times X_a \times X_a \to X_a$ describes the κ-scheme or the limiter scheme. Notice that the κ-schemes are recovered by applying certain linear 'limiter' functions. However a κ-scheme does not necessarily satisfy monotonicity conditions, see [17].

Locally non-uniform grids

On a locally refined grid, one or more of the mean states in (3.11) or (3.12) are not available, because the cells with which the states should be associated, do not exist. For this, we introduce the concept of *virtual cell* and associated *virtual state*.

With each integer coordinate pair $(2^n i + r, 2^n j + s) \notin I^{l+n}$, $0 \le r, s < 2^n$, $n \ge 1$ and $(i,j) \in I^l$, $l_b \le l \le L-n$, we associate the virtual cell $\widetilde{\omega}^{l+n}_{2^n i+r, 2^n j+s} \subset \widetilde{\Omega}$, given by

$$\widetilde{\omega}^{l+n}_{2^n i+r, 2^n j+s} = 2^{-(l+n)} (2^n i + r, 2^n i + r + 1) \\ \times 2^{-(l+n)} (2^n j + s, 2^n j + s + 1).$$

In the physical space the virtual cell $\omega^{l+n}_{2^n i+r, 2^n j+s} \subset \widehat{\Omega}$ is defined as

$$\omega^{l+n}_{2^n i+r, 2^n j+s} = M^{l+n}(\widetilde{\omega}^{l+n}_{2^n i+r, 2^n j+s}). \tag{3.13}$$

Note that $\omega^{l+n}_{r,s}$ is exactly $\Omega^{l+n}_{r,s}$, if the grid would be sufficiently refined.

With the virtual cell $\omega^l_{i,j}$ we associate a virtual state $v^l_{i,j} \in X_a$, which can be interpreted as an approximation of the mean of the state vector function on $\omega^l_{i,j}$. In general a virtual state $v^l_{i,j}$ depends on $\{q^m\}_{m=l_b,...,l}$.

Virtual cells and virtual states are used for the discretisation in the neighbourhood of green boundaries. To a large extent the virtual states determine the accuracy of the algebraic equations associated with the locally non-uniform grid.

The concept of virtual states allows us to compute left and right states in a locally non-uniform grid situation, in a way similar to (3.11) and (3.12). The requirements to be satisfied for the proper computation of virtual states, are discussed in Section 4. Regardless of the way how the virtual states are computed, for first-order consistency we take for the eastern cell face of $\Omega_{i,j}^l$, similarly to (3.11), and $(i+1,j) \notin I^l$

$$(q^L)_{i,j,E}^l = q_{i,j}^l, \tag{3.14a}$$
$$(q^R)_{i,j,E}^l = v_{i+1,j}^l. \tag{3.14b}$$

Similar to (3.12), we take for second-order consistency, if $(i+1,j), (i+2,j) \notin I^l$

$$(q^L)_{i,j,E}^l = C\left(q_{i-1,j}^l, q_{i,j}^l, v_{i+1,j}^l\right), \tag{3.15a}$$
$$(q^R)_{i,j,E}^l = C\left(v_{i+2,j}^l, v_{i+1,j}^l, q_{i,j}^l\right), \tag{3.15b}$$

and if $(i+1,j) \in I^l$, but $(i+2,j) \notin I^l$

$$(q^L)_{i,j,E}^l = C\left(q_{i-1,j}^l, q_{i,j}^l, q_{i+1,j}^l\right), \tag{3.16a}$$
$$(q^R)_{i,j,E}^l = C\left(v_{i+2,j}^l, q_{i+1,j}^l, q_{i,j}^l\right), \tag{3.16b}$$

Formulae similar to (3.14)–(3.16) are used for the cell faces, $\partial \Omega_{i,j,k}^l$, $k = N, S, W$.

4 Error analysis of the discretisation

In this section we study the local discretisation error and the consistency of the discretisation described in Section 3. In the discretisation we distinguish three approximations, each of which have a contribution to the local discretisation error. These contribution are:

- approximation of the mapping from the computational space into the physical space; in equations this error is denoted by $\tau_m^l(q)$;

- approximation of the mean flux on a cell face by the flux evaluated at the mean state along the cell face (quadrature rule); in equations this error is denoted by $\tau_q^l(q)$;

- approximation of the mean state on a cell face by biased reconstructions; in equations this error is denoted by $\tau_r^l(q)$.

Often the approximation of M by M^l is not essential, since the change from M to M^l, merely changes the partitioning in the physical domain slightly, without affecting the accuracy of the resulting set of algebraic equations. However, it does affect the approximation of the domain of definition of the problem. Hence, for the interior cells of $\widetilde{\Omega}$ we can assume $M = M^l = M^{l_0}$, for some constant l_0 and $l_0 \leq l \leq L$, without affecting the accuracy. At the domain-boundary, $\partial \widetilde{\Omega}$, the error of this approximation M^l can be important. Here, in general we do not have $M = M^l = M^{l_0}$, because it results in an approximation of the boundary, independent of the level l. Hence, with increasing levels of refinement, the geometry of the discrete problem would not converge to the geometry of the continuous problem. A more important reason to study this approximation is related to a-posteriori estimation of the local discretisation error and the application of τ-extrapolation. This uses the so-called relative local discretisation errors of two consecutive levels of refinement (cf. [3, 20]).

The two errors $\tau_q^l(q)$ and $\tau_r^l(q)$ together make up the local truncation error. For the fluxes approximated with the projection of the true solution, i.e. with $q = \bar{q}$, the exact solution of the problem, they form the error in the system of algebraic equations, for a given partitioning.

The analysis presented in this section leads to requirements to be imposed on the reconstruction of cell-wise smooth functions from the cell-wise constant numerical data. These requirements, depend on the goal set out to be reached in a particular discretisation. We distinguish between the goals (i) obtain a given order of consistency and (ii) obtain a given order of discrete convergence. This distinction is made, since a certain order of the local discretisation error for *all* equations may not be essential to obtain a given order of discrete convergence. Assume a total number of equations of $\mathcal{O}(h^{-2})$. An $\mathcal{O}(h^{-1})$ number of equations with low-order accuracy may not affect the rate of discrete convergence, not even in supremum norm. To study this in detail, we have to redefine the notion of consistency for a non-uniform mesh. In fact, we define a slightly *weaker* form of consistency. This weaker form is the discrete L_1-norm of the local truncation error for a collection of discrete equations. The requirements for consistency both in the weak and in the usual sense are studied.

4.1 Approximation of the mapping

To make an a-posteriori estimation of the local discretisation error, with sufficient accuracy, we first study the consequences of approximating M by M^l. Actually, we are only interested in the relation between the mappings for two consecutive levels of refinement, since we want to study the use of two consecutive levels of refinement in the estimation of the local discretisation error. The relation between the mappings of two consecutive levels can be established through their relation to M. This relation also allows us study the accuracy of the restriction to the coarse grid (cf. Section 4.2).

We consider a surjective mapping $M : \overline{\tilde{\Omega}} \to \overline{\Omega}$, also injective on $\tilde{\Omega}$. We also consider its continuous, piecewise bilinear approximation associated with level of refinement l, $M^l : \overline{\tilde{\Omega}} \to \overline{\tilde{\Omega}^l}$, which is exact in the vertices $\tilde{P}_{i,j,k}^l$. To simplify notations, in the present local analysis, we drop the indices which are constant throughout this part of the analysis. We consider a cell on level of refinement l. In the computational space, the corresponding cell is denoted by $\widehat{\Omega}$. Its images in $\widehat{\Omega}$ are

$$\Omega = M(\tilde{\Omega}),$$
$$\Omega^l = M^l(\tilde{\Omega}),$$

where we drop the subscripts. We use a local Cartesian coordinate system (ξ, η) in the computational space, with its origin in the center of $\tilde{\Omega}$. Similar to (3.2), we use the Cartesian coordinates in the physical space as obtained by M

$$\begin{pmatrix} x(\xi, \eta) \\ y(\xi, \eta) \end{pmatrix} = M(\xi, \eta),$$

where the origin in the physical space is

$$(0,0)^T = M(0,0). \tag{4.1}$$

We assume $\tilde{\Omega}$ being a square with edges $h_l = 2h$, $\tilde{\Omega} = (-h, h)^2$. The area in the computational space is $4h^2$. In the physical space the area is denoted by A for Ω and A^l for Ω^l.

The Jacobians of M and M^l are denoted by J and J^l respectively:

$$\begin{aligned} J(\xi,\eta) &= x_\xi y_\eta - x_\eta y_\xi, \\ J^l(\xi,\eta) &= x^l_\xi y^l_\eta - x^l_\eta y^l_\xi, \end{aligned}$$

where the subscripts denote differentiation with respect to ξ and η respectively. Further, we assume that M is sufficiently smooth and for $(\xi,\eta) \in \overline{\tilde\Omega}$ it can be written

$$\begin{aligned} M &= m_0 + m_1\xi + m_2\eta + m_3\xi^2 + m_4\xi\eta + m_5\eta^2 \\ &\quad + m_6\xi^3 + m_7\xi^2\eta + m_8\xi\eta^2 + m_9\eta^3 + \ldots . \end{aligned} \qquad (4.2)$$

Note that by (4.1) $m_0 = (0,0)^T$. The mapping M^l is piecewise bilinear:

$$M^l = m^l_0 + m^l_1\xi + m^l_2\eta + m^l_4\xi\eta.$$

We define $(x_i, y_i)^T$, $(x^l_i, y^l_i)^T$ by

$$\begin{aligned} m_i &= (x_i, y_i)^T, & i &= 0,1,2,\ldots, \\ m^l_i &= (x^l_i, y^l_i)^T, & i &= 0,1,2,4. \end{aligned} \qquad (4.3)$$

From the exactness of M^l at the cell vertices, given by (3.1), we can express m^l_i, $i = 0, 1, 2, 4$ in terms of m_i, $i = 0, 1, \ldots$. This gives

$$\begin{aligned} m^l_0 &= m_0 + m_3 h^2 + m_5 h^2 + \mathcal{O}(h^4), & (4.4\text{a}) \\ m^l_1 &= m_1 + m_6 h^2 + m_8 h^2 + \mathcal{O}(h^4), & (4.4\text{b}) \\ m^l_2 &= m_2 + m_7 h^2 + m_9 h^2 + \mathcal{O}(h^4), & (4.4\text{c}) \\ m^l_4 &= m_4 + \mathcal{O}(h^4). & (4.4\text{d}) \end{aligned}$$

As a result, we have M^l expressed in terms of M, for $|\xi|, |\eta| \leq h$, and

$$M^l - M = m_3(h^2 - \xi^2) + m_5(h^2 - \eta^2) + \mathcal{O}(h^3).$$

We are now interested in the difference between the weak form (3.5), for $\Omega = M(\tilde\Omega)$ and $\Omega^l = M^l(\tilde\Omega)$, divided by the respective areas A and A^l. This error is denoted by $\tau_m(q)$.

Let a sufficiently smooth, integrable function $w : \Omega \cup \Omega^l \to \mathbb{R}^d$ be defined, and let its Taylor series expansion around the origin $(x,y) = (0,0)$ be given by

$$w(x,y) = w_0 + w_1 x + w_2 y + w_3 x^2 + \ldots . \qquad (4.5)$$

For the error of the weak form on Ω^l with respect to the weak form on Ω, it suffices to consider the difference

$$t(w) \equiv \frac{1}{A^l} \int_{\Omega^l} w\, d\Omega - \frac{1}{A} \int_\Omega w\, d\Omega, \qquad (4.6)$$

for a sufficiently smooth and integrable function w. For M and M^l, assuming from this point that $J, J^l > 0$, this is equal to

$$t(w) = \frac{1}{A^l} \int_{\tilde\Omega} w^l J^l\, d\tilde\Omega - \frac{1}{A} \int_{\tilde\Omega} w J\, d\tilde\Omega,$$

where $w^l = w(x^l, y^l)$. With

$$\delta_w = w^l - w,$$
$$\delta_J = J^l - J,$$

the integrand $w^l J^l$ can be written as

$$w^l J^l = wJ + \delta_w J + w\delta_J + \delta_w \delta_J.$$

An expression for δ_w in terms of m_i and w_i can be obtained by using (4.5) for (x^l, y^l) and (x, y), and by subtraction and substitution of (4.4). From this exercise it appears that $\delta_w = \mathcal{O}(h^2)$. Multiplication with J and integration over $\tilde{\Omega}$ yields

$$\int_{\tilde{\Omega}} \delta_w J \, d\tilde{\Omega} = \frac{8}{3} J_0 h^4 \{w_1(x_3 + x_5) + w_2(y_3 + y_5)\} + \mathcal{O}(h^5), \tag{4.7}$$

where J_0 is $J(0,0)$.

Similarly, based on (4.4), we can also find an expression for δ_J. It appears that generally $\delta_J = \mathcal{O}(h)$, with the $\mathcal{O}(h)$ terms linear in ξ or η. A straightforward calculation yields for the integral of $w\delta_J$

$$\int_{\tilde{\Omega}} w\delta_J \, d\tilde{\Omega} =$$
$$\frac{8}{3} h^4 \{w_0(x_1 y_7 - x_7 y_1 + x_8 y_2 - x_2 y_8 - x_3 y_4 + x_4 y_3 - x_4 y_5 + x_5 y_4) \tag{4.8}$$
$$+ w_1(-x_1 x_3 y_2 + x_1 x_2 y_3 - x_1 x_2 y_5 + x_2 y_5 y_1)$$
$$+ w_2(-x_3 y_1 y_2 + x_2 y_1 y_3 - x_1 y_2 y_5 + x_5 y_1 y_2)\} + \mathcal{O}(h^5).$$

The term $\delta_w \delta_J$ gives only an $\mathcal{O}(h^5)$ contribution to the integral over $\tilde{\Omega}$.

By (4.7) and (4.8) we can define $C(w)$ as the coefficient in the first term of an asymptotic expansion, given by the sum of (4.7) and (4.8), i.e.

$$\int_{\tilde{\Omega}} (w^l J^l - wJ) \, d\tilde{\Omega} = h^4 C(w) + \mathcal{O}(h^5),$$

with C independent of h. Now we can write

$$A^l - A = h^4 C(1) + \mathcal{O}(h^5). \tag{4.9}$$

For the area A we have

$$A = 4h^2 J_0 + \mathcal{O}(h^4), \tag{4.10}$$

combination of (4.9) and (4.10) yields after some manipulation

$$\frac{1}{A^l} - \frac{1}{A} = -\frac{C(1)}{16 J_0^2} + \mathcal{O}(h).$$

Using this, it can be easily shown that the difference $t(w)$ satisfies

$$t(w) = \frac{h^2}{4 J_0} (C(w) - w_0 C(1)) + \mathcal{O}(h^3). \tag{4.11}$$

With

$$\tau_m^l(q) = t \left(\frac{\partial f(q)}{\partial x} + \frac{\partial g(q)}{\partial y} \right),$$

we establish the asymptotic relation between the approximation of the weak form on two consecutive levels of refinement, caused by the approximation of the grid geometry. Although the difference between the partitioning obtained with M and the partitioning obtained with M^l itself is not essential for the convergence of the local discretisation error, we use the result (4.11) when we show that an a-posteriori error estimation is sufficiently accurate. Further, we use (4.11) to show that on a composite grid, the equations for a refined cell (3.10), in general give at best second-order accuracy on the coarse grid. Assume for (4.11) that w is the differential operator, applied to the solution of the continuous problem. According to the result in (4.11), in general the discrete equations derived for boundary cells are at most second-order accurate, because (in general) we cannot use $M = M^l = M^{l_0}$, l_0 a constant, $l_0 \leq l \leq L$, in cells along the boundary of the domain. A piecewise bilinear mapping $M = M^l$ would yield $\tau_m^l(q) = 0$.

4.2 Accuracy of the coarse-grid restriction

We consider the restriction $\overline{R}_{l+1}^l$, as defined by (3.3). We study the difference between the restrictions $\overline{R}_{l+1}^l \overline{R}^{l+1} q$ (cf. (3.3)) and $\overline{R}^{l,l+1} q$. In the computation of the virtual states we use as the coarse-grid discrete function on Ω_f^l, the restricted function $q_{i,j}^l = \{\overline{R}_{l+1}^l q^{l+1}\}_{i,j}^l$, $(i,j) \in I_f^l$, where $q^{l+1} = \overline{R}^{l+1} q$, while we assume that $q_{i,j}^l$ is the mean of q on $\Omega_{i,j}^l$, given by $\overline{R}^{l,l+1} q$. The error we make with this assumption is studied here.

In our analysis we again drop the unnecessary indices. We consider a cell Ω^l on level l, which has refinements on level $l+1$. A kid of Ω^l is identified by the subscript $m \in K$, where K is the set of indices (cf. (2.2)), associated with the cell Ω^l. For level l we have the approximation M^l of M and for level $l+1$ we have M^{l+1}. Similar to the notations in the previous subsection, we use the superscripts in our notations to distinguish between variables for $\Omega = M(\tilde{\Omega})$ (e.g. A, no superscript) and $\Omega^l = M^l(\tilde{\Omega})$ (e.g. A^l). Hence, e.g. A_m^{l+1}, $m \in K$, denotes the area of the mth kid $M^{l+1}(\tilde{\Omega}_m^{l+1})$.

Consider $\overline{R}_{l+1}^l q^{l+1}$, where $q^{l+1} = \overline{R}^{l+1} q$. By definition of $\overline{R}_{l+1}^l$ we have

$$\overline{R}_{l+1}^l \overline{R}^{l+1} q = \frac{\sum_{m \in K} A_m^{l+1} q_m^{l+1}}{\sum_{m \in K} A_m^{l+1}}. \tag{4.12}$$

With the results from the previous subsection we find

$$\frac{1}{\sum_{m \in K} A_m^{l+1}} = \frac{1}{A} - \frac{h_{l+1}^4 C(1)}{4 A^2} + \mathcal{O}(h_{l+1}). \tag{4.13}$$

For the numerator in (4.12) we find

$$\sum_{m \in K} A_m^{l+1} q_m^{l+1} = \int_{\tilde{\Omega}} q J \, d\tilde{\Omega} + \frac{1}{4} h_{l+1}^4 C(q) + \mathcal{O}(h_{l+1}^5). \tag{4.14}$$

Multiplication of (4.13) and (4.14) gives for (4.12):

$$\overline{R}^{l,l+1} \overline{R}^{l+1} q = \frac{1}{A} \int_{\tilde{\Omega}} q J \, d\tilde{\Omega} + \frac{h_{l+1}^2}{16 J_0}(C(q) - q_0 C(1)) + \mathcal{O}(h_{l+1}^3).$$

With (4.6) and (4.11) we obtain

$$\overline{R}^l q = \frac{1}{A} \int_{\tilde{\Omega}} q J \, d\tilde{\Omega} + \frac{h_{l+1}^2}{4 J_0}(C(q) - q_0 C(1)) + \mathcal{O}(h_l^3).$$

Hence, the difference between the two restrictions is

$$\overline{R}^{l,l+1}q - \overline{R}^{l}_{l+1}\overline{R}^{l+1}q = \frac{3}{16}\frac{h_{l+1}^2}{J_0}(C(q) - q_0 C(1)) + \mathcal{O}(h_l^3).$$

4.3 Consistency and weak consistency

Now we consider the approximation of the weak form on a given partitioning by the set of discrete equations. We assume that $M = M^l = M^{l_0}$, where $l_0 < l$. With (2.1) this implies $\Omega = M^l(\tilde{\Omega})$. This in turn implies that we do not consider the differences between the discretisation on different levels, due to the differences in mappings from the computational into the physical space, on each level.

Local discretisation error and consistency

We use the notation as defined in Section 2 and 3. Let $N(q) = r$ denote the continuous equations (3.5). We denote the solution of this continuous problem by $\bar{q}$:

$$N(\bar{q}) = r.$$

Let the non-linear operator N be approximated by the discrete finite-volume operator N^l, on a level of refinement $l \in \{0, \ldots, L\}$, of a sufficiently smooth grid, consisting of quadrilaterals. Further, let projections R^l and $\overline{R}^l$ be defined properly, for example by the definitions in Section 3.2, and related to each other through A^l, as in (3.4). Assume that the right-hand side of (1.1a) is exactly integrated, as denoted by (3.7). The local discretisation error for $N^l(q^l; q^{l-1}) = r^l$ is $\tau^l(\bar{q})$, where τ^l is given by

$$\tau^l(q) = (A^l)^{-1}\left(N^l(\overline{R}^l q; \overline{R}^{l-1} q) - R^l N(q)\right).$$

We define $\tau^l_{i,j}(q) \equiv \{\tau^l(q)\}^l_{i,j}$. The discretisation N^l is called an approximation to N of order of consistency p, if for all $(i, j, l) \in I$

$$\tau^l_{i,j}(\bar{q}) = \mathcal{O}(h_l^p),$$

for $h_l \to 0$. Notice that $h_l \to 0$, if $l \to \infty$.

Weak consistency

We also introduce a weak form of consistency, related to the above mentioned consistency in the usual sense. The new definition of consistency is weaker than the usual definition, because it considers the collective behaviour of the local discretisation error for a set of equations, rather than the behaviour of each equation separately. Consider a partitioning of the domain Ω, obtained by refining n times a previously obtained locally refined composite grid. Let a discretisation as described in Section 3 be defined for this new system. Then the *collective local truncation error*, $T_n^l(q)$, for an n times completely refined system is defined by

$$T_n^l(q) = A^{-1} R^l_{l+1} \ldots R^{l+n-1}_{l+n}\left(N^{l+n}(\overline{R}^{l+n} q) - R^{l+n} N(q)\right).$$

In absolute value, this is the discrete L_1-norm of the local truncation error for the set of equations for all descendants of each cell $\Omega^l_{i,j}$. Note that

$$T_0^l(q) = \tau^l(q).$$

A discretisation N^l is called *weakly consistent of order p* if

$$\left\{T_n^l(\bar{q})\right\}_{i,j}^l = \mathcal{O}(h_{l+n}^p), \quad \forall (i,j,l) \in I,$$

for $n \to \infty$.

4.4 Analysis of local truncation error

In this section we study the local truncation error of the discretisation introduced in Section 3. We use the same assumptions as formulated at the beginning of Section 4.3. From the analysis we obtain requirements for the computation of the virtual states, both for a consistent and weakly consistent discretisation.

As described at the beginning of Section 4, the local discretisation error consists of two contributions. The first contribution, denoted by $\tau_q^l(q)$, is a result of the quadrature rule that is used to approximate the mean value along a cell face, of the flux across that cell face. This mean value is approximated by the flux evaluated at the mean value of the state function along that face. Further, the mean value of the state function q along a cell face is approximated by reconstruction of piecewise polynomial functions from cell-wise constant functions which represents the average of q in each cell. The reconstruction is done twice for each cell face, i.e. for each of the two sides, with a bias in both opposite directions. These reconstructed approximations of the mean value along a cell face is then used in an approximate Riemann solver. If q is a solution of the continuous equation (1.1a), this procedure gives a contribution to the local discretisation error, denoted by $\tau_r^l(q)$.

Quadrature rule

Similar as before, we drop the indices which are superfluous here. As noted, the mapping M is assumed to be bilinear on $\tilde{\Omega}$. It is given by

$$M(\xi, \eta) = m_0 + m_1\xi + m_2\eta + m_4\xi\eta, \qquad (4.15)$$

where m_0, m_1, m_2 and m_4 are defined by (4.3). Note that the cell-wise constant parameters m_i are fully determined by the coordinates of the vertices of that cell. The area of cell Ω is

$$A = \int_\Omega d\Omega = \int_{\tilde{\Omega}} J \, d\tilde{\Omega} = 4h^2 J_0,$$

where $J_0 = x_1 y_2 - x_2 y_1 > 0$, the Jacobian of M at the origin.

We denote the mean value of the flux across the kth cell face by f_k and since M is linear at $\partial\tilde{\Omega}_k$, we have

$$\begin{aligned} f_k(q) &= \frac{1}{s_k} \int_{\partial\Omega_k} f(q)n_x + g(q)n_y \, ds \\ &= \frac{1}{2h} \int_{\partial\tilde{\Omega}_k} f(q)n_x + g(q)n_y \, d\tilde{s}. \end{aligned} \qquad (4.16)$$

The mean of $q(x,y)$ along $\partial\Omega_k$ is denoted by

$$q_k = \frac{1}{s_k} \int_{\partial\Omega_k} q \, ds.$$

The unit normal along a cell face is constant, and we use the notation

$$(n_x)_k = n_x|_{\partial\Omega_k}, \qquad (n_y)_k = n_y|_{\partial\Omega_k}.$$

By q^* and q_k^* we denote

$$q^* = (q, n_x, n_y)^T,$$
$$q_k^* = (q_k, (n_x)_k, (n_y)_k)^T,$$

and by f^* we denote the flux in the direction of $n = (n_x, n_y)^T$, given by

$$f^*(q^*) = f(q)n_x + g(q)n_y.$$

The contribution $\tau_q(q)$ of the quadrature rule to the local truncation error is given by

$$\tau_q(q) = \frac{1}{A} \sum_{k \in D} (f^*(q_k^*) - f_k(q)) s_k.$$

An expansion of $f^*(q^*)$ around q_k^* yields

$$\begin{aligned} f^*(q^*) &= f^*(q_k^*) + \left.\frac{\partial f^*}{\partial q^{*\alpha}}\right|_{q_k^*} (q^* - q_k^*)^\alpha \\ &\quad + \tfrac{1}{2} \left.\frac{\partial^2 f^*}{\partial q^{*\alpha} \partial q^{*\beta}}\right|_{q_k^*} (q^* - q_k^*)^\alpha (q^* - q_k^*)^\beta + \ldots, \end{aligned} \quad (4.17)$$

where the superscripts denote the components of q^* and q_k^*, and the summation convention is used for $\alpha, \beta = 1, \ldots, d+2$. We assume that the function $q(x(\xi, \eta), y(\xi, \eta))$ can be written as a Taylor series expansion around the origin of the local Cartesian coordinate system in the *computational* space. Now some computation, the details of which are found in [20] shows that the contribution of the quadrature rule to the local discretisation error is $\mathcal{O}(h^2)$.

Reconstruction

Here we study the role of the reconstruction step, in which the left and right states are computed. This yields requirements which should be satisfied by the reconstruction, in order to obtain a consistent or a weakly consistent discretisation. In the previous subsections we found that the local truncation error is limited to second-order by the choice of the mapping and by the choice of quadrature rule. Hence, we are interested in first and second-order consistency only.

In our notations we again drop indices if convenient. As we are interested in a local discretisation error, which is $\mathcal{O}(h^p)$, $p = 1, 2$, we consider the equation for a cell Ω on some level in the geometric structure. We assume that the solution of the problem is sufficiently smooth and that the numerical flux function is sufficiently differentiable.

The outward pointing unit normals on $\partial \Omega_k$ are given by

$$n_E = \begin{pmatrix} y_\eta \\ -x_\eta \end{pmatrix} \frac{1}{\sqrt{x_\eta^2 + y_\eta^2}}\bigg|_{\partial\Omega_E}, \quad n_S = \begin{pmatrix} y_\xi \\ -x_\xi \end{pmatrix} \frac{1}{\sqrt{x_\xi^2 + y_\xi^2}}\bigg|_{\partial\Omega_S},$$

$$n_W = \begin{pmatrix} -y_\eta \\ x_\eta \end{pmatrix} \frac{1}{\sqrt{x_\eta^2 + y_\eta^2}}\bigg|_{\partial\Omega_W}, \quad n_N = \begin{pmatrix} -y_\xi \\ x_\xi \end{pmatrix} \frac{1}{\sqrt{x_\xi^2 + y_\xi^2}}\bigg|_{\partial\Omega_N}.$$

We introduce the unit normal on the kth cell face $\bar{n}_k$ in the physical space, and $\bar{s}_k$, in absolute value equal to the length of $\partial\Omega_k$. The $\bar{n}_k$ and $\bar{s}_k$ are defined by

$$\bar{n}_k = \begin{cases} n_k, & k = E, N, \\ -n_k, & k = W, S, \end{cases} \quad \bar{s}_k = \begin{cases} s_k, & k = E, N, \\ -s_k, & k = W, S. \end{cases} \quad (4.18)$$

Note that this gives $\bar{n}_k \bar{s}_k = n_k s_k$, $\forall k \in D$. We also introduce the vector $w \in X_a^2 \times E$, of length $2d + 2$, given by
$$w(q, q', n) = (q, q', \bar{n}),$$
where $q, q' \in X_a$ and $\bar{n} \in E$. For the kth cell face we define
$$w_k = (q_k, q_k, \bar{n}_k), \tag{4.19}$$
$$w_k^{LR} = (q_k^L, q_k^R, \bar{n}_k). \tag{4.20}$$

Assume that the following, usual conditions hold
$$\begin{aligned} F(q, q', -n) &= -F(q, q', n), & \forall q, q' \in X_a, \quad n \in E, \\ F(q, q, n) &= f^*(q^*), & \forall q \in X_a, \quad n \in E. \end{aligned} \tag{4.21}$$

The local discretisation error due to the reconstruction can now be written as
$$\tau_r = \frac{1}{A} \sum_{k \in D} \left(F(q_k^L, q_k^R, n_k) - f^*(q_k^*) \right) s_k. \tag{4.22}$$

With (4.19) and (4.21) we now have
$$F(w_k) = \begin{cases} f^*(q_k^*), & k = E, N, \\ -f^*(q_k^*), & k = W, S, \end{cases}$$

and with (4.18), (4.22) can be written as
$$\tau_r = \frac{1}{A} \sum_{k \in D} \left(F(w_k^{LR}) \bar{s}_k - f^*(q_k^*) s_k \right). \tag{4.23}$$

We denote the *reconstruction error* r_k, for the kth cell face by
$$r_k = w_k^{LR} - w_k. \tag{4.24a}$$

We denote the difference between $(w_0)_k$ and the mean w_k on $\partial \Omega_k$ by
$$\Delta_k = w_k - (w_0)_k, \tag{4.24b}$$

where $(w_0)_k$ is defined by
$$(w_0)_k = (q_0, q_0, (\bar{n}_0)_k),$$

and $(\bar{n}_0)_k$ defined by
$$(\bar{n}_0)_k = \begin{cases} \begin{pmatrix} y_2 \\ -x_2 \end{pmatrix} \frac{1}{\sqrt{x_2^2 + y_2^2}}, & k = E, W, \\ \begin{pmatrix} -y_1 \\ x_1 \end{pmatrix} \frac{1}{\sqrt{x_1^2 + y_1^2}}, & k = N, S. \end{cases}$$

Now, first making a Taylor series expansion of $F(w)$ around the mean w_k along cell face $\partial \Omega_k$ and substitution of it in (4.23) and next, substitution of an expansion of $F(w)$ around $(w_0)_k$, gives
$$\begin{aligned} \tau_r = \frac{1}{A} \sum_{k \in D} \Bigg\{ & \left. \frac{\partial F}{\partial w^\alpha} \right|_{(w_0)_k} r_k^\alpha + \left. \frac{\partial^2 F}{\partial w^\alpha \partial w^\beta} \right|_{(w_0)_k} r_k^\alpha (\Delta_k^\beta + \frac{1}{2} r_k^\beta) \\ & + \frac{1}{2} \left. \frac{\partial^3 F}{\partial w^\alpha \partial w^\beta \partial w^\gamma} \right|_{(w_0)_k} r_k^\alpha (\Delta_k^\beta \Delta_k^\gamma + r_k^\beta \Delta_k^\gamma + \frac{1}{3} r_k^\beta r_k^\gamma + \ldots) + \ldots \Bigg\} \bar{s}_k. \end{aligned} \tag{4.25}$$

Here, again the summation convention is used for α, β and γ. We use (4.25) to study the requirements for consistency.

4.5 Consistency requirements

Now we can formulate the requirements to be satisfied in the reconstruction phase in order to obtain a pth-order consistent or weakly consistent discretisation.

Consistency

We consider the contributions due to the cell faces $\partial\Omega_k$, $k = E, W$. Assume that the following asymptotic relations hold for $p, q, s = 1, 2$, and for r_k and δ_k as defined by (4.24):

$$r_E = \mathcal{O}(h^q), \tag{4.26a}$$
$$r_W = r_E + \mathcal{O}(h^r), \quad r \geq q, \tag{4.26b}$$
$$\Delta_E = \mathcal{O}(h^s), \tag{4.26c}$$
$$\Delta_W = \Delta_E + \mathcal{O}(h^t), \quad t \geq s. \tag{4.26d}$$

For the mapping M we also have

$$s_E = \mathcal{O}(h), \tag{4.27a}$$
$$s_W = s_E + \mathcal{O}(h^2). \tag{4.27b}$$

With (4.26) and (4.27), and by changing the order of summation, we find for the first term of (4.25),

$$r_E \bar{s}_E + r_W \bar{s}_W = r_E s_E + (r_E + \mathcal{O}(h^r))(-\bar{s}_E + \mathcal{O}(h^2))$$
$$= \mathcal{O}(h^{q+2}) + \mathcal{O}(h^{r+1}).$$

Hence, a pth-order consistent discretisation requires

$$q \geq p, \tag{4.28a}$$
$$r - 1 \geq p, \tag{4.28b}$$

It can easily be shown that all other terms in (4.25) give an $\mathcal{O}(h^p)$ or smaller contributions to τ_r, provided $p, q, s \geq 1$ and (4.28) are satisfied.

Weak consistency

For weak consistency and its requirements with respect to reconstruction, we consider a cell $\Omega^l_{i,j}$ and all its descendants when the composite grid is n times refined. The definitions of the previous subsections hold, but a superscript specifying the level and a subscript for the cell number are added. The superscript n denotes level $l + n$, subscript r denotes cell index $2^n i + r$ and subscript s denotes index $2^n j + s$. The collective local discretisation error is now given by

$$\{T_n^l(q)\}_{i,j}^l = \frac{1}{A_{i,j}^l} \sum_{r,s=0}^{2^n-1} A_{r,s}^n \tau_{r,s}^n(q). \tag{4.29}$$

With $\tau^l(q) = \tau_q^l(q) + \tau_r^l(q)$, we have

$$A_{r,s}^n \tau_{r,s}^n(q) = \sum_{k \in D} \left(F((w^{LR})_{r,s,k}^n) \bar{s}_{r,s,k}^n - f_{r,s,k}^{*n}(q) s_{r,s,k}^n \right) \tag{4.30}$$

Assume the numerical flux $F((w^{LR})^n_{r,s,k})$ across cell face $\partial\Omega^n_{r,s,k}$ to be an $\mathcal{O}(h^q_{l+n})$ accurate approximation of the mean flux $f^{*n}_{r,s,k}(q)$. We have

$$F((w^{LR})^n_{r,s,k})\bar{s}^n_{r,s,k} = f^{*n}_{r,s,k}(q)s^n_{r,s,k} + \mathcal{O}(h^{q+1}_{l+n}). \tag{4.31}$$

Substitution of (4.31) and (4.30) into (4.29) yields a summation over all cell faces on the 'outer' boundary $\partial\Omega^n_{r,s,k} \subset \partial\Omega^l_{i,j,k}, \forall k \in D$. We consider $k = W$ and $n \to \infty$. For the contribution of the approximations on these cell faces to $\{T^l_n\}^l_{i,j}$ we find with (4.31)

$$\frac{1}{A^l_{i,j}} \sum_{s=0}^{2^n-1} \left(F((w^{LR})^n_{0,s,W})\bar{s}^n_{0,s,W} - f^{*n}_{0,s,W}(q)s^n_{0,s,W} \right)$$
$$= \frac{1}{A^l_{i,j}} \sum_{s=0}^{2^n-1} \left(C_1 h^{q+1}_{l+n} + \mathcal{O}(h^{q+2}_{l+n}) \right)$$
$$= C_2 h^q_{l+n} + \mathcal{O}(h^{q+1}_{l+n}),$$

where C_1 and C_2 are constants independent of the level n. Similar results hold for $k = N, E, S$. Hence, if we have an $\mathcal{O}(h^q_{l+n})$ accurate approximation of the mean value along a cell face of the flux across the cell face, then the collective local discretisation error is $\mathcal{O}(h^q_{l+n})$ and hence, the approximation is qth-order weakly consistent. So, a weakly consistent approximation of order $p = 1, 2$, is obtained by pth-order accurate approximation of the mean flux. Let the reconstruction error be defined by (4.24). By the fact that the quadrature rule yields a second-order contribution, and by expansion of $F(w)$ around the mean $w^{l+n}_{r,s,k}$, it follows that a reconstruction error of order $p = 1, 2$, yields a pth-order weakly consistent discretisation.

Note that consistency of order $p = 1, 2$, implies weak consistency, since a necessary but not sufficient condition for consistency is a reconstruction error of order p, as given by (4.28a).

5 Interpolations for virtual states

In this section we study a procedure for the computation of the virtual states. We consider a grid in the physical domain obtained by an affine mapping $M^l = M$. An analysis for more general mappings would probably require an invariant description in general coordinates, similar as e.g. in [11]. We derive expressions for the error of the discretisation due to the computation of the virtual states. For an affine mapping M and a function q we have

$$q^l_{i,j} = \{\bar{R}^l q\}^l_{i,j} = \frac{1}{h^2_l} \int_{\tilde{\Omega}^l_{i,j}} q\, d\tilde{\Omega},$$

This simplifies the analysis, since we can express all in computational (ξ, η) coordinates. We first establish the accuracy of the virtual states for first and second-order accurate discretisations.

We assume a sufficiently smooth function q, and a Taylor series expansion, given by

$$\begin{aligned} q &= q_0 + q_1\xi + q_2\eta + q_3\xi^2 + q_4\xi\eta + q_5\eta^2 \\ &+ q_6\xi^3 + q_7\xi^2\eta + q_8\xi\eta^2 + q_9\eta^3 + \ldots, \end{aligned}$$

around the centre of the cell $\Omega^l_{i,j}$. We consider the equations for a cell $\Omega^{l+1}_{2i+1,2j+1}$, where $\partial\Omega^{l+1}_{2i+1,2j+1,E}$ is part of the green boundary on level $l+1$. The virtual state $v^{l+1}_{2i+2,2j+1}$

required for these equations, is an approximation of the mean of q on $\omega_{2i+2,2j+1}^{l+1}$. This mean can be expressed as

$$\frac{1}{h_{l+1}^2} \int_{\omega_{2i+2,2j+1}^{l+1}} q \, d\tilde{\Omega} = q_0 + \tfrac{3}{2}h_{l+1}q_1 + \tfrac{1}{2}h_{l+1}q_2 \\ + \tfrac{7}{3}h_{l+1}^2 q_3 + \tfrac{3}{4}h_{l+1}^2 q_4 + \tfrac{1}{3}h_{l+1}^2 q_5 + \tfrac{15}{4}h_{l+1}^3 q_6 \\ + \tfrac{7}{6}h_{l+1}^3 q_7 + \tfrac{1}{2}h_{l+1}^3 q_8 + \tfrac{1}{4}h_{l+1}^3 q_9 + \mathcal{O}(h_{l+1}^4).$$ (5.1)

This equals $q_{2i+2,2j+1}^{l+1}$ in the situation $(2i+2, 2j+1) \in I^l$, since $\omega_{2i+2,2j+1}^{l+1}$ is the part of $\hat{\Omega}$ that would be $\Omega_{2i+2,2j+1}^{l+1}$ if the grid would be sufficiently refined (cf. (3.13)). In [17] it is shown that $q_{2i+2,2j+1}^{l+1}$ (and hence (5.1)) satisfies the requirements for consistency (4.28), for the reconstructions discussed in Section 3.4. For any virtual state $v_{2i+2,2j+1}^{l+1}$ that differs $\mathcal{O}(h_{l+1}^p), p = 1, 2$ from (5.1), the reconstructions discussed in Section 3.4 do not satisfy the consistency requirements and introduce an $\mathcal{O}(h_{l+1}^{p-1})$ error in the equations for $\Omega_{2i+1,2j+1}^{l+1}$. However, the numerical flux will be a pth-order accurate approximation of the mean value of the exact flux, and hence pth-order consistency in the weak sense is obtained (cf. 4.5).

5.1 Virtual states for weak consistency

For pth-order weak consistency we only require a pth-order accurate approximation of the mean flux across a cell face. We consider virtual states for both first and second-order weak consistency.

First-order

The formula for a first-order accurate virtual state $v_{2i+2,2j+1}^{l+1}$ is

$$v_{2i+2,2j+1}^{l+1} = q_{i+1,j}^l.$$ (5.2)

For this virtual state we find with the Taylor series expansion

$$v_{2i+2,2j+1}^{l+1} = q_0 + 2h_{l+1}q_1 + \mathcal{O}(h_{l+1}^2).$$ (5.3)

This differs $\mathcal{O}(h_{l+1})$ from (5.1). Hence, it yields a zeroth-order error for the equations. However, since the virtual state is $\mathcal{O}(h_l)$ accurate, the reconstruction gives a first-order accurate virtual state, which yields a flux computation which is first-order accurate. This implies first-order weak consistency.

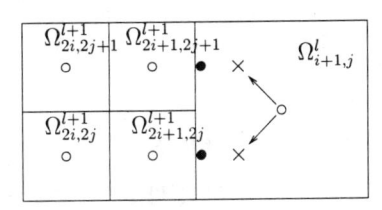

Figure 5.1: Virtual state for first-order weak consistency on a locally refined grid; $\circ$: available state; $\bullet$: left or right state; $\times$: virtual state.

Second-order

A similar situation for second-order weak consistency is found, if the virtual states are computed with the second-order accurate formulas

$$v^{l+1}_{2i+2,2j+1} = \tfrac{3}{4}q^{l+1}_{i+1,j} + \tfrac{1}{4}q^{l+1}_{i,j+1}, \quad (5.4a)$$

$$v^{l+1}_{2i+3,2j+1} = \tfrac{3}{4}q^{l+1}_{i+1,j} + \tfrac{1}{4}q^{l+1}_{i+2,j+1}. \quad (5.4b)$$

Again, the virtual state $v^{l+1}_{2i+2,2j+1}$ is an approximation of the mean of q on $\omega^{l+1}_{2i+2,2j+1}$, given in (5.1). For the virtual state computed in this way, we have

$$\begin{aligned}v^{l+1}_{2i+2,2j+1} &= q_0 + \tfrac{3}{2}h_{l+1}q_1 + \tfrac{1}{2}h_{l+1}q_2 + \tfrac{10}{3}h^2_{l+1}q_3 \\ &+ \tfrac{4}{3}h^2_{l+1}q_5 + \mathcal{O}(h^3_{l+1}).\end{aligned} \quad (5.5)$$

A similar result is obtained for $v^{l+1}_{2i+3,2j+1}$. Apparently, this is a second-order accurate approximation of (5.1). Similar to the first-order weakly consistent situation, the consistency requirements (4.28) are not satisfied for $p = 2$. However, the flux is second-order accurate, hence (5.4) yield second-order weak consistency. These formulas are two of a number of possible choices for second-order accurate virtual states. We have chosen these for their relative *compactness* and their *symmetry*. They are symmetric with respect to the diagonal through the centres of $\Omega^l_{i,j+1}$ and $\Omega^l_{i+1,j}$. A virtual state required for e.g. $\Omega^{l+1}_{2i+2,2j+2}$ with $\partial\Omega_{2i+2,2j+2,S}$, which is a part of the green boundary on level $l+1$, exactly results in (5.4a).

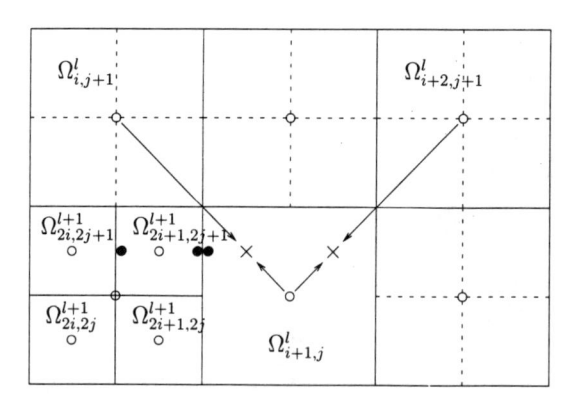

Figure 5.2: Virtual states for second-order weak consistency and first-order consistency on a locally refined grid; ○: available state; ●: left or right state; ×: virtual state.

5.2 Virtual states for consistency

The requirements to be satisfied for a consistent discretisation, are given by (4.28). We consider both first and second-order consistency. A pth-order consistent discretisation for equations for cells near green boundaries requires a $(p+1)$st accurate computation of virtual states.

First-order

First-order consistency can be obtained by second-order accurate computation of virtual states. For this the formulas (5.4) are applied. From (5.1) and (5.5) it is clear that the requirements (4.28) are satisfied for $p = 1$, if they are satisfied by $q^{l+1}_{2i+2,2j+1}$ when $\Omega^{l+1}_{2i+2,2j+1}$ would exist. This is shown to be the case in [17].

Second-order

Similarly a third-order accurate computation of the virtual state is required for second-order consistency. This causes no extra second-order error with respect to the situation where $\Omega^{l+1}_{2i+2,2j+1}$ would exist. As shown in [17], in that situation (i.e. $\Omega^{l+1}_{2i+2,2j+1}$ exists) a second-order accurate discretisation is obtained.

A third-order accurate computation of virtual states is given by

$$v^{l+1}_{2i+2,2j+1} = \tfrac{17}{16}q^l_{i+1,j} + \tfrac{1}{16}\left(q^l_{i,j} + q^l_{i,j+1} + q^l_{i+1,j+1}\right) \\ - \tfrac{2}{16}\left(q^l_{i+2,j} + q^l_{i+1,j-1}\right), \quad (5.6a)$$

$$v^{l+1}_{2i+3,2j+1} = \tfrac{17}{16}q^l_{i+1,j} + \tfrac{1}{16}\left(q^l_{i+2,j} + q^l_{i+2,j+1} + q^l_{i+1,j+1}\right) \\ - \tfrac{2}{16}\left(q^l_{i,j} + q^l_{i+1,j-1}\right). \quad (5.6b)$$

This is schematically represented in Fig. 5.3.

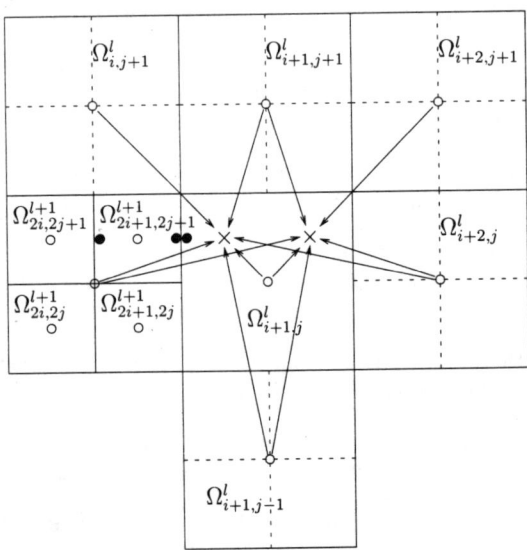

Figure 5.3: Virtual states for second-order consistency on a locally refined grid; ○: available state; ●: left or right state; ×: virtual state.

These are also chosen from a number of possible alternatives. Apart from compactness and symmetry, this choice is based on the size of the in absolute value largest negative coefficients. For the present choice, the negative coefficients are smaller in absolute value than for possible alternatives with similar compactness and symmetry. From the Taylor series expansion of q it can be shown that the virtual state $v_{2i+2,2j+1}^{l+1}$, obtained by (5.6a) can be expressed as

$$v_{2i+2,2j+1}^{l+1} = q_0 + \tfrac{3}{2}h_{l+1}q_1 + \tfrac{1}{2}h_{l+1}q_2 + \tfrac{7}{3}h_{l+1}^2 q_3 + \tfrac{3}{4}h_{l+1}^2 q_4 \\ + \tfrac{1}{3}h_{l+1}^2 q_5 + \tfrac{3}{2}h_{l+1}^3 q_6 + \tfrac{5}{3}h_{l+1}^3 q_7 + 3h_{l+1}^3 q_9 + \mathcal{O}(h_{l+1}^4).$$

Clearly this is a third-order accurate approximation of the mean value of q on $\omega_{2i+2,2j+1}^{l+1}$. Hence, since this virtual state does not introduce additional second-order errors in the reconstruction, the requirements for consistency (4.28) are satisfied for $p = 2$ and the discretisation is second-order consistent.

References

[1] T.J. Barth and P.O. Frederickson. Higher order solution of the Euler equations on unstructered grids, using quadratic reconstruction. In *AIAA-90-0013*, 1990.

[2] T.J. Barth and D.C. Jespersen. The design and application of upwind schemes on unstructured meshes. In *AIAA-89-0366*, 1989.

[3] A. Brandt. Multi-level adaptive techniques (MLAT) for singular perturbation-problems. In P.W. Hemker and J.J.H. Miller, editors, *Numerical Analysis of Singular Perturbation Problems*, pages 53–142. Academic Press, 1979.

[4] S.K. Godunov. Finite difference method for numerical computation of discontinuous solutions of the equations of fluid dynamics. *Math. Sbornik*, 47:271–306, 1959. Translated from Russian: Cornell Aeronautical Laboratory.

[5] W. Hackbusch. *Multi-Grid Methods and Applications*, volume 4 of *Springer Series in Computational Mathematics*. Springer-Verlag, Berlin, 1985.

[6] W. Hackbusch and U. Trottenberg, editors. *Multigrid Methods II, Proc. of the 2nd European Conference on Multigrid Methods, held in Cologne, 1985, Lecture Notes in Mathematics, vol. 1228*. Springer-Verlag, 1986.

[7] P.W. Hemker. Defect correction and higher order schemes for the multi grid solution of the steady Euler equations. In Hackbusch and Trottenberg [6], pages 149–165.

[8] P.W. Hemker and S.P. Spekreijse. Multiple grid and Osher's scheme for the efficient solution of the steady Euler equations. *Appl. Num. Math.*, 2:475–493, 1986.

[9] B. Koren. Defect correction and multigrid for the efficient and accurate computation of airfoil flows. *J. Comput. Phys.*, 77:183–206, 1988.

[10] P.D. Lax. *Hyperbolic Systems of Conservation Laws and the Mathematical Theory of Shock Waves*, volume 11 of *Regional Conference Series in Applied Mathematics*. Society for Industrial and Applied Mathematics, Philadelphia, 1973.

[11] C.W. Oosterlee and P. Wesseling. A multigrid method for an invariant formulation of the incompressible Navier-Stokes equations in general coordinates. *Comm. Appl. Numer. Methods*, 8:721–734, 1992.

[12] S. Osher and F. Solomon. Upwind difference schemes for hyperbolic systems of conservation laws. *Math. Comp.*, 38(158):339–374, 1982.

[13] P.L. Roe. Approximate Riemann solvers, parameter vectors and difference schemes. *J. Comput. Phys.*, 43:357–372, 1981.

[14] G.H. Schmidt and F.J. Jacobs. Adaptive local grid refinement and multi-grid in numerical reservoir simulation. *J. Comput. Phys.*, 77:140–165, 1988.

[15] S.P. Spekreijse. Multigrid solution of monotone second-order discretizations of hyperbolic conservation laws. *Math. Comp.*, 49(179):135–155, 1986.

[16] S.P. Spekreijse. Second order accurate upwind solutions of the 2D steady Euler equations by the use of a defect correction method. In Hackbusch and Trottenberg [6], pages 285–300.

[17] S.P. Spekreijse. *Multigrid Solution of the Steady Euler Equations*. CWI, Amsterdam, 1988. CWI-tract 46.

[18] J.L. Steger and R.F. Warming. Flux vector splitting in the inviscid gasdynamic equations with application to finite-difference methods. *J. Comput. Phys.*, 40:263–293, 1981.

[19] P.K. Sweby. High resolution schemes using flux limiters for hyperbolic conservation laws. *SIAM J. Numer. Anal.*, 21:995–1011, 1984.

[20] H.T.M. van der Maarel. *A Local Grid Refinement Method for the Euler Equations*. PhD thesis, University of Amsterdam, 1993. To appear as CWI Tract, CWI, Amsterdam, 1994.

[21] B. van Leer. Towards the ultimate conservative difference scheme. IV. A new approach to numerical convection. *J. Comput. Phys.*, 23:276–299, 1977.

[22] B. van Leer. Flux-vector splitting for the Euler equations. In E. Krause, editor, *Proceedings of the Eight International Conference on Numerical Methods in Fluid Dynamics, Aachen, 1982, Lecture Notes in Physics, vol. 170*, pages 507–512. Springer-Verlag, 1982.

[23] B. van Leer. Upwind difference methods for aerodynamic problems governed by the Euler equations. In B.E. Engquist, S.J. Osher, and R.C.J. Sommerville, editors, *Large Scale Computations, Lectures in Applied Mathematics part 2*, pages 327–336. AMS, 1985.

[24] B. van Leer. Towards the ultimate conservative difference scheme. V. A second order sequel to Godunov's method. *J. Comput. Phys.*, 32:101–136, 79.

On the solution of nonlinear equations for nondifferentiable mappings

Owe Axelsson
Department of Mathematics
University of Nijmegen
NL-6525 ED Nijmegen
The Netherlands

Igor E. Kaporin
Department of Computational Mathematics
and Cybernetics
Moscow State University
119899 Moscow, Russia

Summary

An efficient version of an inexact Newton method to solve nonlinear systems of equations with automatic choice of steplength is presented. The method can be combined with minimization over subspaces and an optimal scaling of the approximate Newton direction vector is achieved. The method converges globally even if the Frechet derivative of the nonlinear mapping is singular or does not exist. The method is applied for the solution of the driven cavity problem on a square domain and the onset of turbulence is discussed.

Introduction

Let $\|\cdot\| = (\cdot,\cdot)^{\frac{1}{2}}$ where $(\cdot,\cdot)$ is an innerproduct in $\mathbf{R}^n$ and consider the mapping $F: \mathbf{R}^n \to \mathbf{R}^n$, which is assumed to the Lipschitz continuous:

$$\|F(x) - F(y)\| \leq \Gamma \|x-y\|$$

for all x, y in some sufficiently large balls about the iteration points. In general, F is not differentiable. We want to solve $F(x) = 0$. Such systems arise in many contexts, for instance for certain partial differential equations, modelling the displacement of bodies in contact through friction and discretized by finite difference or finite element methods. The solution of $F(x) = 0$ can be embedded in an evolutionary problem.

If F is a strongly monotone mapping, that is, if there exists $m > 0$ such that

$$\bigl(F(x) - F(y), x-y\bigr) \geq m(x-y, x-y) \qquad (1.1)$$

for all x, y in some sufficiently large ball S, then all solutions of the evolution equation

$$x'(t) = -F\bigl(x(t)\bigr), \quad t > 0, \qquad (1.2)$$

where $x(0)$ is given, converges as $t \to \infty$ to a stationary solution, i.e. a solution of $F(x) = 0$, which in this case is unique.

For the numerical computation, equation (1.2) must be discretized by some numerical integration method, such as the Euler forward method:

$$x^{k+1} = x^k - \tau_k F(x^k), \quad k = 0, 1, \ldots \tag{1.3}$$

where $x^0 = x(0)$, x^k is the corresponding approximation of $x(t_k)$, and $t_{k+1} = t_k + \tau_k$, $\tau_k > 0$, where $\{\tau_k\}$ is a sequence of stepsizes. If $\tau_k = \tau$ is fixed, then (1.3) takes the form of a fixed point mapping

$$x^{k+1} = G_\tau(x^k), \quad k = 0, 1, \ldots \tag{1.4}$$

where $G_\tau(x) = x - \tau F(x)$. (In fact, x is a solution of $F(x) = 0$ if and only if x is a fixed point of G_τ for some $\tau \neq 0$.)
It is well known and readily seen that (1.4) converges if G_τ is a contractive mapping, that is,

$$\|G_\tau(x) - G_\tau(y)\| \leq \alpha \|x-y\|, \quad \alpha < 1,$$

for all $x, y \in S$. In such a case,

$$\|x^{k+1} - x^k\| \leq \alpha \|x^k - x^{k-1}\|,$$

and it follows that $\{x^k\}$ is a Cauchy sequence, a limit point $\widehat{x}$ exists and

$$\|\widehat{x} - x^k\| \leq \frac{\alpha^k}{1-\alpha} \|x^1 - x^0\|.$$

Given a strongly monotone and Lipschitz continuous mapping F and using the relation $G_\tau = x - \tau F(x)$, we find by an elementary computation,

$$\|G_\tau(x) - G_\tau(y)\|^2 \leq (1 - 2\tau m + \tau^2 \Gamma^2) \|x-y\|^2, \tag{1.5}$$

that is, G_τ is contractive if $\tau < \frac{2m}{\Gamma^2}$. (This gives an alternative proof of existence and unicity of solutions of $F(x) = 0$, for such mappings.) The upper bound in (1.5) is minimized if $\tau = \frac{m}{\Gamma^2}$, in which case the contraction factor $\alpha = \sqrt{1-(m/\Gamma)^2}$. In general m/Γ is a small number, so the convergence can be very slow. Furthermore, the assumption (1.1) is quite restrictive.

A faster rate of convergence can be expected when F is differentiable and $F'(x(t))$ is nonsingular and one solves

$$x'(t) = -F'(x(t))^{-1} F(x(t)), \quad t > 0, \tag{1.6}$$

and $x(0)$ is given. Here $F'(x)$ is the Frechet derivative, i.e., in our case, an $n \times n$ matrix, frequently called the Jacobian. The solution to the differential equation satisfies $F(x(t)) = e^{-t} F(x(0))$. It can be shown further that if F is twice continuously differentiable with bounded derivatives and if $F'(x(t))^{-1}$ is bounded and $C_1 \|x(t_0) - x_\infty\| < 1$, where C_1 depends on the above bounds and $x(t) \to x_\infty$, $t \to \infty$ then

$$\|x(t) - x_\infty\| \leq C\epsilon^{-t}/(1 + C\epsilon^{-t}), \quad t > t_0,$$

for some constant C.

Hence (1.6) can be an efficient continuation method for the solution of $F(x) = 0$. However, for the numerical computation, the differential equation in (1.6) must be discretized. Using the Euler forward method, one finds

$$x^{k+1} = x^k - \tau_k F'(x^k)^{-1} F(x^k), \quad k = 0, 1, \ldots \tag{1.7}$$

This is the classical Newton-Kantorovich method with stepsizes $\{\tau_k\}$. When $\tau_k = 1$, the method can be shown to converge, but only locally in general, if $\|F'(x^k)^{-1}\|$ is bounded and F' is Hölder continuous. In this case the initial vector must already be sufficiently close to the solution. The rate of convergence is superlinear, $\|\hat{x} - x^{k+1}\| = O(\|\hat{x} - x^k\|^{1+\gamma})$, where $\gamma > 0$ is the exponent in the Hölder continuity condition.

There are many papers devoted to this topic. For references, see [O,R; D.S; Y] for instance. When x^0 is not sufficiently close to the solution, the stepsizes must be chosen sufficiently small initially to get convergence. The initial rate of convergence is linear. Eventually, when the approximations reaches some sufficiently small ball about the solution, the superlinear rate of convergence starts.

Normally, the computational effort in solving the systems of the form $F'(x^k)p^k = -F(x^k)$, which appears in (1.7) is large and when the approximation x^k is far from the solution, it is not cost efficient to solve this system exactly. Instead, one can compute a vector p^k for which

$$\|F'(x^k)p^k + F(x^k)\| \leq \rho_k \|F(x^k)\|, \tag{1.8}$$

where $0 < \rho_k < 1$, i.e. ρ_k controls the relative error in solving $F'(x^k)p^k = -F(x^k)$ approximately.

In addition to saving computational labor, this formulation enables one to treat singular, or arbitrarily nearly singular, matrices $F'(x^k)$.

Such methods have been presented and analysed in [D.E.S.; B.R.; Y] and in [A,a]. Here the method is presented in a more general form where the choices of search vectors and steplengths are clarified. It is based mainly on results in [A,b] and [K.A.] but with a simplified presentation and with the method extended to the case of nondifferentiable mappings.

The remainder of the paper is organized as follows. In the next section the algorithm is presented and some alternative implementations are discussed. In the following section, the global convergence of the method is shown. An application of the method to solve the driven cavity problem for Navier Stokes equations is given in the final section.

The presentation here is based to a major part on [K.A.].

2. The damped inexact Newton algorithm for nondifferentiable mappings

Consider the solution of $F(x) = 0$, $F : \mathbf{R}^n \to \mathbf{R}^n$, where F is not differentiable in general. Let $K : \mathbf{R}^n \to \mathbf{R}^n$ be another, but differentiable, mapping such that (cf. [Z])

$$\|F(x) - K(x) - F(y) + K(y)\| \leq \delta \|x - y\|, \tag{2.1}$$

where $\delta > 0$ must be sufficiently small (see below), for all x, y in sufficiently large balls around the iteration points x^k.

Consider first the following prototype of the damped inexact Newton (DIN) algorithm:

1a. Compute a vector p^k s.t.

$$\|F(x^k) + K'(x^k)p^k\| \le \rho_k \|F(x^k)\|, \qquad (2.2)$$

for some ρ_k, $0 \le \rho_k < 1$.

1b. Normalize p^k:

Let

$$c_k = -\frac{F(x^k)^T K'(x^k) p^k}{\|F(x^k)\|\,\|K'(x^k)p^k\|}, \qquad (2.3)$$

$$t_k = c_k \frac{\|F(x^k)\|}{\|K'(x^k)p^k\|}$$

and let

$$\widehat{p}^k = t_k p^k.$$

2a. Let $x^{k+1} = x^k + \sum_{j=0}^{r_k} \alpha_j^{(k)} \widehat{p}^{k-j}$, $0 \le r_k \le \min(r,k)$, where $r \ge 0$ and the real coefficients $\alpha_j^{(k)}$ are computed such that

$$\|F(x^{k+1})\| \le \sqrt{1 - \tau c_k^2}\,\|F(x^k)\| \qquad (2.4)$$

for some $\tau > 0$.

2b. The search for an approximation x^{k+1} in 2a. takes normally place as a line search along the line between $\widehat{x}^k$ and $\widehat{x}^{k+1} + \widehat{p}^k$. However, it is possible to improve the approximation found in 2a. using a sequence of successive line searches along $\widehat{p}^k$, $\widehat{p}^{k-1}$, $\widehat{p}^{k-2}$, etc, each search improving the current value of x^{k+1}, by making the norm of the current residual smaller.

We now comment on each of the above steps.

Step 1a. In general, the vector p^k in (2.1) is not scaling invariant. However, the vector can be computed using a (preconditioned) minimum residual conjugate gradient method, for instance. Step 1a. can then be deleted as it can, in general, if one computes the vector $\widehat{p}_k$ from a least square minimum residual method on subspaces, as will be shown later.

Step 1b. Note that the number c_k is the cosine of the acute angle between the vectors $-F(x^k)$ and $K'(x^k)p^k$.

The normalization, $\widehat{p}^k = t_k p^k$ is done such that

$$\|K'(x^k)\widehat{p}^k\| = c_k \|F(x^k)\|, \qquad (2.5)$$

i.e.

$$t_k = \frac{c_k \|F(x^k)\|}{\|K'(x^k)p^k\|}.$$

This is illustrated in Figure 1, where $c_k = \cos(\alpha_k)$.

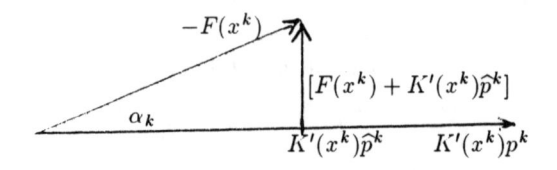

Figure 1. The optimality of $\widehat{p}^k$.

In this way
$$\|F(x^k) + K'(x^k)\widehat{p}^k\| \le \min_t \|F(x^k) + tK'(x^k)p^k\|.$$

Also
$$\|\widehat{p}^k\| = t_k\|p^k\| = c_k\beta_k\|F(x^k)\| \tag{2.6}$$

where
$$\beta_k = \frac{\|p^k\|}{\|K'(x^k)p^k\|}.$$

Since
$$\beta_k = \frac{\|K'(x^k)^{-1}K(x^k)p^k\|}{\|K'(x^k)p^k\|}$$

if $K'(x^k)$ is nonsingular, it holds

$$\beta_k \le \|K'(x^k)^{-1}\|.$$

However, a finite value of β_k may exist even if $K'(x^k)$ is singular. If $K'(x^k)$ is singular, or close to singular, we assume that the direction p^k has been computed such that $\|K'(x^k)p^k\|$ is not very small, i.e. p^k should not have dominating components along the eigenvectors corresponding to the smallest eigenvalues of $K'(x^k)$. Problems with singular or close to singular Jacobians occur in many contexts, such as for turning point and bifurcation problems. Methods to solve certain singular or nearly singular systems efficiently have been discussed in [A.N.P.].

Finally, let
$$\widehat{\rho}_k = \|F(x^k) + K'(x^k)\widehat{p}^k\|/\|F(x^k)\|.$$

Then
$$\widehat{\rho}_k = \sqrt{1 - c_k^2},$$

i.e. $\widehat{\rho}_k$ is the sine for the acute angle α_k in Figure 1. Note that any method which minimizes the residual $\|F(x^k) + K'(x^k)p^k\|$ on some subspace will automatically give a vector $\widehat{p}^k$, normalized as in 1b.

For instance, if we seek p^k in a subspace V of dimension m, $m \ll n$, we let

$$U = K'(x^k)V$$

and compute
$$p^k = -V(U^T U)^{-1} U^T F(x^k).$$

(Here we have assumed that U has full column rank. Otherwise we must use a generalized inverse of $U^T U$.) Note that $K'(x^k)p^k = -QF(x^k)$, where $Q = U(U^T U)^{-1} U$ is the orthogonal projector onto the subspace spanned by the columns of U. Such a choice was made in [B.S.] and [K.A.] and is identical to the Gauss-Newton algorithm for the solution of, in general low-dimensional, least squares problems (see e.g. [P.], [D.H.]),

$$\min_{y \in \mathbb{R}^m} \|F(x^k + Vy)\|.$$

For such a vector p^k we have

$$c_k = \cos(-F(x^k), K'(x^k)p^k) = \frac{\|QF(x^k)\|}{\|F(x^k)\|} =$$
$$= \sqrt{1 - \frac{\|(I-Q)F(x^k)\|^2}{\|F(x^k)\|^2}} = \sqrt{1 - \min_{p \in \text{span} V} \frac{\|F(x^k) + K'(x^k)p\|}{\|F(x^k)\|^2}}.$$

Hence $p^k = \hat{p}^k$, i.e. is automatically normalized as in Step 1b.
Possible choices of V will be discussed later.

Step 2a. If $\hat{p}^k$ is computed as an orthogonal projector of $-F(x^k)$ on a subspace as shown above, then there is no need to form a linear combination of vectors as in 2a. Instead we simply search for a steplength τ for which $\|F(x^k + \tau \hat{p}^k)\|$ is sufficiently small. The above mentioned subspace can include also some of the previous search directions. It will later (Theorem 2.1) be shown that there exists $\tau > 0$ such that

$$\|F(x^k + \tau \hat{p}^k)\| \le \sqrt{1 - \tau(1 - \varepsilon/c_k) c_k^2} \|F(x^k)\|, \qquad (2.7)$$

for some ε, $0 < \varepsilon < \delta_k$. Assume, for simplicity that $\varepsilon/c_k \le \frac{1}{2}$. The computation of a vector x^{k+1} can therefore take place using backtracking in the following way (see [K.A.]):

Let $\tau_{i,k} = 2^{-i}$, $i = 0, 1, \ldots$ and let s be the smallest integer such that

$$\|F(x^k + \tau_{s,k} \hat{p}^k)\|^2 \le (1 - \tau_{s,k} c_k^2) \|F(x^k)\|^2 \qquad (2.7a)$$

but

$$\|(F(x^k + \tau_{s-1,k} \hat{p}^k)\|^2 > (1 - \tau_{s-1,k} c_k^2) \|F(x^k)\|^2. \qquad (2.7b)$$

The number of backtracking steps is bounded by $s < \log_2 \frac{2}{\tau_{s,k}}$.

Step 2b. This step can be performed in several ways. Using the values of $\|F(x^k + \tau_{i,k} \hat{p}^k)\|$, one can use interpolation and find the value τ_k' for which the minimal value of the interpolant is taken. If $\|F(x^k + \tau_k' \hat{p}^k)\|$ is smaller than $\|F(x^k + \tau_{s,k} \hat{p}^k)\|$, the new argument replaces x^{k+1}, i.e. one lets $x^{k+1} = x^k + \tau_k' \hat{p}^k$. Similar approximate minimizations can take place along some of the directions $\hat{p}^{k-1}$, $\hat{p}^{k-2}$, etc.
It will now be shown that there exists $\tau > 0$ such that (2.4) and hence (2.7), holds if δ in (2.1) is sufficiently small. It will also be shown that we can get a lower bound on

c_k if the vector $\widehat{p}^k$ is computed from a Krylov subspace, possibly extended with some previous search direction vectors.

Theorem 2.1. Let $F, K : \mathbb{R}^n \to \mathbb{R}^n$, let K be differentiable and assume that (2.1) holds for some sufficiently small $\delta = \delta_k$ (see below) in a sufficiently large ball about x^k.

Assume also that K' is Lipschitz continuous with relative Lipschitz constant Γ_k,

$$\|\int_0^1 [K'(x^k + t\tau \widehat{p}^k) - K'(x^k)]K'(x^k)^{-1} dt\| \le \frac{1}{2}\Gamma_k \tau \|\widehat{p}^k\| \tag{2.8}$$

for all τ, $0 < \tau \le 1$. Let $\widehat{p}^k$ be a vector such that

$$\|K'(x^k)\widehat{p}^k\| = c_k \|F(x^k)\|,$$

where c_k is the acute angle between $-F(x^k)$ and $K'(x^k)\widehat{p}^k$. Assume further that β_k exists such that

$$\frac{\|\widehat{p}^k\|}{\|K'(x^k)\widehat{p}^k\|} \le \beta_k.$$

Then if (2.1), or more precisely (2.11a) below, holds with $\delta_k \le \varepsilon \beta_k^{-1}$, $0 < \varepsilon < c_k$ there exists $\tau > 0$ such that

$$\|F(x^k + \tau \widehat{p}^k)\| \le \sqrt{1 - (1 - \frac{\varepsilon}{c_k})\tau c_k^2} \|F(x^k)\|. \tag{2.9}$$

In particular, (2.9) holds for $0 < \tau \le \tau_k^0 = \frac{1 - \varepsilon/c_k}{1 + a_k}$ where $a_k = \Gamma_k \beta_k \|F(x^k)\|$.

Proof. To prove the global convergence estimate (2.9) we write

$$\begin{aligned} F(x^k + \tau \widehat{p}^k) =& [F(x^k) + \tau K'(x^k)\widehat{p}^k] \\ &+ [F(x^k + \tau \widehat{p}^k) - K(x^k + \tau \widehat{p}^k) - F(x^k) + K(x^k)] \\ &+ [K(x^k + \tau \widehat{p}^k) - K(x^k) - \tau K'(x^k)\widehat{p}^k]. \end{aligned} \tag{2.10}$$

The three terms within brackets are estimated as follows:

$$\begin{aligned} \|F(x^k) + \tau K'(x^k)\widehat{p}^k\| =& \{\|F(x^k)\|^2 + 2\tau F(x^k)^T K'(x^k)\widehat{p}^k \\ &+ \tau^2 \|K'(x^k)\widehat{p}^k\|^2\}^{\frac{1}{2}} \\ =& \sqrt{1 - 2\tau c_k^2 + \tau^2 c_k^2} \|F(x^k)\|, \end{aligned} \tag{2.11a}$$

where we have used (2.5) and (2.3);

$$\begin{aligned} \|F(x^k + \tau \widehat{p}^k) - K(x^k + \tau \widehat{p}^k) - F(x^k) + K(x^k)\| & \\ \le \tau \delta_k \|\widehat{p}^k\| \le \tau_k c_k \delta_k \beta_k \|F(x^k)\| \le \tau_k c_k \varepsilon \|F(x^k)\| & \end{aligned} \tag{2.11b}$$

where we have used (2.1), (2.6) and the assumed bound on δ_k;

$$\|K(x^k + \tau \widehat{p}^k) - K(x^k) - \tau K'(x^k)\widehat{p}^k\| \tag{2.11c}$$

$$= \|\int_0^1 [K'(x^k + t\tau \widehat{p}^k) - K'(x^k)]\tau \widehat{p}^k dt\|$$

$$= \|\int_0^1 [K'(x^k + t\tau \widehat{p}^k) - K'(x^k)]K'(x^k)^{-1} K'(x^k)\tau \widehat{p}^k dt\|$$

$$\le \frac{1}{2}\Gamma_k \tau^2 \|\widehat{p}_k\| \|K'(x^k)\widehat{p}^k\| \le \frac{1}{2}\tau^2 c_k^2 \Gamma_k \beta_k \|F(x^k)\|^2,$$

where (2.8), (2.5) and (2.6) have been used.
By (2.10) and (2.11a,b,c) we find then, using the triangle inequality,

$$\|F(x^k+\tau\hat{p}^k)\| \leq \{\sqrt{1-2\tau c_k^2+\tau^2 c_k^2} + \tau c_k \varepsilon + \frac{1}{2}\tau^2 c_k^2 a_k\}\|F(x^k)\|, \quad (2.12)$$

where $a_k = \Gamma_k \beta_k \|F(x^k)\|$. We minimize now the convergence factor in the r.h.s.,

$$\varphi(\tau) = \sqrt{1-2\tau c_k^2+\tau^2 c_k^2} + \tau c_k \varepsilon + \frac{1}{2}\tau^2 c_k^2 a_k.$$

If $c_k < 1$ we find

$$\min_{0 \leq \tau \leq 1} \varphi(\tau) = \varphi(\hat{\tau}_k),$$

where $\hat{\tau}_k$ is the solution of

$$\tau = g(\tau) \equiv \frac{1 - \varepsilon/c_k \sqrt{1 - 2\tau c_k^2 + \tau^2 c_k^2}}{1 + a_k \sqrt{1-2\tau c_k^2+\tau^2 c_k^2}}.$$

It can be seen that $g'(\tau)$ is positive in the interval $[0, 1)$ and decreases monotonically with $g'(1) = 0$ (see figure 2). If $c_k = 1$ then $\varphi(\tau) = 1 - \tau + \tau\varepsilon + \frac{1}{2}\tau^2 a_k$, so $\hat{\tau}_k = \min(1, (1-\varepsilon)/a_k)$.

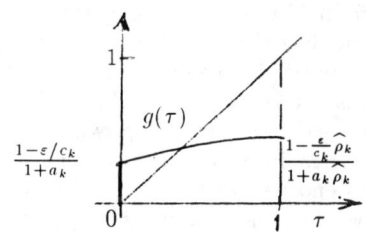

Figure 2. Existence of a fixed point

When a_k is large, we can use $\tau_k^0 = \frac{1-\varepsilon/c_k}{1+a_k}$ as a sufficiently accurate lower bound of $\hat{\tau}_k$ and the above shows that

$$\min_{0 \leq \tau \leq 1} \varphi(\tau) \leq \varphi(\tau_k^0) =$$

$$\sqrt{1-2\frac{(1-\varepsilon/c_k)}{1+a_k}c_k^2 + \left(\frac{(1-\varepsilon/c_k)c_k}{1+a_k}\right)^2} + \frac{1-\varepsilon/c_k}{1+a_k}\varepsilon c_k + \frac{1}{2}\left(\frac{(1-\varepsilon/c_k)c_k}{1+a_k}\right)^2 a_k. \quad (2.13)$$

We use now the elementary inequality

$$\sqrt{1-\zeta} + \frac{1}{2}\eta \leq \sqrt{1-\zeta+\eta},$$

which holds if $0 \leq \frac{1}{2}\eta \leq \zeta \leq 1$, to find

$$\varphi(\tau_k^0) \leq \sqrt{1 - \frac{(1-\varepsilon/c_k)^2}{1+a_k}c_k^2}.$$

Similarly, it follows from (2.12)

$$\|F(x^k+\tau\hat{p}^k)\| \leq \sqrt{1 - (1-\frac{\varepsilon}{c_k})\tau c_k^2}\,\|F(x^k)\|$$

if $0 < \tau \leq \tau_k^0$. $\square$

Asymptotic convergence factor:

Recall that $a_k = \Gamma_k \beta_k \|F(x^k)\|$. If $\Gamma_k \beta_k$ is bounded for all k, the convergence factor $\varphi(\tau_k^0)$ becomes eventually $\approx \sqrt{2\varepsilon - \varepsilon^2}$, as $\|F(x^k)\| \to 0$ and $c_k \to 1$. It is seen from (2.13) that if $\varepsilon = 0$ ($\delta = 0$, i.e. $K(x) = F(x)$) and $\hat{\rho}_k^2 = 1 - c_k^2 = O(a_k^2)$, then the convergence factor is

$$\varphi(\tau_k^0) \leq \sqrt{a_k^2 + \hat{\rho}_k^2} + \frac{1}{2}a_k = O(a_k) = O(\|F(x^k)\|), \quad \|F(x^k)\| \to 0.$$

This means that, under the above conditions, the convergence rate becomes eventually quadratic.

The theorem shows that the steplength is limited by $\hat{\tau}_k \simeq \tau_k^0 = \frac{1-\varepsilon/c_k}{1+a_k}$. If there is a good estimate available of this number, the backtracking in step 2a can start with a stepsize, say $4\tau_{k,}^0$, in order to save some evaluations of $F(x^k + \tau_{i,k}\hat{p}^k)$. If (2.7a) is satisfied already for this value, one can use some forward tracking steps, i.e. increase the current value of $\tau_{i,k}$ until a step is found where (2.7b) holds.

If F is differentiable we can let $K = F$ in which case $\varepsilon = 0$. If F is singular we may perturb F slightly to form K which is nonsingular. The mapping K may change from step to step. It must only be differentiable along each of the lines between x^k and $x^k + \hat{p}^k$, $k = 0, 1, \ldots$. If F corresponds to a discretization of $\nabla(a(\nabla u)^2)\nabla u)$, where the coefficient a is a discontinuous function for some value of $|\nabla u|$, we can let K correspond to $\nabla(\tilde{a}(|\nabla u|^2)\nabla u)$, where $\tilde{a}$ is some smoothed approximation of a.

The final problem which remains is to show that one can guarantee a lower bound on the numbers c_k. To this end we use a Krylov subspace and constant p^k so that $K'(x^k)p^k$ becomes an orthogonal projection of $-F(x^k)$ (see comments on Step 1b). It turns out that it can be efficient to add to the Krylov subspace also some previous search direction vectors, which was suggested in [A.b]. Hence, following [K.A.] let

$$m_k = 1 + \min(k, l),$$

where l is the maximum number of previous directions used and

$$V_k = [q_k, x_k - v_{k-1}, \ldots, x_{k-m_k+2} - x_{k-m_k+1}]. \tag{2.14}$$

Hence q_k is an approximate Newton direction, for instance obtained by using s preconditioned (conjugate gradient minimal residual) iterations to solve

$$K'(x^k)q = -F(x^k).$$

Hence $q_k \in \text{span}\{[C_k K'(x_k)]^{j-1} C_k F(x_k)\}_{j=1}^s$ where C_k is a preconditioner to $K'(x_k)$. (If we let $C_k = A_k K'(x_k)^T$, then A_k^{-1} should be an approximation of $K'(x_k)^T K'(x_k)$. The latter choice can be efficient if $K(x_k)$ is nonsymmetric.) Alternatively, we can let

$$V_k = \{C_k F(x_k), \ldots, [C_k F(x_k)]^{s-1} C_k F(x_k), x_k - x_{k-1}, \ldots x_{k-m_k+2} - x_{k-m_k+1}\}.$$

We derive now a lower bound of c_k. For any $p \in \text{span}(V_k)$ we have

$$F(x^k) + K'(x^k)p = \pi_s(M_k)F(x^k) + Wu$$

where
$$M_k = K'(x^k)C_k,$$

π_s is a polynomial of degree s, normalized at the origin as $\pi_s(0) = 1$. W is the $n \times (m_k-1)$ matrix
$$W = [K'(x^k)(x_k - x_{k-1}), \ldots, K'(x^k)(x_{k-m_k+2} - x_{k-m_k+1})],$$

and u is a $\mathbb{R}^{m_k-1}$ vector.

By construction
$$\begin{aligned}
c_k^2 &= \cos(-F(x^k), K'(x^k)\widehat{p}^k) \\
&= 1 - \min_{p \in \text{span}(V_k)} \frac{\|F(x^k) + K'(x^k)p\|^2}{\|F(x^k)\|^2} \\
&= 1 - \min_{p \in \pi_k, q \in \mathbb{R}^{m_k-1}} \frac{\|\pi(M_k)F(x^k) + Wq\|}{\|F(x^k)\|^2}.
\end{aligned}$$

As shown in [K.A.], this yields
$$c_k^2 \geq 1 - \min_{p \in \pi_k} \|P\pi(M_k)\|^2 \geq 1 - \min_{p \in \pi_k} \|\pi(M_k)\|^2,$$

where $P = I - W(W^T W)^{-1} W^T$, a projection matrix.

As was proposed in [A.a] and [B.S.], a proper choice of q in (2.14) is the s'th iterate of the minimum residual algorithm. Then, using well known bounds for Chebyshev polynomials, we find
$$c_k^2 \geq 1 - \frac{4}{\left[\left(\frac{\sqrt{\kappa}+1}{\sqrt{\kappa}-1}\right)^s + \left(\frac{\sqrt{\kappa}-1}{\sqrt{\kappa}+1}\right)^s\right]^2}$$

where $\kappa_k = \text{cond}(M_k) = \text{cond}(K'(x^k)C_k K'(x^k)^T)$. Hence the sequence $\{c_k\}$ is bounded from below if there exists an upper bound of the condition number K_k of M_k.

3. The driven cavity flow problem

We consider now the problem of steady flow of an incompressible viscous fluid in a square cavity. The flow is described by the Navier-Stokes equations and using the streamfunction ψ and vorticity ζ, the equations take the form

$$\left. \begin{aligned} -\nu \Delta \zeta + u\zeta_x + v\zeta_y &= 0 \\ \zeta + \Delta \psi &= 0 \end{aligned} \right\} \qquad (3.1)$$

where $u = \psi_y$, $v = -\psi_x$ are the velocity components. The flow region Ω is defined by $0 < x < 1$, $0 < y < 1$. The flow is induced by the sliding motion of the top wall ($y = 1$) from left to right. The boundary conditions on the boundary are $u = 1$, $v = 0$ on the sliding wall and the no-slip conditions $\psi = 0$, $\frac{\partial \psi}{\partial n} = 0$, on the side and bottom walls.

Equation (3.1) can be written as a fourth order problem in the stream function alone,
$$\Delta^2\psi + c(\psi_y(\Delta\psi)_x - \psi_x(\Delta\psi)_y) = 0, \quad (x,y) \in \Omega \tag{3.2}$$
where $c = \frac{1}{\nu}$, the non-dimensional Reynold's number.

A uniform staggered grid with the stepsize $h = \frac{1}{m+2}$, that is, the set of points
$$\{(\xi_i, \eta_j) = (ih + \frac{h}{2}, jh + \frac{h}{2}) : -1 \le i \le m+2, \; -1 \le j \le m+2\}, \tag{3.3}$$
was used and standard central differences were employed in deriving the following finite difference equations:

$$\begin{aligned}
& 20 y_{i,j} - 8(y_{i-1,j} + y_{i+1,j} + y_{i,j-1} + y_{i,j+1}) \\
& + 2(y_{i-1,j} + y_{i+1,j-1} + y_{i-1,j-1} + y_{i+1,j+1}) + y_{i-2,j} + y_{i+2,j} + y_{i,j-2} + y_{i,j+2} \\
& + \frac{c}{4}(y_{i,j+1} - y_{i,j-1})(y_{i-2,j} + y_{i-1,j-1} + y_{i-1,j+1} - 4y_{i-1,j} + 4y_{i+1,j} \\
& \qquad - y_{i+1,j-1} - y_{i+1,j+1} - y_{i+2,j}) \\
& - \frac{c}{4}(y_{i+1,j} - y_{i-1,j})(y_{i,j-2} + y_{i-1,j-1} + y_{i+1,j-1} - 4y_{i,j-1} + 4y_{i,j+1} \\
& \qquad - y_{i-1,j+1} - y_{i+1,j+1} - y_{i,j+2}) = 0, \quad (i,j) \in \Omega_h,
\end{aligned} \tag{3.4}$$

with boundary conditions
$$y_{i,j} = 0, \quad (i,j) \notin \Omega_h \text{ and } j \le m,$$
$$-y_{i,m+1} = y_{i,m+2} = \frac{h}{2}, \quad 0 \le i \le m+1. \tag{3.5}$$

Here the index set Ω_h is defined by (3.3).

The discrete boundary conditions were obtained from difference approximations of the type, say, for the top boundary
$$\frac{y_{i,m+2} + y_{i,m+1}}{2} = 0, \qquad \frac{y_{i,m+2} - y_{i,m+1}}{h} = 1,$$
where $i = 0, 1, \ldots, m+1$; each such pair of equations can be easily solved, which gives (3.5). The advantage of the above discretization is that the diagonal part of the Jacobian of the system, resulting after elimination of (3.5) from (3.4), is not affected by any terms equal to the Reynolds number multiplied by some quantity of unpredictable sign, as is the case when using standard uniform grid and boundary conditions (for the same boundary line)
$$y_{i,m+1} = 0, \qquad \frac{y_{i,m+2} - y_{i,m}}{2h} = 1,$$
as was done in [S.K,b]. The set of nonlinear equations of order $n = m \times m$ was then generated in the same way as above.

No spurious solutions, that is, non-physical ones but delivering very small residuals to the equation $f(x) = 0$, as in [S.K,a] were revealed during the present test runs most likely due to the use of different boundary conditions.

The initial guess was chosen to be zero for all test runs, and, following [B.S.], the exact inverse of the Jacobian corresponding to $c = 0$ was chosen as the preconditioner. For that purpose, the exact Cholesky factorization using the nested dissection reordering of the resulting discrete biharmonic operator was computed before starting the iterations and only forward and back substitutions were further employed.

The following table gives the performance measures of some testruns. Here $\varepsilon = 10^{-8}$ was used as iterative residual norm tolerance. s denotes the dimension of the Krylov subspace and l is the number of previously computed directions used.

Table 3.1. Performance measures for $c = 5000$, $n = 961$

s, l	# outer iterations	total # of mult. of precond. by a vector	# Jacobian evaluations	# function evaluations	linear algebra operations per unknowns (excl. for K' and F)
16,2	135	2160	397	412	213484
24,2	93	2232	285	295	266129
24,4	89	2136	274	285	274097

In figures 3.1 and 3.2 the contour plots of the stream function for $c = 5000$ and $c = 10000$, respectively, on a 127×127 mesh are shown.

Using a linearized model, in [G.] it has been shown that the corner singularities causes an infinite number of steadily smaller vortices (eddies) in the lower two corners of the cavity. The intensities of these decay with a factor of the order 10^{-4}. Hence, in practice only the first few of them have any practical significance. As argued in [G.H.L.] a physical flow may not behave as a truly infinite dimensional flow. Rather what is sought is a very large finite dimensional system.

Similar to [S.K.] we consider now the appearance of turning points when the flow was studied by continuation in the Reynolds number parameter. For $h = 1/16$ the value $\psi(\frac{1}{2}, \frac{1}{2})$ of the stream function at the center point was followed which revealed the typical S-shaped graph.. The results indicate a laminar flow for $c < c_1$ for $c_1 \simeq 5502$, $\psi_1 \simeq -0.317$ and a 'supersonic' behaviour for $c > c_2$ for $c_2 \simeq 5136$, $\psi_2 \simeq -0.823$. The approximate limit for value of $\psi(\frac{1}{2}, \frac{1}{2})$ very large values of c was found to be -3.7. For $c_1 < c < c_2$ there exists three solutions and for this region of Reynold numbers, the flow can be referred to as transonic or turbulent.

However, for finer meshes, the values of c_1, c_2 increased fast, for a 32×32 mesh we found $c_1 = 19010$, $\psi_1 = -0.0744$. This indicates a strong dependence on the second (mesh) parameter also. It is claimed in [S.K.b] that all limit points dissappeared when $h < 1/100$. (In their paper, the vortex center of the primary vortex was traced.) The dissappearance of limit points for sufficiently fine meshes would show the existence of spurious (i.e. nonphysical) solutions for coarser meshes and a unique solution for sufficiently fine meshes. We intend to make a further study with our approximation of this phenomena.

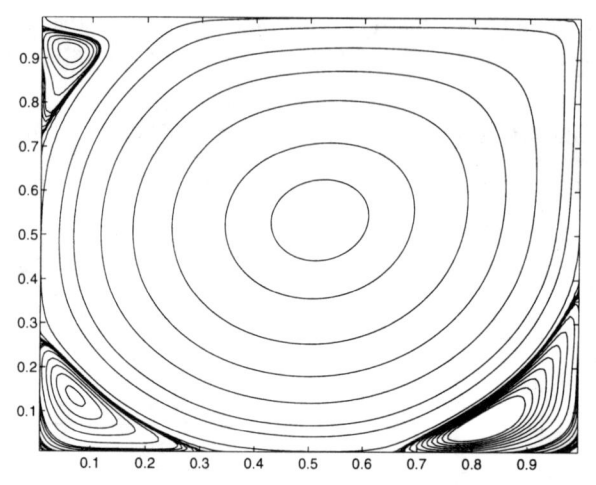

Figure 3.2 Stream function contour plot for the driven cavity problem: $Re = 5000$, 127×127 mesh, 139 iterations

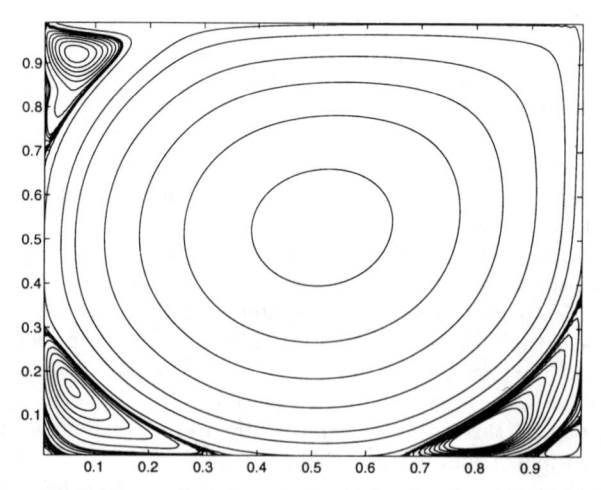

Figure 3.2 Stream function contour plot for the driven cavity problem: $Re = 10000$, 127×127 mesh, 191 iterations

Acknowledgement.

The help given by Maya Neytcheva with some testruns and use of the graphic plotter is gratefully appreciated.

References

[A.a] O. Axelsson, *On global convergence of iterative methods*, in Iterative Solution of Nonlinear Systems of Equations, Lect. Notes in Math., 953, (R. Ansorge et al., eds.), Springer-Verlag, 1981, 1-15.

[A.b] O. Axelsson, *On mesh independence and Newton-type methods*. Applications of Mathematics, 38 (1993), 249-265. [Proceedings of ISNA'92.]

[A.N.P.] O. Axelsson, M. Neytcheva and B. Polman, *An application of the bordering method for solving nearly singular systems*, Seminar "Numerical Mathematics in Theory and Practice", Plzen, Czech Republic, January 25-26, 1993, 28-54, University of Bohemia.

[B.R.] R.E. Bank and D.J. Rose, *Global approximate Newton methods*, Numer. Math., 37 (1980), 279-295.

[B.S.] P.N. Brown and Y. Saad, *Hybrid Krylov methods for nonlinear systems of equations*, SIAM J. Sci. Statist. Comput., 11 (1990), 450-481.

[D.H.] P. Deuflhard and G. Heindl, *Affine invariant convergence theorems for Newton's method and extensions to related methods*, SIAM J. Numer. Anal., 16 (1979), 1-10.

[D.E.S.] R.S. Dembo, S.C. Eisenstat and T. Steihaug, *Inexact Newton methods*, SIAM J. Numer. Anal., 19 (1982), 400-408.

[D.S.] J.E. Dennis and R.B. Schnabel, *Numerical Methods for Unconstrained Optimization and Nonlinear Equations*, Prentice-Hall, Englewood Cliffs, N.J., 1983.

[G.] K. Gustafson, *Vortex separation and fine structure dynamics*, Applied Numerical Mathematics 3 (1987), 167-182.

[G.H.L.] K. Gustafson, K. Halasi and R. Leben, *Controversies concerning finite/infinite sequences of fluid corner vortices*, Contemporary Mathematics, Vol. 99, 1989, American Mathematical Society, pp. 351-357.

[K.A.] I.E. Kaporin and O. Axelsson, *On a class of nonlinear equation solvers based on the residual norm reduction over a sequence of affine subspaces*, SIAM J. Scientific Computing, to appear.

[O.R.] J.M. Ortega and W.C. Rheinboldt, *Iterative solution of nonlinear equations in several variables*, Academic Press, New York, 1970.

[P.] V. Pereyra, *Iterative methods for solving nonlinear least squares problems*, SIAM J. Numer. Anal., 4 (1967), 27-36.

[S.K,a] R. Schreiber and H.B. Keller, *Spurious solutions in driven cavity calculations*, J. Comput. Phys., 49 (1983), 165-172.

[S.K,b] R. Schreiber and H.B. Keller, *Driven cavity flows by efficient numerical techniques*, J. Comput. Phys., 49 (1983), 310-333.

[Y.] T.J. Ypma, *Local convergence of inexact Newton methods*, SIAM J, Numer. Anal., 21 (1984), 583-590.

[Z.] D. Zinćello, *A class of approximate methods of solving operator equations with non-differentiable operators*, (Ukrainian), Dopovidi Akad. Nauk, Ukrain, R.S.R., 1963, 852-856.

Finite Element Solution of the Incompressible Navier–Stokes Equations on Anisotropically Refined Meshes

Roland Becker, Rolf Rannacher
Institut für Angewandte Mathematik, Universität Heidelberg
INF 294, D–69120 Heidelberg, Germany

Abstract

We consider the solution of the incompressible Navier–Stokes equations by the finite element method on strongly anisotropic meshes which typically occur in resolving boundary or interior layers. By theoretical analysis and numerical examples we demonstrate how the nonconforming "rotated" $\widetilde{Q}_1/P_0$-element and the stabilized Q_1/Q_1-element can be modified to achieve approximation and stability properties, as well as multigrid performance, which are essentially independent of the cell aspect ratios.

1 Introduction

In the theoretical analysis of finite element discretizations of viscous flow problems it is commonly assumed that the underlying mesh is "quasi-uniform", i.e., satisfies the uniform shape and size conditions, or that it satisfies at least the uniform shape condition which allows certain local mesh refinements. This restriction appears to be essential as it makes available so-called local "inverse inequalities" for finite elements in the stability proofs for the discretizations (Babuska-Brezzi condition). However, in practice, one frequently wants to use meshes which are locally anisotropically refined, e.g., to properly resolve boundary layers or edge singularities. This situation is not covered by the available theory and it is not clear whether the known stability and convergence results can be extended to such type meshes. Another important problem is that of the efficient solution of the resulting algebraic systems, e.g., by multigrid methods. Numerical experience shows that on irregular meshes the convergence properties of standard multigrid methods deteriorate and may even turn to divergence depending on the aspect ratio of the mesh cells. It is generally not clear whether this phenomenon has to be blamed to stability defects of the underlying discretizations or to problems of the smoother used in the multigrid algorithm.

This note aims to contribute to answer these questions by showing that some of the finite element methods presently in use for solving the incompressible Navier-Stokes equations actually possess stability and approximation properties which are independent of the mesh aspect ratio. The corresponding problem of robustness of multigrid solvers is only briefly addressed and will be treated in more detail elsewhere. For sake of simplicity, we concentrate on the special case of cartesian tensor-product meshes as shown in Figure 1 ("backward facing step"-mesh). Extensions to more general types of meshes in 2D and 3D including also local refinement are possible and will be mentioned below.

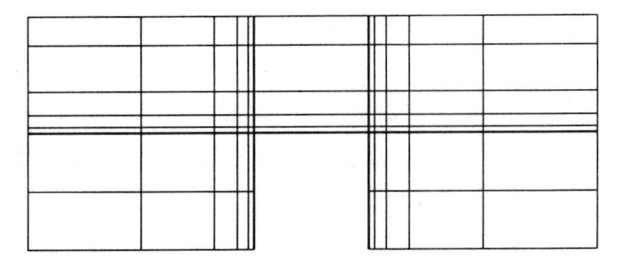

Figure 1: Typical anisotropically refined mesh

2 Finite Elements for the Stokes Problem

As a model problem, we consider the stationary Stokes system

$$-\Delta u + \nabla p = f \quad \text{in } \Omega, \quad u|_{\partial\Omega} = 0, \tag{1}$$

$$\nabla \cdot u = 0 \quad \text{in } \Omega, \tag{2}$$

on a bounded domain Ω with sufficiently regular boundary. The finite element discretization of this problem is based on its variational formulation

$$(\nabla u, \nabla v) - (p, \nabla \cdot v) = (f, v) \quad \forall v \in H, \tag{3}$$

$$(\nabla \cdot u, q) = 0 \quad \forall q \in L, \tag{4}$$

where $L = L^2(\Omega)$ and $H = H_0^1(\Omega)$ are the usual Lebesgue and Sobolev spaces, and $(\cdot,\cdot)_\Omega$ and $\|\cdot\|_\Omega$ denote the inner product and norm of $L^2(\Omega)$ the subscript Ω being usually omitted. This notation is simultaneously used for spaces and norms of scalar– as well as vector–valued functions.

For approximating problem (3), (4) one chooses finite–dimensional spaces $H_h \text{“}\subset\text{”} H$ and $L_h \subset L$ consisting of functions which are piecewise polynomial with respect to a (regular) decomposition $\Pi_h = \{K\}$ of Ω into simple elements K (triangles, quadrilaterals, etc.). The symbol h is used as a parameter characterizing the maximum width of the elements of Π_h, and $\partial\Pi_h = \{\Gamma\}$ denotes the set of all edges (faces) Γ of the elements of Π_h. The velocity spaces H_h may be non–conforming, i.e., $H_h \not\subset H$. The discrete Stokes problem reads: Find $\{u_h, p_h\} \in H_h \times L_h/R$, such that

$$(\nabla u_h, \nabla v_h)_h - (p_h, \nabla \cdot v_h)_h = (f, v_h) \quad \forall v_h \in H_h, \tag{5}$$

$$(\nabla \cdot u_h, q_h)_h = 0 \quad \forall q_h \in L_h, \tag{6}$$

where, in the nonconforming case, the subscript h refers to differentiation and integration in the piecewise sense with respect to the mesh Π_h.

The necessary and sufficient conditions for the existence and convergence of the approximations $\{u_h, p_h\}$ are the approximability properties

$$\inf_{v_h \in H_h} \|\nabla(v - v_h)\|_h \leq c\, h^{m-1} \|v\|_{H^m}, \quad v \in H \cap H^m(\Omega), \tag{7}$$

$$\inf_{q_h \in L_h} \|q - q_h\| \leq c\, h^{m-1} \|q\|_{H^{m-1}}, \quad q \in L \cap H^{m-1}(\Omega), \tag{8}$$

for some integer $m \geq 2$, and the stability estimate (Babuska–Brezzi condition)

$$\inf_{q_h \in L_h/R} \sup_{v_h \in H_h} \frac{(q_h, \nabla \cdot v_h)_h}{\|\nabla v_h\|_h \|q_h\|} \geq \gamma , \qquad (9)$$

for some fixed constant $\gamma > 0$ independent of h (see, e.g., [5]). Under these conditions one has the following approximation result

$$\|\nabla(u - u_h)\|_h + \|p - p_h\| \leq c \, h^{m-1} \{\|u\|_{H^m} + \|p\|_{H^{m-1}}\} . \qquad (10)$$

There are various admissible pairs of finite element spaces ("Stokes elements") proposed in the literature, of which we consider here two particularly simple examples.

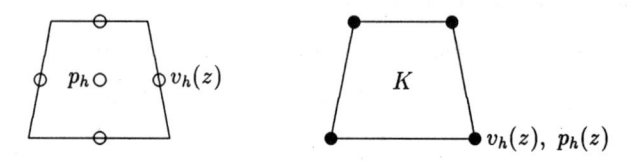

Figure 2: Nodal points of the nonconforming "rotated" $\widetilde{Q}_1/P_0$- and the (stabilized) conforming Q_1/Q_1–Stokes element

1. *The nonconforming $\widetilde{Q}_1/P_0$–Stokes element*
This Stokes element uses rotated bi-linear (tri-linear) shape functions for the velocities and piecewise constants for the pressure. It was introduced and analysed in [10] and may be viewed as the natural quadrilateral analogue of the well-known triangular Crouxeiz-Raviart element (see [5]). In numerical tests it has shown satisfactory stability and approximation properties and its 2D and 3D versions have been implemented in recent Navier-Stokes codes (see [12], [7], and [3]). In 2D, the local velocity space is given by

$$\widetilde{Q}_1(K) := \{q \circ \psi_K^{-1} : q \in \text{span}(1, x, y, x^2 - y^2)\} ,$$

where ψ_K denotes the transformation of K to a certain reference unit element. This ansatz is unisolvent with the nodal functionals ("fluxes")

$$F_\Gamma(v) := \int_\Gamma v \, ds, \quad \Gamma \in \partial \Pi_h .$$

Accordingly, the global finite element spaces for velocities and pressures are defined as

$$H_h := \{v_h \in H : v_h \in \widetilde{Q}_1(K) \quad \forall K \in \Pi_h,$$
$$F_\Gamma(v_h) \text{ continuous}, F_\Gamma(v_h) = 0, \Gamma \in \partial \Pi_h \cap \partial \Omega\} ,$$
$$L_h := \{p_h \in L : p_h \in P_0(K) \quad \forall K \in T_h\}.$$

Instead of the fluxes over the element edges, one may also use the corresponding function values at the midpoints of the edges as nodal values. leading to a different finite element space for the quadrilateral element (in contrast to the triangular case). The above is the fully parametric version of the $\widetilde{Q}_1/P_0$–element as it is defined via transformation (bi-

or tri-linear) to a fixed reference element. It was shown in [10] that the stability and approproximation properties of this ansatz deteriorate on strongly distorted meshes, i.e., depend very sensitively on the amount of deviation of the elements K from parallelogram shape. This defect can be neglected if the computational mesh is generated through a systematic refinement process starting from a fixed macro-decomposition (as commonly used in the static multigrid approach). Alternatively, one may also use a "nonparametric" version of the element where the reference space $\widetilde{Q}_1(K) := \{q \in \text{span}(1, \xi, \eta, \xi^2 - \eta^2)\}$ is defined for each element K independently with respect to the coordinate system (ξ, η) spanned by the directions connecting the midpoints of sides of K. This ansatz turns out to be robust with respect to the shape of the elements K, and the error estimate (10) holds true. But is of course still depending on the regularity of the mesh (defined in the usual way). Below, we will further relax this requirement by allowing the elements to be stretched in one or more (in 3D) directions.

2. The stabilized Q_1/Q_1-Stokes element

For several practical reasons, it is desirable to work with equal degree trial functions for velocity and pressure, particularly allowing the use of continuous pressure approximations. The simplest element of this type is the Q_1/Q_1-Stokes element which uses globally continuous (isoparametric) bi- or tri-linear shape functions for both the velocity and the pressure. This ansatz would be unstable if used with the variational formulation (5), (6). It can be stabilized by adding certain terms in the continuity equation (6) (see [8]):

$$(\nabla \cdot u_h, q_h) + \alpha \sum_{K \in \mathcal{T}} h_K^2 (\nabla p, \nabla q)_K = \text{"correction terms"} \quad \forall q_h \in L_h. \qquad (11)$$

The "correction terms" are chosen in such a way that the solution of the continuous problem exactly satisfies the discrete equation. Thus, this formulation introduces a stabilizing term for the pressure in a consistent way. It is also a well-known fact that for the triangular analogue of this element, the corresponding system can be derived from the stable MINI-element (see [4]). We mention that a similar construction is possible for the Q_1/P_0-Stokes element which will not be considered here.

The stabilizing effect of the modification (11) is based on the relation

$$\sup_{v \in H} \frac{(p, \nabla \cdot v)}{\|\nabla v\|} \geq \gamma \|p\| - c \left(\alpha \sum_{K \in \mathcal{T}} h_K^2 \|\nabla p\|_K^2 \right)^{\frac{1}{2}}, \qquad (12)$$

with γ and c independent of h. On quasi-uniform meshes, the equations (5) and (11) define a stable and consistent approximation of the Stokes problem (1), (2), for which the convergence estimate (10) holds again with $m = 2$. This ansatz has been implemented in recent 2D and 3D Navier-Stokes codes (see, e.g., [7] and [3]). However, it was already reported in [7] that the convergence properties of this element sensitively depend on the parameter α and may deteriorate on strongly stretched meshes.

3 The Stokes Elements on Anisotropic Meshes

On strongly stretched meshes, i.e., those with large element aspect ratios, two different effects may occur: 1) the approximation properties (7) and (8) may deteriorate, and 2) the stability estimate (9) may brake down. It is known (see, e.g., [1] and [9]) that for most of the lower-order finite elements (including the elements considered here) the local interpolation estimates remain valid even on highly stretched elements ("maximum angle condition" versus "minimum angle condition"). Accordingly, the failure of the considered Stokes elements on stretched meshes is not so much a problem of consistency but rather one of stability. Hence, we will discuss the stability in more detail and leave the approximation aspects for a more complete treatment in [9] and [3].

In analysing the stability of the schemes (5), (6) and (5), (11), we have to show that the Babuska–Brezzi stability inequalities (9) and (12) hold independently of the mesh aspect ratio. To this end, let us first introduce some notation. We consider rectangular tensor-product meshes where a typical element K looks as shown in Figure 3.

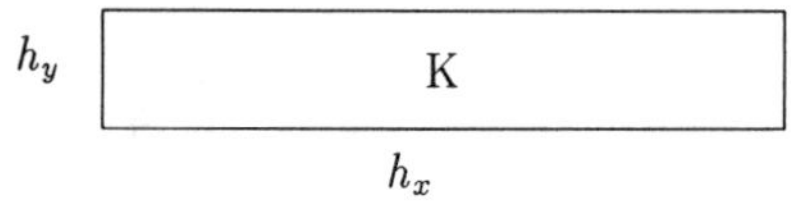

$$h_y \quad \boxed{\quad K \quad} \quad h_x \qquad \sigma_K = \frac{h_{max}}{h_{min}} = \frac{h_x}{h_y}$$

Figure 3: Element of a rectangular tensor-product mesh

where h_x and h_y denote the extensions in the cartesian coordinate directions and $\sigma = max_{K \in \Pi_h} \sigma_K$ is the mesh aspect ratio.

We begin with the nonconforming $\widetilde{Q}_1/P_0$–Stokes element which is considered in its non–parametric form, i.e., the local trial spaces for the velocities are $\widetilde{Q}_1(K) := \{q \in \text{span}(1, x, y, x^2 - y^2)\}$.

Proposition 1. *The non-parametric version of the nonconforming $\widetilde{Q}_1/P_0$–Stokes element is stable on arbitrary rectangular tensor-product meshes, i.e., there holds:*

$$\sup_{v_h \in V_h} \frac{(p_h, \nabla \cdot v_h)}{\|\nabla v_h\|_h} \geq \gamma \|p_h\|, \quad p_h \in L_h,$$

with a constant γ independent of the mesh width and the mesh aspect ratio.

Proof: For any given $p_h \in L_h$, there exists a $v \in H$, such that (see [5])

$$(p_h, \nabla \cdot v) \geq \gamma \|p_h\| \|\nabla v\|.$$

Then, the discrete function $v_h \in H_h$ defined by

$$\int_\Gamma v_h ds = \int_\Gamma v ds \quad \forall \Gamma \in \partial \Pi_h$$

satisfies $(p_h, \nabla \cdot v_h) = (p_h, \nabla \cdot v) \geq \gamma \|p_h\| \|\nabla v\|$. It remains to proof the H^1–stability of this interpolation process. To this end, we write

$$\int_K \nabla v_h \nabla v_h \, dx = -\int_K \Delta v_h v_h \, dx + \int_\Gamma v_h \nabla v_h \cdot n \, ds$$
$$= -\int_K \Delta v_h v \, dx + \int_\Gamma v \nabla v_h \cdot n \, ds$$
$$= \int_K \nabla v \nabla v_h \, dx ,$$

where we have used that Δv_h vanishes on K and that $\nabla v_h \cdot n$ is constant on the edges. (Notice that this argument requires some technical modifications for general non-rectangular meshes.) We conclude the proof by the Cauchy–Schwarz inequality. □

Remark: The essential point in the above proof is that $\Delta v_h|_K = 0$. This property is generally not satisfied for the parametric version of this element even on a rectangular mesh. We can however use a properly scaled ansatz:

$$\widetilde{Q_1}(K) := \{q \circ \psi_K^{-1} : q \in \text{span}(1, \sigma_K x, y, (\sigma_K x)^2) - y^2\},$$

where $\sigma_K = h_x/h_y$. To demonstrate that this scaling is actually necessary for the stability of this element, we compare in Figure 4 the results of a standard "driven cavity" calculation on meshes with $\sigma = 32$.

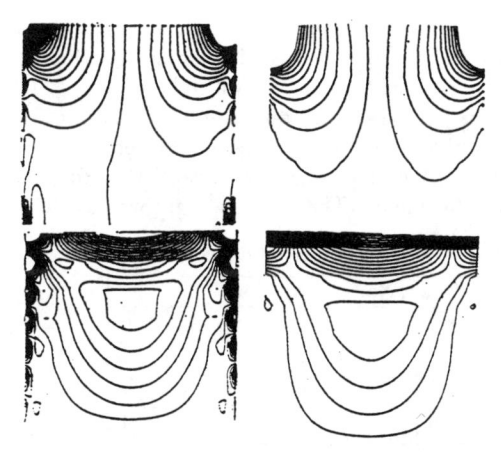

Figure 4: Pressure (above) and velocity (below) isolines for the unscaled (left) and the scaled (right) version of the $\widetilde{Q_1}/P_0$-Stokes element

Next, we consider the stabilized Q_1/Q_1-Stokes element, also on rectangular tensor-product meshes. On such anisotropic meshes the question arises: What meaning should we give to the parameter h_K in the stabilising term in (11)? The right choice of the parameter α is crucial for the accuracy of solutions as well as for the convergence of iterative solvers (see [7], [14]). We consider the following three different choices for the stabilizing bilinear form:

$$c(p,q) = \begin{cases} c_1(p,q) &= \sum_K \text{measure}(K)(\nabla p, \nabla q)_K \\ c_2(p,q) &= \sum_K \text{diam}(K)^2(\nabla p, \nabla q)_K \\ c_3(p,q) &= \sum_K h_x^2(\partial_x p, \partial_x q)_K + h_y^2(\partial_y p, \partial_y q)_K \,. \end{cases}$$

The form c_1 is built in analogy to the so-called MINI-element, since condensation of the bubble functions leads directly to "$h_K^2 = \text{measure}(K)$". We see that c_1 gets smaller with increasing σ, an undesirable effect which is avoided by c_2. Finally, c_3 distinguishes between the different coordinate directions which requires the use of a local coordinate system in the definition of the stabilization. By a local "inverse estimate" for bi-linear functions on rectangles we get the stability relation

$$c_3(p_h, p_h) \leq \|p_h\|^2$$

which appears necessary to achieve uniformity with respect to the mesh aspect ratio. This may be seen by writing the discrete system (5), (6) in matrix notation

$$\begin{bmatrix} A & B \\ -B^T & \alpha C_i \end{bmatrix} \begin{bmatrix} \hat{u} \\ \hat{p} \end{bmatrix} = \begin{bmatrix} b \\ c \end{bmatrix},$$

where C_i corresponds to the stabilizing bilinear form c_i. The Schur complement of the main diagonal block A is $S = \alpha C_i - B^T A^{-1} B$. Then, the stability constant γ in (11) is given by

$$\gamma = \lambda_{min}(M^{-1}S)^2,$$

where M denotes the mass matrix of the pressure ansatz. The relation (11) implies that S is spectrally equivalent to M. The question is whether this remains valid for $\sigma \to \infty$. In view of the relation (5), the continuity constant, which bounds $\lambda_{max}(B^T A^{-1} B)$, can be estimated independently of σ. Hence, we may estimate γ by counting the number of cg–iterations (preconditioned by the mass matrix) needed to invert the Schur complement. For these test calculations we have choosen a sequence of uniformly anisotropic grids obtained by one-directional refinements. The results are given in Table 1.

Table 1: Number of cg–iterations

	No. of Itarations						
$C_i\|\sigma$	2	4	8	16	32	64	128
1	8	18	39	98	559	*	*
2	8	18	39	88	193	*	*
3	8	16	29	31	29	27	24

The increase of the stability constant for c_1 reflects the fact that $\lambda_{min}(M^{-1}S) \sim \sigma^{-1}$, whereas the bad behaviour of c_2 can be explained by $\lambda_{max}(M^{-1}S) \sim \sigma$ (overstabilization), as we only have $c_2(p,q) \leq \sigma \|p\| \|q\|$. We also see that the anisotropic stabilization using c_3 leads to an aspect ratio independent behavior.

Proposition 2. *The anisotropically stabilized Q_1/Q_1–Stokes element is stable on arbitrary rectangular tensor-product meshes, i.e., there holds:*

(13) $$\sup_{v_h \in H_h} \frac{(p_h, \nabla \cdot v_h)}{\|\nabla v_h\|} + \alpha\, c_3(p_h, p_h)^{\frac{1}{2}} \geq \gamma \|p_h\|, \quad p_h \in L_h,$$

with a constant γ independent of the mesh width and the mesh aspect ratio.

The proof of this proposition is more involved than that of Proposition 1, since it requires the construction of an H^1–stable interpolation operator on anisotropic meshes. The conventional Clément–operator which can be used in the quasi-uniform case loses its stability for $\sigma \to \infty$ and has to be replaced by some more elaborate construction. The details of this argument will be given in ([3]).

For illustrating the relevance of the above result, we consider the accuracy of the different stabilizations. For this we construct a sequence of grids with a fixed number of nodes and increasing aspect ratio. This is done by pulling nodes exponentially to the boundary. The results of this test are shown in Table 2, where L denotes the number of refinement steps used in the generation of the computational mesh.

Table 2: Accuracy of the various versions of the stabilized Q_1/Q_1-Stokes element

c_i	L	$\sigma = 100$	$\sigma = 500$	$\sigma = 1000$	
1	3	2.10 − 2	2.84 − 2	6.17 − 2	
1	4	1.37 − 2	7.84 − 3	5.77 − 2	$O(?)$
1	5	1.12 − 2	4.69 − 3	5.82 − 2	
2	3	3.57 − 2	4.33 − 2	1.80 − 1	
2	4	1.27 − 2	1.63 − 2	1.70 − 2	$O(1)$
2	5	5.13 − 3	3.60 − 1	2.15 − 1	
3	3	4.95 − 2	5.44 − 2	5.51 − 2	
3	4	1.45 − 2	1.63 − 2	1.70 − 2	$O(h^2)$
3	5	3.90 − 3	4.42 − 3	4.76 − 3	

4 Solution Algorithm

Finally, we briefly describe our concept of the multigrid algorithm used for solving the discrete Stokes problems (5), (6) and also the related discrete Navier-Stokes problems. Here, we concentrate on the stabilized Q_1/Q_1-Stokes element. The nonconforming $\widetilde{Q}_1/P_0$-element can be treated by similar techniques. Within a stationary or nonstationary Navier-Stokes calculation a simple fixed–point iteration is employed for treating the non–linearity, as described for instance in [12]. The resulting linearized systems are then solved in each step by a multigrid algorithm on a sequence of hierarchical meshes. In our test calculations these meshes have been generated by systematic subdivision (bisection of element edges) of a macro-decomposition of the domain Ω.

In this multigrid algorithm the generic restrictions and prolongations are used. Let us describe the smoothing process first for the case of uniform meshes. The smoother is an adaptation of the block–Gauss–Seidel iteration introduced by Vanka [13]. Within an iteration sweep, to each nodal point the patch consisting of the four surrounding elements is associated and the corresponding velocity and pressure unknowns are updated simultaneously. This requires the (direct) solution of local 9 × 9–systems of the form

$$\begin{bmatrix} a_{11} & 0 & \cdots & 0 & d_1 \\ 0 & a_{22} & \cdots & 0 & d_2 \\ \vdots & \vdots & \cdots & \vdots & \vdots \\ 0 & 0 & \cdots & a_{88} & d_8 \\ d_1 & d_2 & \cdots & d_8 & c \end{bmatrix} \begin{bmatrix} u_1 \\ u_2 \\ \vdots \\ u_8 \\ p \end{bmatrix} = \begin{bmatrix} \vdots \\ \vdots \\ \text{defects} \\ \vdots \\ \vdots \end{bmatrix}$$

where for further simplification the off–diagonal entries in the velocity-block are neglected. The good performance of this smoother for the Q_1/Q_1–element on regular meshes will be demonstrated in [3].

From the classical multigrid–theory for elliptic equations we know that point–iterations lose their smoothing property for large aspect ratios. The remedy is the use of a smoother which becomes a direct solver in the limit $\sigma \to \infty$. Consequently, since our smoother is similiar to a point–Gauss-Seidel iteration, we expect problems in the case of strongly stretched grids. Our startegy to overcome this difficulty is as follows:
Since we expect the mesh aspect ratio to be large only in a small part of the computational domain, we should use an *adaptive* smoother. This means that we will combine the point–smoother with a more robust version just where we need it, for instance on elements with a large σ_K. In this approach the nodes are grouped in the direction of the anisotropic mesh refinement and iterated implicitely leading to a process which may be termed "stringwise"-block-Gauss-Seidel method. Our algorithm works as follows:
– Define a set of lines (strings) in the direction of the anisotropic mesh refinement which can be determined from the matrix entries.
– Along these strings, use a direct band–solver for updating the corresponding local velocity and pressure values.
This process is controlled by a switch aspect ratio σ_s. Figure 5 illustrates the process by showing those areas of the computational domain where the point–smoother is adaptively modified into a "string"–smoother for $\sigma_s = 2, 3, 10$. For more details on this procedure, we refer to [3].

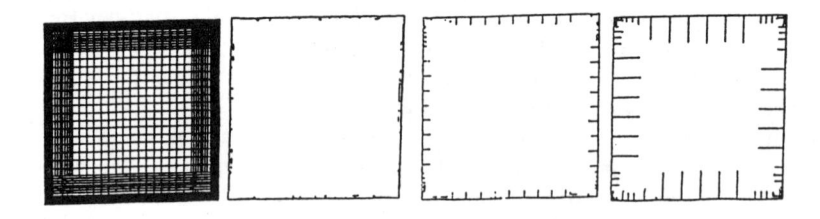

Figure 5: Adaptively choosen "strings" for a "driven cavity" grid

For testing the effectiveness of the above stabilization strategy we have calculated a jet flow in a channel. This more realistic example shows that the discussed stability proporties are also crucial in Navier-Stokes calculations. The coarse grid and the boundary conditions are described by Figure 6.

inflow $\quad\quad\quad\quad\quad\quad\quad\quad\quad\quad\quad\quad\quad\quad\quad\quad\quad\quad\quad\quad$ $\nu\frac{\partial u}{\partial n} + p \cdot n = 0$

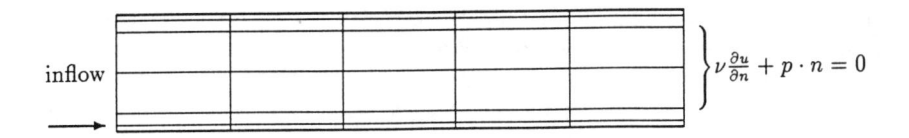

Figure 6: Coarse grid and boundary conditions for the jet flow

Table 4 contains the multigrid convergence rates in dependence of the viscosity ν and the mesh aspect ratio σ, using 2 adaptive ($\sigma_s = 5$) pre- and postsmoothing sweeps.

Table 3: Multigrid convergence rates for the jet problem depending on ν and σ

	ρ_{mg}			
$\nu \mid \sigma$	50	100	500	1000
1	0.30	0.31	0.33	0.33
10^{-2}	0.30	0.33	0.40	0.45
10^{-3}	0.30	0.40	0.42	0.54

To conclude this discussion we present some numerical results for the case with $\nu = 1/3000$, where the upper picture shows the effect of a wrong stabilization.

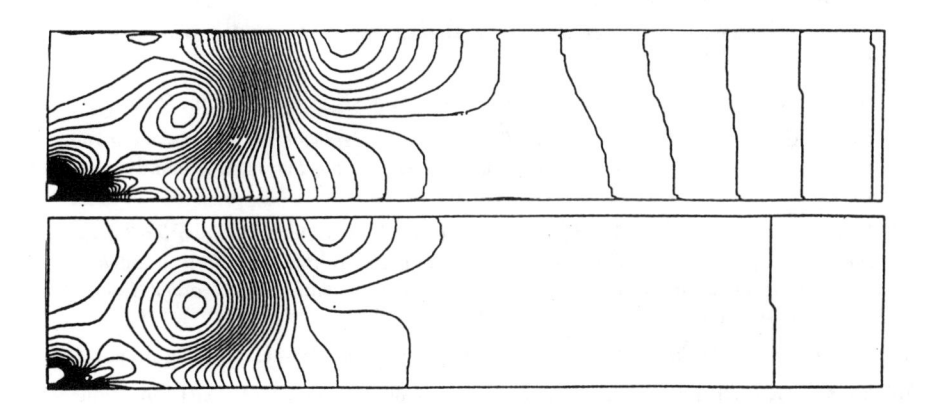

Figure 7: Pressure isolines for the jet problem calculated with isotropic (upper) and anisotropic (lower) stabilization

Let us finally mention an important problem especially in the use of non–uniform grids. The use of iterative solvers makes it necessary to define a stopping criterion. To this end we need to measure the residual in the right norm. Clearly, the common weighting by the number of unknowns is not appropriate for non–uniform grids. For a recent approach towards a solution of this problem based on the Galerkin orthogonality inherent in the multigrid process, we refer to [2].

Acknowledgement

The authors thank the Deutsche Forschungsgemeinschaft for the financial support of this research within the Graduiertenkolleg "Modellierung und wissenschaftliches Rechnen" at the IWR, Universität Heidelberg.

References

[1] Apel, T., Dobrowolski, M.: *Anisotropic interpolation with applications to the finite element method*, Computing, 47, 277–293 (1992).

[2] Becker, R., Johnson, C., Rannacher, R. : *Adaptive error control for multigrid finite element methods*, Preprint, April 1994.

[3] Becker, R.: *Finite Element Solution of Flow Problems on Domains with Moving Boundaries*, Thesis, Universität Heidelberg, 1994, in preparation.

[4] Brezzi, R.E., Fortin, T.: *Mixed and Hybrid Finite Element Methods*, Springer, Berlin, 1991.

[5] Girault, V., Raviart, P.A.: *Finite Element Methods for Navier–Stokes Equations*, Springer, Berlin, 1986.

[6] Hackbusch, W.: *Multi-Grid Methods and Applications*, Springer, Berlin, 1985.

[7] Harig, J.: *Eine robuste und effiziente finite Elemente Methode zur Lösung der inkompressiblen 3-D Navier–Stokes-Gleichungen auf Vektorrechnern*, Thesis, Universität Heidelberg, 1991.

[8] Hughes, T.J.R., Franca, L.P., Balestra, M.: *A new finite element formulation for computational flui dynamics: V. Circumventing the Babuska–Brezzi condition: a stable Petrov-Galerkin formulation of the Stokes problem accommodating equal-order interpolation*, Comp. Meth. Appl. Mech. Eng., 59, 85–99 (1986).

[9] Rannacher, R.: *Finite element discretization on irregular meshes*, 1994, in preparation.

[10] Rannacher, R., Turek, S.: *Simple nonconforming quadrilateral Stokes element*, Num. Meth. Part. Diff. Eqns., 8, 97–111 (1992).

[11] Sockol, P.M.: *Multigrid solution of the Navier–Stokes equations on highly stretched grids*, Int. J. Appl. Num. Meth. Fluids, 17, 543–566 (1993).

[12] Turek, S.: *Tools for simulating non-stationary incompressibl flow via discretely divergence-free finite elemen models* Int. J. Numer. Meth. Fluids, 18, 71–105 (1994).

[13] Vanka, S.P.: *Block-implicit multigrid solution of Navier Stokes equations in primitive variables*, J. Comp. Phys., 65, 138–158 (1986).

[14] Wathen, A., Sylvester, D.: *Fast iterative solution of stabilized Stokes systems, Part I. Using simple diagonal preconditioners*, SIAM J. Numer. Anal., 30, 630–649 (1993).

Downwind Numbering : A Robust Multigrid Method for Convection-Diffusion Problems on Unstructured Grids

Jürgen Bey
Mathematisches Institut, Univ. Tübingen,
Auf der Morgenstelle 10, 72076 Tübingen, Germany

Gabriel Wittum
ICA III, Univ. Stuttgart,
Pfaffenwaldring 27, 70569 Stuttgart, Germany

Summary

In the present paper we introduce and investigate a robust smoothing strategy for convection-diffusion problems in two and three space dimensions without any assumption on the grid structure. The main tool to obtain such a robust smoother is an ordering strategy for the grid points called "*downwind numbering*", which follows the flow direction and – combined with a Gauß-Seidel type smoother – yields robust multi-grid convergence for adaptively refined grids, provided the convection field is cycle-free. The algorithms are of nearly optimal complexity and the corresponding smoothers are shown to be robust in numerical tests.

1. Introduction

In recent years the research interest in robust multigrid methods has grown significantly. This is due to the fact that modelling challenging application problems like problems from fluid mechanics typically results in singularly perturbed partial differential equations.

Since early multigrid days the basic problems with singularly perturbed equations are known and several strategies to work around them have been developed. One of the most powerful of these is the robust smoothing strategy, introduced by Wesseling and Kettler in 1982 (cf. [13], [9]). The basic idea is to choose a smoother which is a fast solver for the singular perturbation and thus allows the coarse-grid correction cycle to handle the rest. This strategy has been used and theoretically analyzed for many practical applications (see [2], [11], [12], [14] and the references therein).

Unfortunately up to now robust multigrid methods usually required structured grids, which is in direct contradiction to the requirement for adaptive refinement caused by local phenomena which are typical for singularly perturbed problems, such as boundary and shear layers, transition regions between phases etc. Thus we need robust multigrid methods which can be applied to locally refined grids.

The topic of this paper is to introduce and investigate a robust smoothing strategy for convection-diffusion problems in two and three space dimensions without any assumption on the grid structure. The main tool to obtain such a robust smoother for these problems is an ordering strategy for the grid points which follows the flow direction and – combined with a Gauß-Seidel type smoother – yields robust multigrid convergence for adaptively refined grids, provided the convection field is cycle-free.

The idea to combine a Gauß-Seidel method with a downwind ordering process to obtain a very fast solver for the convective terms is an old one. But usually combinations of this idea with multigrid methods were given for special problems and structured grids only (see e.g. [9]), where numbering is very easy.

In the present paper we give an ordering strategy called downwind numbering which is of nearly optimal complexity. This downwind numbering proofs to be robust not only for globally convection dominated problems but also for problems which are convection dominated only in a part of the region. There multigrid correction shows to be really necessary.

We base our investigations on the scalar equation

$$-\varepsilon \Delta u + c \cdot \nabla u = f \quad \text{in} \quad \Omega, \tag{1}$$

with $\operatorname{div} c = 0$ and Dirichlet boundary conditions. Discretization grids are constructed by combining triangles and quadrilaterals in 2d and by tetrahedra in 3d. In the latter case stable refinements are obtained using a strategy that was proposed by one of the authors in [3]. On these grids we discretize (1) by means of a finite volume method and a corresponding first order upwinding of the convective terms. These strategies are described in section 2. After reviewing the basic multigrid schemes in section 3, we describe a downwind numbering algorithm and some of it's basic properties in section 4. Section 5 contains the numerical tests validating the robustness of the downwind numbered smoother.

2. Finite Volume Discretization

Our discretization grids consist of triangles resp. quadrilaterals in 2d and of tetrahedra in 3d. Any such triangulation $\mathcal{T}$ of Ω is required to be "*closed*" in the sense that there are no so called "*hanging nodes*". For the discretization of the convection diffusion equation (1) – which can be interpreted as a conservation law – we use a finite volume method (also called box method). See for example the paper of W. Hackbusch [7] for a detailed description in the 2d case.

The finite volume method is based on a usual finite element triangulation $\mathcal{T}$ of the solution domain and a dual box mesh $\mathcal{B}$ consisting of boxes b_i for every node x_i of $\mathcal{T}$, in such a way, that $x_i \in b_i$ and the boxes $b_i \in \mathcal{B}$ cover Ω and have pairwise empty inner intersection.

Using with respect to $\mathcal{T}$ piecewise linear ansatz functions represented in terms of the corresponding nodal basis $\{\varphi_j\}$, integration of (1) over the box b_i and the divergence theorem lead us to the discrete linear system

$$\sum_j u_j \int_{\partial b_i} (-\varepsilon \nabla \varphi_j + c \varphi_j) \cdot d\sigma = \int_{b_i} f\, dx \qquad \forall b_i \in \mathcal{B}. \qquad (2)$$

The discrete solution u is then given by $u = \sum_j u_j \varphi_j$. Following [7], the boxes b_i in 2d are constructed by connecting the center of mass of any element of $\mathcal{T}$ to the midpoints of it's edges. This method is extended here to three dimensions : The center of mass of every tetrahedral element is connected with the centers of mass of all it's faces, and these again to the midpoints of their edges.

Thus we obtain four portions of the element each of the same volume, and one for each corner. Figure 1 shows the corresponding part of a tetrahedron for one of it's corners. The box b_i is then defined to be just the union of all portions of x_i in any element with corner x_i. In this way and under usual finite element assumptions we obtain an $O(h^2)$ discretization [7].

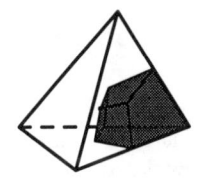

Figure 1: Box portion of an elements corner

Unfortunately finite volume as well as finite element discretizations are known to be unstable in the case of dominating convection, which is observable by strong oscillations in the discrete solution. These are explained by the fact, that the stiffness matrix is not an M-matrix, but is skew symmetric up to terms of order $O(\varepsilon)$. In special cases the linear system can even be singular.

To avoid those stability problems we use a first order upwind sheme (see for example [8]), resulting in an $O(h)$-discretization. In one dimension this upwind method can be interpreted as the substitution of central by backward differences, or as addition of some artificial diffusion. We want to give some short explanation here formulated in the box terminology :

Due to the symmetry of the nodal basis functions φ_i and φ_j the usual discretizations treat the convective flux from box b_i to the adjacent box b_j half as outflow from box b_i and half as inflow to box b_j, which leads to the nearly skew symmetric form of the stiffness matrix. In contrast to that the upwind sheme attributes the convective flux completely to the downstream box b_j, leading to a diagonal dominant stiffness matrix and a stable discretization.

Furthermore – if there are no cycles in the convection graph – with a proper ordering of the unknowns we can obtain a nearly lower triangular stiffness matrix up to entries of order $O(\varepsilon)$. For such linear systems Gauß-Seidel is a very efficient smoother.

Up to now we were only concerned with the discretization of (1) with respect to a certain triangulation $\mathcal{T}$. In order to apply a multigrid method to the solution of the arising linear system we have to define a whole sequence $\mathcal{T}_0, \mathcal{T}_1, \ldots, \mathcal{T}_J = \mathcal{T}$ of hierarchically nested triangulations, where $\mathcal{T}_k$ is obtained from $\mathcal{T}_{k-1}$ by a set of certain refinement rules.

These rules make sure that the triangulations $\mathcal{T}_k$ are "*closed*" and that the generated sequence is "*stable*", i.e. the elements do not deteriorate for $J \to \infty$. In 2d the well known "*red*" and "*green*" refinements from [1] are used for the triangles and the quadrilaterals are subdivided in a straightforward way.

Stable refinement and closure of tetrahedral meshes is a somewhat more complicated task, but it is known that the usual 2d refinement strategies can be generalized to three dimensions. We here use the refinement method proposed by the first author in [3], which produces elements of at most six similarity classes for any level 0 tetrahedron. For the green closure we use three additional refinement rules which are given for example in [5]. See also [4] for the implementaion details.

3. Multigrid Methods for Convection-Diffusion Problems

Multigrid methods were originally developed for the efficient solution of elliptic PDE's (see [6] for an introduction). In this case multigrid has been proven to be of optimal complexity, that is the amount of work to solve the arising linear system up to a fixed accuracy grows only linearly with the number of unknowns, provided the triangulations used are refined uniformly.

One of the main difficulties applying multigrid to convection-diffusion equations is the singularly perturbed character of these problems. Discretizations mix up convection and diffusion operators specifically with respect to the grid scale. This makes the coarse grid approximation within the multigrid process deteriorate. Further convection-diffusion equations show local phenomena as boundary or interior layers which have to be resolved by adaptive gridding. Thus we need a robust, adaptive multigrid method to treat this type of equations.

An overview of robust and adaptive multigrid is given in [2]. As pointed out there, one typical approach is to use a robust smoother and a standard coarse-grid correction cycle. In this case the smoother has to eliminate the influence of the singular perturbation part such that the coarse-grid correction is able to do it's work on the elliptic part. That means that the smoother should be exact or very fast for the limit case. This idea has been introduced quite early by Wesseling [13] and investigated and extended by a couple of authors in the meantime ([9], [14], [11], [12]).

In case of convection-diffusion problems discretized by upwinding, it is quite easy to obtain an iterative scheme which solves the case of pure convection exactly. We can just choose a Gauß-Seidel like method and order the unknowns in flow direction. This trivial fact is well known in literature. If the diffusion part does not vanish, however, such a scheme is not sufficient as solver for the whole equation. Nevertheless it is an ideal candidate as robust smoother within a multigrid cycle, since it satisfies the main requirement for a robust smoother explained above. Thus Gauß-Seidel and ILU schemes have been used in literature as robust smoothers (see [9], [13]) for two dimensional convection-diffusion problems discretized on structured grids, where ordering is quite easy.

For adaptively refined grids we have to be aware of the complexity problem as well as of the robustness as described in [2]. I. e. we have to choose a multigrid method allowing for robust smoothing on the one hand and having optimal complexity in case of strongly local refinements. There exist several classes of adaptive multigrid methods, additive and multiplicative ones with global or local smoothing. As described in [2], these can be classified by their basic additive or multiplicative structure and the pattern which is used for smoothing.

As pointed out there, only local smoothing yields multigrid methods of optimal complexity for adaptively refined grids. Local smoothing means, that only the unknowns of the locally refined regions are smoothed on a particular grid level. A similar strategy has already been used by M. C. Rivara in 1984 [10]. Since local multigrid is twice as efficient as the BPX preconditioner, we base our numerical tests on this algorithm. The kernel of our robust smoother is the ordering algorithm which is described in the next chapter.

4. Downwind Numbering

We assume the convection-diffusion equation (1) to be discretized either according to section 2 or by a similar method. Let $\mathcal{N}$ be the set of nodes of the underlying triangulation $\mathcal{T}$, and $A = (a_{xy})_{x,y \in \mathcal{N}}$ be the corresponding stiffness matrix. Since every level of the grid hierarchy is numbered separately, we can skip the level index k at this place.

Using the upwinded finite volume discretization of section 2, we know A to be weakly diagonal-dominant and the off-diagonal entries of A to be non-positive. Splitting any edge of $\mathcal{T}$ into two directed links, we define the link from node x to node y to be a "*downwind link*" if $|a_{yx}| > |a_{xy}|$, that is the link follows the flow direction. These downwind links form the directed convection graph. Futhermore, for any node $x \in \mathcal{N}$ we denote by $I(x)$ the number of it's inflow edges, i.e. the number of downwind links pointing to x.

The main aim of the ordering process is to bring the stiffness matrix for the convective terms in triangular form. This can be performed easily, provided the flow field is cycle-free. In this case the convection graph defines a partial ordering of the unknowns. This partial ordering can be used to number the unknowns in streamwise direction. For this purpose we developed algorithm *Downwind_Numbering* which is shown in Figure 2.

The main feature of this algorithm is a kind of wavefront technique. It works as follows : First for every node $x \in \mathcal{N}$ we compute the number of inflow edges and denote it by $I(x)$. Our first wavefront $\mathcal{L}_0$ consists of all nodes x with $I(x) = 0$. It is nonempty if we assume the convection graph to be cycle-free. Now we step downstream across the remaining graph in a wavefront manner : The partial list $\mathcal{L}_{k+1}$ consists of all nodes y which are downstream neighbours of $\mathcal{L}_k$-nodes and which are not in the ordered list $\mathcal{L}$ of already processed nodes (i.e. $I(y) > 0$ in (8)) and for which all inflow edges have been already taken into account this way. The last condition is checked by decrementing $I(y)$ for every visit. Then all inflow edges are taken into account if $I(y) = 0$ in line (10) is true.

67

```
Algorithm  Downwind_Numbering

(1)     for ( $x \in \mathcal{N}$ ) compute $I(x)$;
(2)     let $\mathcal{L} = \mathcal{L}_0$ be the list of all nodes $x$ with $I(x) = 0$;
(3)     $k = 0$;

(4)     while ( $\mathcal{L}_k \neq \emptyset$ )
        {
(5)         $\mathcal{L}_{k+1} = \emptyset$;
(6)         for ( $x \in \mathcal{L}_k$ )
(7)             for all neigbours $y$ of $x$ in downwind direction
                {
(8)                 if ( $I(y) \neq 0$ )
                    {
(9)                     $I(y) = I(y) - 1$;
(10)                    if ( $I(y) = 0$ ) then $\mathcal{L}_{k+1} = \mathcal{L}_{k+1} \cup \{y\}$;
                    }
                }
(11)        sort $\mathcal{L}_{k+1}$ and append it to the list $\mathcal{L}$;
(12)        $k = k + 1$;
        }

(13)    if ( $|\mathcal{L}| = |\mathcal{N}|$ ) then stop.
(14)    else handle cycles.
```

Figure 2: Downwind numbering algorithm

Finally our current wavefront $\mathcal{L}_{k+1}$ is sorted (e.g. by mergesort) in downwind direction and appended to the list $\mathcal{L}$ of already processed nodes.

The algorithm terminates if $\mathcal{L}_{k+1}$ is empty. If the convection graph is cycle-free, $|\mathcal{L}| = |\mathcal{N}|$ will hold in (13), that is all nodes of $\mathcal{N}$ are taken into consideration by the ordering process. $\mathcal{L}_{k+1} = \emptyset$ but $|\mathcal{L}| < |\mathcal{N}|$ indicates that there is at least one cycle between the remaining nodes.

Remark 1 : If mergesort is used in line (11) to order the partial lists $\mathcal{L}_{k+1}$, the complexity of the algorithm is typically $O(N \log N)$, with N denoting the number of nodes in $\mathcal{N}$. Of course we get the same complexity applying mergesort directly to the whole grid, but in principle we have reduced the sort problem by one dimension. Applying Algorithm *Downwind_Numbering* recursively to sort the grid-points in the wavefront itself, we obtain an algorithm of linear complexity.

Remark 2 : As indicated in the last statement (14), the ordering algorithm can also be applied to problems containing convection cycles. In this case it terminates with $|\mathcal{L}| < |\mathcal{N}|$ and all cycles are contained in the set $\mathcal{N} \setminus \mathcal{L}$. In addition we know, that at least one of the downwind neighbours of the last nonempty wavefront $\mathcal{L}_k$ belongs to a cycle, so the detection of cycles is simplified. After isolation of a cycle our algorithm can continue until the next one is found.

Remark 3 : Cycles in the convection graph will typically occur as soon as the flow field contains vortices. Even in case of vortex-free flow fields, however, it is not always guaranteed, that on the discrete level the matrix graph produced by the finite volume upwinding is free of cycles. It is even possible that those discretization cycles extend over the edges of a single element, thus we call them "*inner element cycles*". These are generated by the use of obtuse triangles (tetrahedra). The treatment of cycles will be subject of a forthcoming paper.

5. Numerical Experiments

We show results for the convection-diffusion equation

$$-\varepsilon \Delta u + c \cdot \nabla u = f \quad \text{in} \quad \Omega, \tag{3}$$

with Dirichlet boundary conditions in two and three space dimensions on adaptively refined grids. For all experiments (3) was discretized by means of a finite volume method with first order upwinding according to section 2. To solve the resulting linear systems we used a standard multigrid V-cycle with local smoother, namely a Gauß-Seidel scheme with downwind numbering. The grids were adaptively refined using a simple gradient refinement criterion, which resolves the boundary and shear layers introduced by the boundary conditions. The refinement process itself was imbedded in a nested iteration (cf. [6]).

5.1. Two-dimensional results

We solve equation (3) on the unit square $\Omega = (0,1) \times (0,1)$ under the following Dirichlet boundary conditions :

$$u(x,y) = \begin{cases} 1, & x \in [0.5, 1], \ y = 0, \\ 0, & (x,y) \in \partial\Omega \quad \text{otherwise.} \end{cases} \tag{4}$$

In Experiment 1 the convection c is chosen to be constant, thus we have a purely convection dominated problem for $\varepsilon \to 0$. In Experiment 2 the convection varies over the domain Ω. Consequently there are convection dominated as well as diffusion dominated regions.

Experiment 1 : We solve equation (3) on the unit square with boundary condition (4). The convection parameter c is defined by

$$c(x,y) = (1,1)^T \quad \text{constant on } \Omega. \tag{5}$$

Thus we have a purely convection dominated problem for $\varepsilon \to 0$. In this case it is no surprise that the local multigrid (1,2,V)-cycle with downwind numbered Gauß-Seidel smoother shows robustness for $\varepsilon \to 0$, as shown in Table 1.

Table 1 : Convergence rate averaged over 10 steps

Level	$\varepsilon = 10^{-2}$	$\varepsilon = 10^{-4}$	$\varepsilon = 10^{-6}$
1	$8.47 \cdot 10^{-3}$	$1.05 \cdot 10^{-3}$	$1.00 \cdot 10^{-3}$
2	$3.10 \cdot 10^{-2}$	$4.92 \cdot 10^{-2}$	$4.88 \cdot 10^{-2}$
3	$5.10 \cdot 10^{-2}$	$8.58 \cdot 10^{-2}$	$9.91 \cdot 10^{-2}$
4	$8.10 \cdot 10^{-2}$	$1.53 \cdot 10^{-2}$	$1.43 \cdot 10^{-2}$
5	$1.36 \cdot 10^{-1}$	$2.01 \cdot 10^{-2}$	$1.77 \cdot 10^{-2}$
6	$1.82 \cdot 10^{-1}$	$2.90 \cdot 10^{-2}$	$2.23 \cdot 10^{-2}$
7	$1.96 \cdot 10^{-1}$	$5.38 \cdot 10^{-2}$	$3.60 \cdot 10^{-2}$

Experiment 2 : For the second experiment we consider a problem which is convection dominated only in parts of the unit square. Therefore we use a convection parameter c of (3) which is given by

$$c(x,y) = [\,1 - (2x - 0.5)\sin\alpha + (2y - 0.5)\cos\alpha\,]^4 \cdot (\cos\alpha, \sin\alpha)^T, \qquad (6)$$

where α is the angle of attack. The boundary conditions are given by (4). The boundary condition jump is propagated in direction α. We have $\text{div}\, c = 0$ and c varies strongly on Ω such that the problem is convection dominated in one part of the domain and diffusion dominated in another part. The absolute value $|c|$ of the convection field for $\alpha = \pi/2$ is shown in Figure 3.

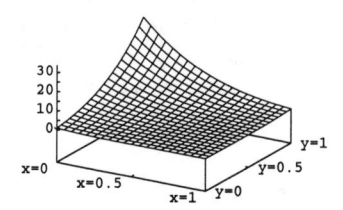

Figure 3: Convection field $|c|$ from Experiment 2

Again we used a gradient refinement criterion to produce adaptively refined triangular grids. The discrete problem on the highest level – that is level 8 – used more than 10.000 unknowns. As smoother we took a Gauß-Seidel scheme with downwind numbering in a local multigrid (1,1,V)-cycle. It is important to note that the smoother itself is not an exact solver. Thus we should see the benefit of multigrid in the diffusion dominated part and of the robust smoother in the convection dominated one. This is confirmed by the results given in Figure 4, where the residual convergence rate averaged over 10 steps for the highest level problem is shown for $\alpha = \pi/2$ and different values of ε.

For the same problem but without downwind numbering the (1,1,V) local multigrid cycle shows a convergence rate of 0.95 averaged over 40 steps. Taking the smoother with downwind numbering but without coarse-grid correction as a solver, we end up with a convergence rate of 0.949 as well. This confirms the outlined concept of robust multigrid.

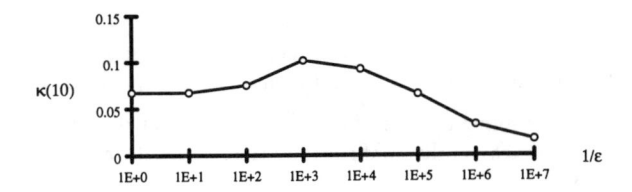

Figure 4: Convergence rate $\kappa(10)$ for Experiment 2

5.2. Three-dimensional results

All three dimensional tests were performed on the unit cube $\Omega = (0,1)^3$. The coarsest triangulation used 48 tetrahedra obtained by dividing the cube into eight subcubes, which are further divided into 6 tetrahedra canonically. Since we use Dirichlet boundary conditions, there is exactly one unkonwn on the coarsest level $\mathcal{T}_0$. This grid is uniformly refined twice leading to triangulations $\mathcal{T}_1$ and $\mathcal{T}_2$. Further refinement steps are performed adaptively by means of a residual based error indicator. The Dirichlet boundary conditions are given by

$$u(x,y,z) = \begin{cases} 1, & 0 \leq x,y \leq 0.5,\ z = 0, \\ 0, & (x,y,z) \in \partial\Omega \quad \text{otherwise.} \end{cases} \quad (7)$$

Again we tested the robustness of our method first by means of a purely convection dominated and than by means of a mixed problem.

Experiment 3 : We solve equation (3) on the unit cube with boundary condition (7). The convection parameter c is defined by

$$c(x,y,z) = (1,1,1)^T \quad \text{constant on } \Omega. \quad (8)$$

This is a purely convection dominated problem for $\varepsilon \to 0$. The convergence results versus ε on different levels are shown in Table 2.

Table 2 : Averaged convergence rates for different levels and ε

Level	$\varepsilon = 10^{-2}$	$\varepsilon = 10^{-4}$	$\varepsilon = 10^{-6}$
3	0.127	0.00046	0.0006
4	0.137	0.00705	0.00084
5	0.148	0.0088	0.001

Figure 5 shows grid and solution of level 5 for the case $\varepsilon = 10^{-4}$. Both plots show a cut of the unit cube along its main diagonal.

Experiment 4 : Finally we consider a mixed problem with both convection and diffusion dominated parts of the domain in analogy to Experiment 2. To this end we choose $c = q \cdot (1,1,1)^T$ and define the weight function q by

$$q(x,y,z) = \begin{cases} 0, & x+y-2z \leq 0, \\ 1, & x+y-2z \geq 1, \\ (x+y-2z)^2, & \text{otherwise.} \end{cases} \quad (9)$$

c vanishes above the main diagonal of the cube, is constant and equal to $(1,1,1)^T$ below the line from $(0.5, 0.5, 0)$ to $(1, 1, 0.5)$, and is interpolated quadratically in between.

Table 3 shows the convergence rates for this problem and different values of ε. The robustness of our approach is fully confirmed by the results. In Figure 6 grid and solution on level 4 are shown for $\varepsilon = 10^{-4}$.

Table 3 : Averaged convergence rates for the mixed problem

Level	$\varepsilon = 10^{-2}$	$\varepsilon = 10^{-4}$
1	0.0687	0.021
2	0.0645	0.028
3	0.1268	0.052
4	0.1367	0.059

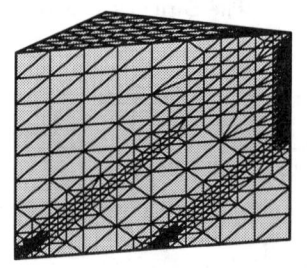

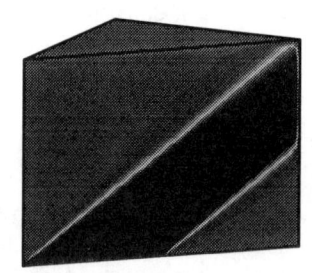

Figure 5: Grid and solution of Experiment 3 on level 5 ($\varepsilon = 10^{-4}$)

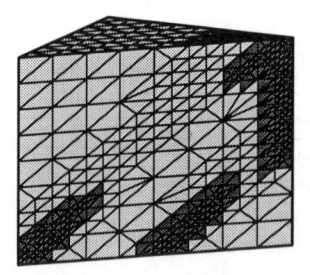

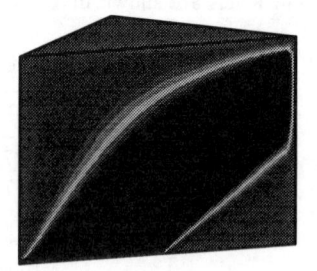

Figure 6: Grid and solution of Experiment 4 on level 4 ($\varepsilon = 10^{-4}$)

References

[1] R. E. BANK, A. H. SHERMAN, AND A. WEISER, *Refinement algorithms and data structures for regular local mesh refinement*, in Scientific Computing, R. Stepleman, ed., Amsterdam: IMACS/North Holland, 1983.

[2] P. BASTIAN AND G. WITTUM, *Adaptivity and robustness*, in Adaptive Methods : Algorithms, Theory and Applications, NNFM, W. Hackbusch and G. Wittum, eds., Vieweg, Braunschweig, 1994.

[3] J. BEY, *Der BPX-Vorkonditionierer in drei Dimensionen : Gitterverfeinerung, Parallelisierung und Simulation*, Preprint no. 92-03, IWR, Univ. Heidelberg, 1992.

[4] ——, AGM^{3D} *Manual*, tech. rep., Univ. Tübingen, 1994.

[5] F. A. BORNEMANN, B. ERDMANN, AND R. KORNHUBER, *Adaptive multilevelmethods in three space dimensions*, Int. J. Numer. Methods Eng., (1993). (to appear).

[6] W. HACKBUSCH, *Multigrid Methods and Applications*, Springer, 1985.

[7] ——, *On first and second order box schemes*, Computing, 41 (1989), pp. 277–296.

[8] C. JOHNSON, *Numerical Solution of Partial Differential Equations by the Finite Element Method*, Cambridge University Press, 1987.

[9] R. KETTLER, *Analysis and comparison of relaxation schemes in robust multigrid and preconditioned conjugate gradient methods*, in Multigrid Methods, Lecture Notes in Mathematics, Vol. 960, W. Hackbusch and U. Trottenberg, eds., Springer, Heidelberg, 1982.

[10] M. C. RIVARA, *Design and data structure of a fully adaptive multigrid finite element software*, ACM Trans. on Math. Software, 10 (1984), pp. 242–264.

[11] R. STEVENSON, *On the robustness of multigrid applied to anisotropic equations : Smoothing and approximation-properties*. Preprint Rijksuniversität Utrecht, Wiskunde, Netherlands, 1992.

[12] ——, *New estimates of the contraction number of V-cycle multigrid with applications to anisotropic equations*, in Incomplete Decompositions : Algorithms, Theory and Applications, NNFM Vol. 41, W. Hackbusch and G. Wittum, eds., Vieweg, Braunschweig, 1993.

[13] P. WESSELING, *A robust and efficient multigrid method*, in Multigrid Methods, Lecture Notes in Mathematics, Vol. 960, W. Hackbusch and U. Trottenberg, eds., Springer, Heidelberg, 1982.

[14] G. WITTUM, *On the robustness of ILU-smoothing*, SIAM J. Sci. Stat. Comp., 10 (1989), pp. 699–717.

A fast solver for gas flow networks

Gabriele Engl
Mathematisches Institut
Technische Universität München
Arcisstrasse 21
D - 80290 München, Germany

Abstract

A network formulation is introduced for the modeling and numerical simulation of complex gas transmission systems like a multi-cylinder internal combustion engine. Basic elements of a network are chambers of finite volume, e.g. cylinders of an internal combustion engine, pipes and connections like valves or nozzles. Chamber states are described by ODE systems in time which include measurement results for heat losses and the combustion energy. The pipe flow is modeled by the unsteady, one-dimensional Euler equations of gas dynamics. A semi-empirical approach for the connections, which are characterized by throttling effects, yields boundary (coupling) conditions – a system of differential algebraic equations (DAEs) in time. The numerical solution is based on a TVD scheme for the pipe flow and a predictor-corrector method for the remaining DAE system. Different extensions of the TVD method to the inhomogeneous Euler equations are discussed. Simulation results for an internal combustion engine show that a fast solver is obtained for complex gas flow problems which is applied in industrial research.

1 Introduction

The simulation of compressible flow in complex gas transmission systems plays an important role in industrial applications. As an example a multi-cylinder internal combustion engine is introduced. The system consists of different components like cylinders, chambers, pipes, valves, branches etc. and can be described as a gas flow network. Figure 1 shows the network representation of a specific four-cylinder engine.

There are two main interests concerning the flow simulation in a multi-cylinder engine: First, a *fast simulation* of the whole system is important. In order to optimize the charge cycle in a cylinder, for example, the flow has to be simulated during several four-stroke cycles, and parameters like the volumetric efficiency have to be computed within minutes. For this purpose the program package PROMO is used in the car industry [3], [9]. The second main interest are *3D-simulations* of specific parts as a cylinder or a catalyst. Boundary conditions for these simulations can be made available by a program like PROMO.

```
1   2   3       4           5           6   7   8           9           10      11  12  1
```

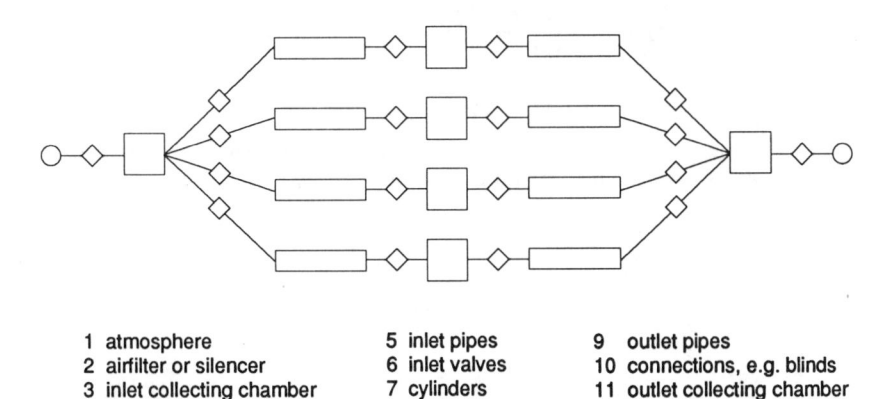

Figure 1: Network representation of a four-cylinder internal combustion engine

1 atmosphere	5 inlet pipes
2 airfilter or silencer	6 inlet valves
3 inlet collecting chamber	7 cylinders
4 connections, e.g. blinds	8 outlet valves

9 outlet pipes	
10 connections, e.g. blinds	
11 outlet collecting chamber	
12 catalyst or silencer	

This work deals with the mathematical model for a fast simulation concept which is based on a semi-empirical approach. The pipe flow is modeled by the one-dimensional Euler equations which are supplemented by appropriate boundary (coupling) conditions. A new solution algorithm is presented which involves a TVD scheme for the Euler equations. The TVD method has originally been developed for homogeneous systems of hyperbolic conservation laws, and its extension to the inhomogeneous Euler equations is discussed.

The work is motivated by deficiencies of the program system PROMO concerning the numerical treatment of the pipe flow and the coupling conditions.

2 Mathematical model

The following network components are distinguished:

- connections, e.g. valves and nozzles
- chambers of finite volume, e.g. cylinders
- straight pipes
- reservoirs of constant state, e.g. atmosphere

Connections

Figure 2 shows a connection which couples two states. Each state is described by density ρ_i, velocity u_i and pressure p_i ($i = 1, 2$) and denotes a chamber state, a pipe boundary state or a constant state. In general the states are functions of time t.

The mass flow $\dot{m}$ through a connection with cross-sectional area $A = A(t)$ is modeled by the flow equation of Saint Venant:

$$\dot{m} = \begin{cases} \alpha \sqrt{2\rho_{10}p_{10}} \, A \, \psi(p_2, p_{10}) , & p_{10} > p_{20} \\ -\alpha \sqrt{2\rho_{20}p_{20}} \, A \, \psi(p_1, p_{20}) , & p_{10} \leq p_{20} \end{cases} \quad (1)$$

where

$$\psi(p_2, p_{10}) = \sqrt{\frac{\kappa}{\kappa - 1} \left[\left(\frac{p_2}{p_{10}}\right)^{\frac{2}{\kappa}} - \left(\frac{p_2}{p_{10}}\right)^{\frac{\kappa+1}{\kappa}} \right]}$$

$\kappa = c_p/c_v$ ratio of the specific heat capacities

and ρ_{i0}, p_{i0} denote density and pressure of the state of rest which is obtained by an isentropic and energy-conserving transformation of ρ_i, u_i, p_i ($i = 1, 2$):

$$\frac{\kappa}{\kappa - 1} \frac{p_{i0}}{\rho_{i0}} = \frac{\kappa}{\kappa - 1} \frac{p_i}{\rho_i} + \frac{u_i^2}{2}$$

$$p_{i0} \rho_{i0}^{-\kappa} = p_i \rho_i^{-\kappa}.$$

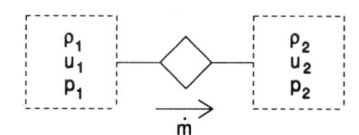

Figure 2: A connection and the coupled states

Throttling effects are taken into account by the empirical flow coefficient α which is defined as the ratio of real mass flow $\dot{m}$ and theoretical (ideal) mass flow $\dot{m}_{th}$ ($0 \leq \alpha \leq 1$). Flow coefficients are determined by measurement and might be functions of time t (e.g. for valves where $A = A(t)$ holds), whereas the dependency on the gas state is negligible.

Chambers

The chamber states are described by ODE systems in time t. As an example, the scheme of a cylinder with two coupled states is depicted in figure 3.

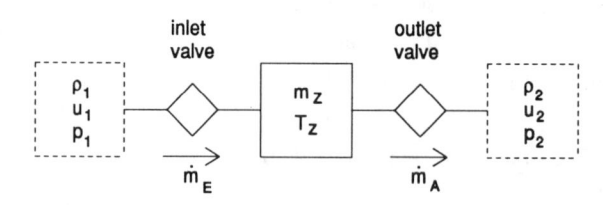

Figure 3: A cylinder with two coupled states

It is assumed that the kinetic energy of the inflowing gas is completely converted into internal energy (velocity $u_Z = 0$). Mass and energy balances yield two ODEs for the mass $m_Z = m_Z(t)$ and the (average) temperature $T_Z = T_Z(t)$ in the cylinder:

$$\begin{aligned} \dot{m}_Z &= \dot{m}_E - \dot{m}_A + \dot{m}_B \\ \dot{T}_Z &= (-p_Z \dot{V}_Z + \dot{m}_E H_E - \dot{m}_A H_A + \dot{Q}_B - \dot{Q}_W - c_v T_Z \dot{m}_Z)/(c_v m_Z) \end{aligned} \qquad (2)$$

where

$\dot{m}_E/\dot{m}_A$	mass flow through inlet/outlet valve
m_B	mass of fuel
V_Z	cylinder volume
H_E/H_A	(specific) enthalpy of the gas flowing through inlet/outlet valve
Q_B	combustion energy
Q_W	heat loss at the cylinder surface .

The (average) pressure $p_Z = p_Z(t)$ is given by the ideal gas law:

$$p_Z = Rm_Z T_Z/V_Z \quad (R: \text{ universal gas constant })\,.$$

The combustion energy Q_B and the heat exchange Q_W are described by empirical functions. A detailed description of the cylinder equations as an example for the chamber modeling can be found in Engl, Rentrop [1]. See also Urlaub [10].

Pipes

Figure 4 shows a pipe with two coupled states. The boundary values at the left and right boundary are denoted by ρ_l, u_l, p_l and ρ_r, u_r, p_r, respectively.

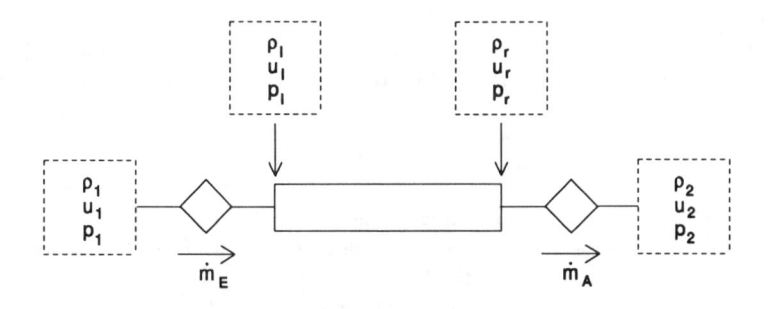

Figure 4: A pipe and the coupled states

The pipe flow is modeled by the 1D Euler equations

$$\bar{U}_t + F(\bar{U})_x = Q(\bar{U}) \quad , \tag{3}$$

a system of hyperbolic conservation laws, where

$$\bar{U} = \begin{pmatrix} \rho A \\ \rho u A \\ e A \end{pmatrix} \quad , \quad F(\bar{U}) = \begin{pmatrix} \rho u A \\ (\rho u^2 + p)A \\ u(e+p)A \end{pmatrix} \quad , \quad Q(\bar{U}) = \begin{pmatrix} 0 \\ p\frac{dA}{dx} \\ 0 \end{pmatrix} \quad ,$$

$$e = \frac{1}{2}\rho u^2 + \frac{1}{\kappa - 1}p \quad \text{(total energy per unit volume)} \quad ,$$

and $A = A(x)$ denotes the cross-sectional area of the pipe.

For $A = const$ the system reduces to

$$U_t + F(U)_x = 0 \quad , \quad U = \begin{pmatrix} \rho \\ \rho u \\ e \end{pmatrix} \quad , \quad F(U) = \begin{pmatrix} \rho u \\ \rho u^2 + p \\ u(e + p) \end{pmatrix} . \quad (4)$$

The characteristic formulation of system (3) is given by

$$p_t - \rho c \, u_t + (u - c)(p_x - \rho c \, u_x) = -\kappa \rho u \frac{1}{A} \frac{dA}{dx} \quad (5)$$

$$(p\rho^{-\kappa})_t + u(p\rho^{-\kappa})_x = 0 \quad (6)$$

$$p_t + \rho c \, u_t + (u + c)(p_x + \rho c \, u_x) = -\kappa \rho u \frac{1}{A} \frac{dA}{dx} \quad (7)$$

where $c = \sqrt{\kappa p / \rho}$ denotes the sound speed. Subsonic flow conditions ($|u| < c$) are assumed in the following.

At each boundary of the pipe a system of three boundary conditions has to be imposed [6]:
In the *inflow case* the continuity equation and the energy equation yield two physical boundary conditions. A numerical boundary condition is given by the characteristic equation which describes a transport of information from the interior of the pipe towards the boundary.
In the *outflow case* the boundary conditions consist of the continuity equation (physical boundary condition) and two characteristic equations (numerical boundary conditions).

The resulting system of equations for the left boundary is summarized in table 1, where the following notations are used:

A_l cross-sectional area of the pipe at the left boundary
A_E and α_E cross-sectional area and flow coefficient of the connection coupled with the left boundary
ρ_{i0}, p_{i0} ($i = 1, l$)   density and pressure of the state of rest corresponding to ρ_i, u_i, p_i.

Table 1: Boundary conditions at the left boundary

	physical boundary condition(s)	numerical boundary condition(s)
inflow ($u \geq 0$)	$\rho_l u_l A_l = \dot{m}_E$ where $\dot{m}_E = \alpha_E \sqrt{2\rho_{10} p_{10}} \, A_E \, \psi(p_l, p_{10})$ $\frac{\kappa}{\kappa - 1} \frac{p_l}{\rho_l} + \frac{u_l^2}{2} = \frac{\kappa}{\kappa - 1} \frac{p_1}{\rho_1} + \frac{u_1^2}{2}$	(5)
outflow ($u < 0$)	$\rho_l u_l A_l = \dot{m}_E$ where $\dot{m}_E = -\alpha_E \sqrt{2\rho_{10} p_{10}} \, A_E \, \psi(p_1, p_{10})$	(5) (6)

3 Numerical solution

A uniform space discretization is introduced for every pipe of the gas flow network. The space derivatives in the numerical boundary conditions are approximated by one-sided difference formulas of second order.
Furthermore the following notations are used:

$Y = Y(t)$ vector of the gas masses and temperatures in all chambers
$U_I = U_I(t)$ vector of the state variables at the internal discretization points of all pipes
$U_R = U_R(t)$ vector of the state variables at the boundaries of all pipes.

The ODE system which describes the chamber states can be written in the form

$$\dot{Y} = B(Y, U_R, t) \quad . \tag{8}$$

Semi-discretization of the boundary conditions yields a system of differential-algebraic equations (DAE system)

$$K(U_R)\dot{U}_R = L(Y, U_I, U_R, t) \tag{9}$$

where the matrix $K(U_R)$ is singular since nonlinear equations (physical boundary conditions) are included.

For a brief description of the solution algorithm it is assumed that approximations of the state variables at a given time level t_n have been computed:

$$Y^n \quad , \quad U_I^n \quad , \quad U_R^n$$

The unknown variables

$$Y^{n+1} \quad , \quad U_I^{n+1} \quad , \quad U_R^{n+1}$$

at the time level $t_{n+1} = t_n + \Delta t$, where Δt denotes the time step, are determined in the following way:

1. An explicit TVD scheme of second order is applied to the Euler equations:

$$U_I^n \, , \, U_R^n \, \rightarrow \, \boxed{U_I^{n+1}} \; .$$

2. The chamber equations (8) and the semi-discretized boundary equations (9) are solved by a second order predictor corrector method.

   Predictor step

   The ODEs of (8) and (9) are discretized by the explicit first order method of Euler. Predictor values $\tilde{Y}^{n+1}$ for Y can be computed directly. Imposing the physical boundary conditions of (9) at the new time level t_{n+1}, a nonlinear system of equations for the predictor values $\tilde{U}_R^{n+1}$ is obtained which is solved by Newton's method:

$$Y^n \, , \, U_R^n \, \rightarrow \, \tilde{Y}^{n+1}$$

$$\tilde{Y}^{n+1} \, , \, U_I^n \, , \, U_R^n \, \rightarrow \, \tilde{U}_R^{n+1} \; .$$

Corrector step

The discretization of the ODEs is based on the implicit trapezoidal rule. Newton's method is applied to the resulting system of nonlinear equations:

$$Y^n, \tilde{Y}^{n+1}, U_R^n, \tilde{U}_R^{n+1} \rightarrow \boxed{Y^{n+1}}$$

$$Y^{n+1}, U_I^n, U_I^{n+1}, U_R^n, \tilde{U}_R^{n+1} \rightarrow \boxed{U_R^{n+1}}$$

The time step Δt is restricted by the CFL-condition for the solution of the Euler equations. Furthermore, discontinuities of the mathematical model have to be taken into account. A discontinuity occurs, for example, when the flow direction through a connection changes or a valve opens or closes. For a detailed description of the algorithm see Engl [2].

The program system PROMO, which is used in the car industry, is based on the Lax-Wendroff method for the numerical solution of the 1D Euler equations. This classical difference scheme produces numerical oscillations near discontinuities and in regions where the solution has steep gradients. Fast changes in temperature are typical for the gas flow in an internal combustion engine, and the Lax-Wendroff method partly causes unacceptable errors.

Therefore the new algorithm involves a TVD scheme which prevents spurious oscillations. TVD schemes have been developed for homogeneous systems of hyperbolic conservation laws. Since the inhomogeneous Euler equations play an important role in practical applications, the extension of a TVD scheme to this case is discussed in the next section.

4 A TVD method for the homogeneous and the inhomogeneous Euler equations

A TVD (*total variation diminishing*) scheme is considered which was introduced by Harten [4]. The scheme is based on the modified flux method which will briefly be described in the following.

Consider the initial value problem for a single conservation law

$$\begin{aligned} u_t + f(u)_x \equiv u_t + a(u)u_x = 0 \quad, \quad x \in \mathbf{R} \quad, \quad 0 < t \in \mathbf{R} \\ u(x,0) = u_0(x) \end{aligned} \quad (10)$$

where $a(u) = df(u)/du$. Let u_i^n, $i \in \mathbb{Z}$, $n \in \mathbb{N}_0$, denote approximations of the exact solution at discretization points (x_i, t_n),

$$x_i = i\,\Delta x \quad, \quad i \in \mathbb{Z} \quad (\Delta x = const) \quad, \quad 0 = t_0 < t_1 < t_2 < \ldots \quad .$$

An explicit 3-point TVD scheme of first order is given by the conservative formulation

$$u_i^{n+1} = u_i^n - \lambda(\tilde{f}_{i+\frac{1}{2}}^n - \tilde{f}_{i-\frac{1}{2}}^n) \quad, \quad \lambda = \frac{\Delta t}{\Delta x} \quad, \quad \Delta t = t_{n+1} - t_n \quad (11)$$

where

$$\tilde{f}_{i+\frac{1}{2}}^n = \tilde{f}(u_{i+1}^n, u_i^n) = \frac{1}{2}[f_i^n + f_{i+1}^n - \frac{1}{\lambda}Q(\lambda a_{i+\frac{1}{2}}^n)(u_{i+1}^n - u_i^n)] \quad (12)$$

denotes the numerical flux function and

$$f_i^n = f(u_i^n)$$

$$a_{i+\frac{1}{2}}^n = \begin{cases} (f_{i+1}^n - f_i^n)/(u_{i+1}^n - u_i^n) & \text{for } u_{i+1}^n - u_i^n \neq 0 \\ a(u_i^n) & \text{for } u_{i+1}^n - u_i^n = 0 \end{cases}$$

$Q(x)$ coefficient of numerical viscosity, $|x| \leq Q(x) \leq 1$ for $0 \leq |x| \leq \mu$, where μ is some real number for which $0 \leq \mu \leq 1$ holds

$\lambda \max_i |a_{i+\frac{1}{2}}^n| \leq \mu$ CFL-condition.

Based on this scheme, an explicit 5-point TVD scheme of second order is derived in the following way:

First, it is observed that the numerical solution (11) is a second order approximation to the modified differential equation

$$u_t + f_x = \frac{1}{\lambda}\tilde{g}_x \quad , \quad \tilde{g} = \tilde{g}(u, \lambda) = \frac{\Delta x}{2}[Q(\lambda a) - (\lambda a)^2]u_x \quad .$$

Thus, if the first order scheme (11), (12) is applied to the modified conservation law

$$u_t + (f + \frac{1}{\lambda}\tilde{g})_x = 0 \quad ,$$

a second order scheme is obtained for the original equation $u_t + f_x = 0$.

Secondly, the TVD property is enforced by a *minmod* limiter function, see e.g. LeVeque [7]. The resulting scheme is given by

$$u_i^{n+1} = u_i^n - \lambda(\bar{f}_{i+\frac{1}{2}}^n - \bar{f}_{i-\frac{1}{2}}^n) \quad , \tag{13}$$

$$\bar{f}_{i+\frac{1}{2}}^n = \bar{f}(u_{i+2}^n, u_{i+1}^n, u_i^n, u_{i-1}^n) = \frac{1}{2}[f_i^n + f_{i+1}^n + \frac{1}{\lambda}(g_i^n + g_{i+1}^n) - \frac{1}{\lambda}Q(\lambda a_{i+\frac{1}{2}}^n + \gamma_{i+\frac{1}{2}}^n)(u_{i+1}^n - u_i^n)] \quad , \tag{14}$$

where the following notations are used:

$$g_i^n = g_i^n(u_{i-1}^n, u_i^n, u_{i+1}^n) = minmod(\tilde{g}_{i+\frac{1}{2}}^n, \tilde{g}_{i-\frac{1}{2}}^n)$$

$$\tilde{g}_{i+\frac{1}{2}}^n = \tfrac{1}{2}(\tilde{g}_i^n + \tilde{g}_{i+1}^n) = \tfrac{1}{2}[Q(\lambda a_{i+\frac{1}{2}}^n) - (\lambda a_{i+\frac{1}{2}}^n)^2](u_{i+1}^n - u_i^n)$$

$$\gamma_{i+\frac{1}{2}}^n = (g_{i+1}^n - g_i^n)/(u_{i+1}^n - u_i^n) \quad .$$

If the CFL-condition $\lambda \max_i |a_{i+\frac{1}{2}}^n| \leq \mu$ is satisfied, the scheme converges to weak solutions of (10) for all initial data of bounded total variation [4].

The TVD method (13), (14) is extended to a system of hyperbolic conservation laws by transforming the system into its characteristic formulation and then applying the scheme to each of the characteristic equations.

In the following, two cases are considered for the Euler equations:

- **Homogeneous case**: cross-sectional area $A = const$

  The TVD scheme for the solution of (4) can be written in the form

  $$U_i^{n+1} = U_i^n - \lambda(\bar{F}_{i+\frac{1}{2}}^n - \bar{F}_{i-\frac{1}{2}}^n) \quad , \quad \bar{F}_{i+\frac{1}{2}}^n = \bar{F}(U_{i+2}^n, U_{i+1}^n, U_i^n, U_{i-1}^n) \quad ,$$

  where $\bar{F}$ denotes the vector-valued flux function.

  Convergence has not been proved in the case of nonlinear systems. However, numerical experiments with the Riemann problem have shown very good results for the Euler equations.

- **Inhomogeneous case**: $A = A(x)$

  In this case the source term $Q(\bar{U})$ of the Euler equations (3) has to be taken into account. Let us consider the following extension of the TVD scheme:

  $$\bar{U}_i^{n+1} = \bar{U}_i^n - \lambda(\overline{\bar{F}}_{i+\frac{1}{2}}^n - \overline{\bar{F}}_{i-\frac{1}{2}}^n) + \frac{\Delta t}{2}(Q_{i+\frac{1}{2}}^n + Q_{i-\frac{1}{2}}^n) \quad , \tag{15}$$

  $$\overline{\bar{F}}_{i+\frac{1}{2}}^n = \bar{F}(\bar{U}_{i+2}^n, \bar{U}_{i+1}^n, \bar{U}_i^n, \bar{U}_{i-1}^n) \quad , \quad Q_{i+\frac{1}{2}}^n = \frac{1}{2}(Q(\bar{U}_i^n) + Q(\bar{U}_{i+1}^n)) \; .$$

Numerical experiments showed an unstable behavior of this method, see Neumeyer [8].

More precisely, consider initial conditions which correspond to a state of rest:

$$\bar{U}(x, t = 0) = const \quad , \quad \text{velocity } u(x, t = 0) = 0$$

The exact solution is equal to the initial state at any time $t > 0$. A moving gas, however, is described by the numerical solution as demonstrated in section 5.

This behavior is not observed, if the TVD scheme is based on the following formulation of the Euler equations which is equivalent to (3):

$$U_t + F(U)_x = -\frac{A_x}{A}\begin{pmatrix} \rho u \\ \rho u^2 \\ u(e+p) \end{pmatrix} \quad , \tag{16}$$

$$U = \begin{pmatrix} \rho \\ \rho u \\ e \end{pmatrix} \quad , \quad F(U) = \begin{pmatrix} \rho u \\ \rho u^2 + p \\ u(e+p) \end{pmatrix} .$$

The reason is still being investigated. Moreover, other high resolution schemes are considered [5].

The next section shows simulation results for an internal combustion engine. Furthermore, the instability of scheme (15) is demonstrated.

5 Simulation results

The new solution algorithm has been applied to the mathematical model of a four-stroke cycle single cylinder internal combustion engine. Figure 5 shows the network representation of the engine which includes a pipe of constant cross-sectional area. A listing of the physical parameters can be found in [2].

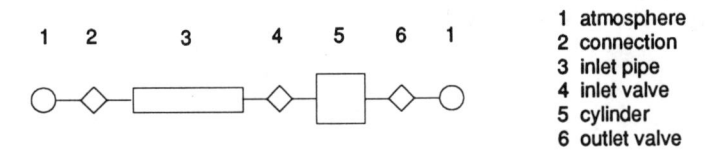

```
     1    2      3       4    5    6    1        1 atmosphere
                                              2 connection
     O—◇—[======]—◇—[====]—◇—O        3 inlet pipe
                                              4 inlet valve
                                              5 cylinder
                                              6 outlet valve
```

Figure 5: Network of a single cylinder internal combustion engine

A complete replacement of the burnt gas during the charge cycle and a high amount of fresh air are important for an efficient performance of the engine. The gas exchange is influenced by pressure waves which are produced in the inlet pipe, and its dependency on the pipe parameters has been studied. Figure 6 shows the volumetric efficiency λ as a function of the pipe length L. The volumetric efficiency represents a measure for the quality of the charge cycle and is defined as the mass of fresh air contained in the cylinder after the charge cycle, divided by a reference mass.

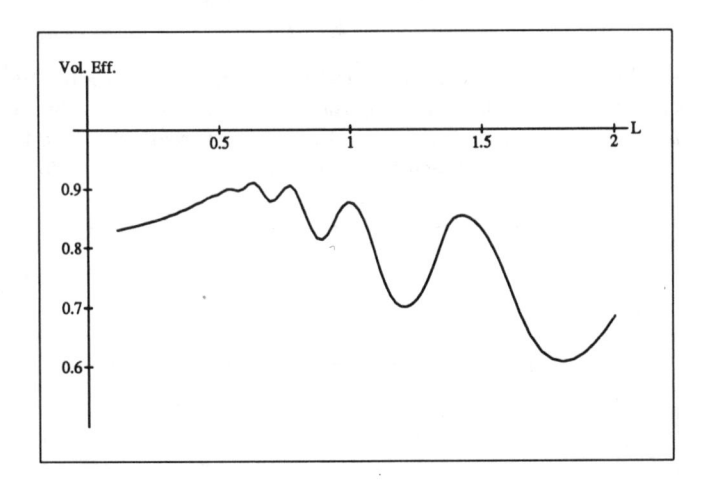

Figure 6: Volumetric efficiency λ versus pipe length L $[m]$

Calculations without pipe yield $\lambda = 0.84$. It is observed that a relatively short pipe does not change this value considerably. The reason are dominating reflections and superposition of pressure waves. If the pipe length is increased, the influence of pressure waves becomes more evident. For $L = 0.62\,[m]$ the maximum value $\lambda = 0.91$ is obtained. This implies an improvement of approximately 8% compared to the calculations without pipe.

The simulation of one four-stroke cycle for $L = 1\,[m]$ and 100 discretization intervals in the pipe takes about 360 CPU-seconds on a workstation HP-APOLLO DN 4500. This shows that a fast solver is obtained which has partly been implemented in an industrial simulation package [8].

Finally, the instability of scheme (15) for the inhomogeneous Euler equations is demonstrated. For this purpose a cone is considered which is filled with a gas at rest. In figure 7 the geometry of the cone is depicted.

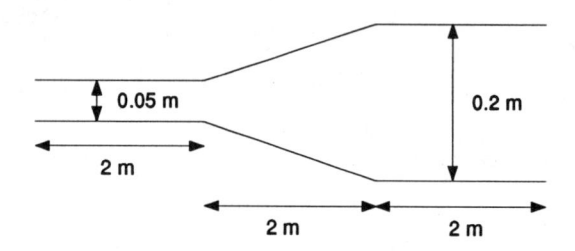

Figure 7: A cone

The initial conditions are given by

$$U(x, t = 0) = \begin{pmatrix} 1.177 \\ 0. \\ 2.5 \end{pmatrix} = const.$$

Figure 8 shows the velocity which is computed by the method (15) after 10 time steps using 300 discretization intervals. Obviously the method produces disturbances at points where the cross-sectional area $A(x)$ is not continuously differentiable. The true solution is computed by the TVD scheme based on formulation (16) of the Euler equations.

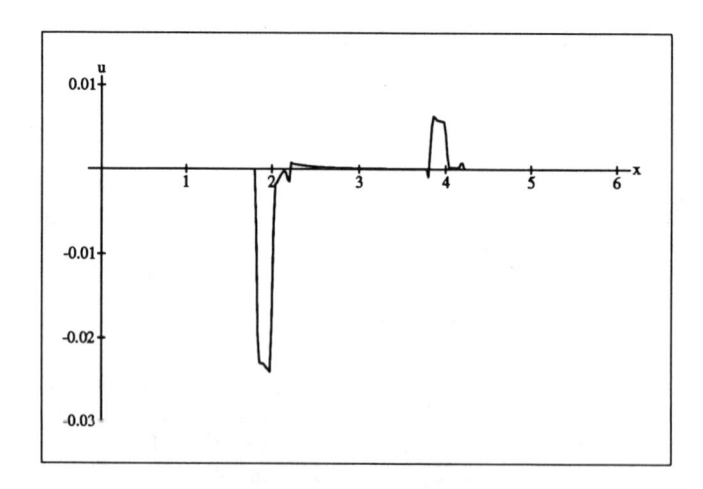

Figure 8: Velocity u versus x after 10 time steps

Acknowledgement

This work has partly been supported within the BMFT-project *Anwendungsorientierte Mathematik*.

References

[1] G. Engl, P. Rentrop: Gas flow in a single cylinder internal combustion engine: A model and its numerical treatment. Int. J. Num. Meth. Heat Fluid Flow 2 (1992) 63–78

[2] G. Engl: Modellierung und numerische Simulation der Gasströmung in Netzwerken am Beispiel des Ladungswechsels im Verbrennungsmotor. Dissertation, Institut für Informatik, Technische Universität München, 1994

[3] K. A. Görg, H.-J. Linnhoff, A. G. Sadek: Optimierung des Programmsystems PROMO zur Berechnung des instationären Ladungswechsels von zündenden Mehrzylinder-Verbrennungsmotoren. FVV (Forschungsvereinigung Verbrennungskraftmaschinen) - Forschungsberichte, Hefte 281-1 bis 281-3, 1981

[4] A. Harten: High resolution schemes for hyperbolic conservation laws. J. Comput. Phys. 49 (1983) 357–393

[5] A. Harten: ENO schemes with subcell resolution. J. Comput. Phys. 83 (1989) 148–184

[6] C. Hirsch: Numerical Computation of Internal and External Flows 1, 2. John Wiley & Sons, Chichester, 1988, 1990

[7] R. J. LeVeque: Numerical Methods for Conservation Laws. Birkhäuser Verlag, Basel, 2nd Edition, 1992

[8] Th. Neumeyer: Ein TVD-Verfahren zur numerischen Simulation des Ladungswechsels im Verbrennungsmotor. Diplomarbeit, Mathematisches Institut, Technische Universität München, 1993

[9] H. Seifert: 20 Jahre erfolgreiche Entwicklung des Programmsystems PROMO. MTZ (Motortechnische Zeitschrift) 51 (1990) 478–488

[10] A. Urlaub: Verbrennungsmotoren 1, 2. Springer-Verlag, Berlin, 1987, 1989

ON THE DESIGN OF PIECEWISE UNIFORM MESHES FOR SOLVING ADVECTION-DOMINATED TRANSPORT EQUATIONS TO A PRESCRIBED ACCURACY

PAUL A. FARRELL

Department of Mathematics and Computer Science, Kent State University, OH 44242, U.S.A..

ALAN F. HEGARTY

Department of Mathematics and Statistics, University of Limerick, Limerick, Ireland.

JOHN J.H. MILLER

Department of Mathematics, Trinity College, Dublin 2, Ireland.

EUGENE O'RIORDAN

Department of Mathematics, Regional Technical College, Tallaght, Dublin 24, Ireland.

AND

G.I. SHISHKIN

Institute of Mathematics and Mechanics, Russian Academy of Sciences, Ekaterinburg, Russia.

SUMMARY

The numerical performance of numerical methods specifically designed for singularly perturbed partial differential equations is examined. Numerical methods whose solutions have an accuracy independent of the small parameter are called ε-uniform methods. In this paper, the advantages of using an ε-uniform numerical method are discussed.

1. Introduction

Partial differential equations with a small parameter, denoted here by ε, multiplying the highest order derivatives are called singularly perturbed problems. Numerical methods whose solutions have an accuracy independent of the small parameter are called ε-uniform methods. The main concern of this paper is an examination of the practical benefits of using existing ε-uniform methods for the numerical solution of singularly perturbed problems, rather than the construction of new methods.

2. Statement of the problem

Consider the linear transport problem

$$Lu \equiv \varepsilon \Delta u + \vec{a}.\nabla u + a_0 u = f \quad \text{on} \quad \Omega = (0,1) \times (0,1) \quad (2.1a)$$

$$u = g \quad \text{on} \quad \partial\Omega \quad (2.1b)$$

$$0 < \varepsilon \leq 1.$$

In the case when $\varepsilon \ll 1$ the problem is singularly perturbed and the problem is said to be advection-dominated. Here u may be thought of as a concentration of some quantity (e.g., heat, a pollutant etc.) that is driven by a known advective velocity field $\vec{a} = (a_1, a_2)$ where ε is the diffusivity. Problem (2.1) may also be regarded as a linear model for the Navier-Stokes flow equations. The functions $\vec{a}$, a_0, f and g are assumed to be sufficiently smooth.

Boundary layers may appear in the solution of (2.1) near certain parts of the boundary for small values of ε. If $\vec{a} \cdot \vec{n} = 0$ on a part of the boundary with a boundary layer, where $\vec{n}$ is the unit normal to the boundary, then the boundary layer is called a parabolic boundary layer; otherwise it is called a regular boundary layer. Here, it is assumed that $\vec{a}$ always satisfies one of the following conditions: either

$$\vec{a}(x,y) = (a_1(x,y), a_2(x,y)) \geq (\alpha_1, \alpha_2) > (0,0), \quad \text{all } (x,y) \in \bar{\Omega} \quad (2.1c)$$

in which case only regular boundary layers can arise, or

$$a_1(x,y) \geq \alpha_1 > 0 \quad a_2(x,y) = 0, \quad \text{all}(x,y) \in \bar{\Omega} \quad (2.1d)$$

in which case both regular and parabolic layers may be present.

A numerical method for solving (2.1) is said to be ε-uniform of order p on the sequence of meshes $\{\Omega_N\}_1^\infty$ where $\Omega_N = \{(x_i, y_j), i,j = 0,1,\ldots,N\}$ if $\exists N_0$, independent of ε, such that for all $N \geq N_0$

$$\sup_{0<\varepsilon\leq 1} \|u - u_N\|_{\Omega_N} \leq CN^{-p} \quad (2.2a)$$

($\|w\|_{\Omega_N} \equiv \max_{x_i \in \Omega_N} |w(x_i)|$) where u is the solution of (2.1), u_N is the numerical approximation to u, C and $p > 0$ are independent of ε and N. In addition, a numerical method for solving (2.1) is said to be globally ε-uniform of order p on the sequence of meshes $\{\Omega_N\}_1^\infty$ if $\exists N_0$, independent of ε, such that for all $N \geq N_0$

$$\sup_{0<\varepsilon\leq 1} \|u - u_N^I\|_\Omega \leq CN^{-p} \quad (2.2b)$$

($\|w\|_\Omega \equiv \max_{x \in \Omega} |w(x)|$), where u_N^I is the bilinear interpolent on Ω of the numerical solution u_N. In these definitions it is clear that the generic parameters C and p depend on N_0.

In this paper, two specific linear transport problems are examined. In the first

$$\vec{a}(x,y) = (2 + x^2 y, 1 + xy), \quad f(x,y) = x^2 + y^3 + \cos(x+2y), \quad a_0 = 0. \quad (2.3a)$$

with the boundary conditions

$$u(x,0) = 0; \quad u(x,1) = 4x(1-x), x < 1/2, u(x,1) = 1, x \geq 1/2; \quad (2.3b)$$

$$u(0,y) = 0; \quad u(1,y) = 8(y - 2y^2), y < 1/4, u(1,y) = 1, y \geq 1/4 . \quad (2.3c)$$

In this case, condition (2.1c) is fulfilled and for small values of ε regular boundary layers of width $O(\varepsilon)$ occur at the outflow boundary $\{(0,y) : 0 \leq y \leq 1\} \cup \{(x,0) : 0 \leq x \leq 1\}$. For the second problem

$$\vec{a}(x,y) = (1 + x^2 + y^2, 0), \quad f(x,y) = 0, \quad a_0 = 0 \quad (2.4a)$$

with the boundary conditions

$$u(x,0) = x^2; \quad u(x,1) = x^3 \quad (2.4b)$$

$$u(0,y) = 0; \quad u(1,y) = 1 . \quad (2.4c)$$

In this case, condition (2.1d) is fulfilled and for small values of ε a regular boundary layer of width $O(\varepsilon)$ occurs at the outflow boundary $\{(0,y) : 0 \leq y \leq 1\}$, while parabolic boundary layers of width $O(\sqrt{\varepsilon})$ occur at the edges $\{(x,0) : 0 \leq x \leq 1\}$ and $\{(x,1) : 0 \leq x \leq 1\}$. Numerical results for (2.1), (2.3) were presented in [1] and in [2] for (2.1), (2.4). Henceforth (2.1),(2.3) is referred to as Problem 1 and (2.1),(2.4) as Problem 2.

3. Construction of the numerical method

It is well known both theoretically and numerically that numerical methods based on upwind operators on a uniform mesh are not ε-uniform. In [1] and [2] it was shown that the use of a piecewise uniform mesh specially adapted to the boundary layers together with a standard upwind difference operator overcomes this difficulty and yields an ε-uniform method.

On an arbitrary mesh $\Omega_N \equiv \{(x_i, y_j); 0 \leq i, j \leq N\}$, the standard upwind operator for (2.1) is defined by

$$[\varepsilon(\delta_x^2 + \delta_y^2) + a_1(x_i, y_j)D_x^+ + a_2(x_i, y_j)D_y^+ + a_0(x_i, y_j)]u_N(x_i, y_j) = f(x_i, y_j) \quad (3.1)$$

where $h_i = x_i - x_{i-1}$, $\bar{h}_i \equiv (h_{i+1} + h_i)/2$, and

$$D_x^+ u_N(x_i, y_j) \equiv (u_N(x_{i+1}, y_j) - u_N(x_i, y_j))/h_{i+1}$$
$$D_x^- u_N(x_i, y_j) \equiv (u_N(x_i, y_j) - u_N(x_{i-1}, y_j))/h_i$$
$$\delta_x^2 u_N(x_i, y_j) \equiv (D_x^+ u_N(x_i, y_j) - D_x^- u_N(x_i, y_j))/\bar{h}_i$$

with analogous definitions for D_y^+, D_y^- and δ_y^2.

Regular boundary layers of width $O(\varepsilon)$ appear in the solution of Problem 1 near the outflow boundary $\{(0,y) : 0 \leq y \leq 1\} \cup \{(x,0) : 0 \leq x \leq 1\}$. The density of mesh points near this outflow boundary is increased by constructing the piecewise-uniform mesh Ω_N^*, condensing near this boundary, where

$$\Omega_N^* \equiv \{(x_i^*, y_j^*) : 0 \leq i, j \leq N\} \quad (3.2)$$

with

$$x_i^* = \begin{cases} ih_1, & \text{for } 0 \leq i \leq N/2, \\ \sigma_x + (i - (N/2))h_2, & \text{for } N/2 < i \leq N \end{cases}$$

and
$$h_1 = 2\sigma_x/N, \quad \text{and} \quad h_2 = 2(1-\sigma_x)/N.$$

The points $\{y_j^*\}$ are defined analogously. The transition point σ_x, which depends on both ε and N, is defined by

$$\sigma_x \equiv \min\{1/2, C_1\varepsilon \ln N\}, \quad \text{where } C_1 \text{ is any constant satisfying } C_1 > \alpha_1^{-1} \quad (3.3)$$

where α_1 is defined in (2.1c). The transition point σ_y is defined analogously. In Problem 1 $\alpha_1 = 2$. It follows that any choice of C_1 satisfying $C_1 > 1/2$ will suffice. In the numerical experiments below the constant C_1 in (3.3) is taken to be 1.

In the solution of Problem 2 a regular boundary layer of width $O(\varepsilon)$ occurs at the outflow boundary $\{(0,y) : 0 \le y \le 1\}$ and parabolic boundary layers of width $O(\sqrt{\varepsilon})$ occur at the edges $\{(x,0) : 0 \le x \le 1\}$ and $\{(x,1) : 0 \le x \le 1\}$. The density of mesh points near each of these boundaries is increased by constructing the piecewise-uniform mesh Ω_N^{**}, condensing on the boundaries, where

$$\Omega_N^{**} \equiv \{(x_i^*, y_j^{**}) : 0 \le i, j \le N\} \quad (3.4)$$

with
$$x_i^* = \begin{cases} ih_1, & \text{for } 0 \le i \le N/2, \\ \sigma_x + (i - (N/2))h_2, & \text{for } N/2 < i \le N \end{cases}$$

$$h_1 = 2\sigma_x/N, \quad h_2 = 2(1-\sigma_x)/N.$$

and
$$y_j^{**} = \begin{cases} jk_1, & \text{for } 0 \le j \le N/4, \\ \sigma_y + (j - (N/4))k_2, & \text{for } N/4 < j < 3N/4, \\ 1 - \sigma_y + (j - (3N/4))k_1, & \text{for } 3N/4 \le j \le N \end{cases}$$

$$k_1 = 4\sigma_y/N, \quad k_2 = 2(1-2\sigma_y)/N.$$

The transition point σ_x satisfies (3.3) and the transition point σ_y is defined by

$$\sigma_y \equiv \min\{1/4, \sqrt{\varepsilon}\ln N\}. \quad (3.5)$$

It should be noted that for parabolic layers no constant analogous to C_1 in (3.3) for regular layers, occurs in the formula for the transition point σ_y.

4. Approximate ε-uniform convergence rate and ε-uniform error constants.

In this section the ε-uniform error constants $C = C(N_0)$ and the ε-uniform convergence rates $p = p(N_0)$ are estimated for various values of N_0 for Problems 1 and 2.

First the method for determining the approximate value of $p(N_0)$ is described. Given N_0 it is assumed that $k+2$ solutions $\{u_{2^l N_0}\}_{l=0}^{k+1}$ have been computed on the meshes $\{\Omega_{2^l N_0}\}_{l=0}^{k+1}$, for $\varepsilon = 2^{-r}, 0 \leq r \leq s$. The double-mesh differences are

$$d_N = \max_{0 \leq r \leq s} d_{N,2^{-r}} \quad \text{and} \quad d_{N,\varepsilon} = \|u_N - u_{2N}^I\|_{\Omega_N}$$

where u_{2N}^I denotes the bilinear interpolant of u_{2N}

In order to bound the double mesh errors, it is necessary to assume that the numerical method is globally ε-uniform (see (2.2b)). From this

$$d_{N,\varepsilon} = \|u_N - u_{2N}^I\|_{\Omega_N} \leq \|u_N - u\|_{\Omega_N} + \|u - u_{2N}^I\|_{\Omega_N}$$

$$\leq \|u_N^I - u\|_{\Omega} + \|u - u_{2N}^I\|_{\Omega} \leq CN^{-p} + C2^{-p}N^{-p} = C(1+2^{-p})N^{-p}.$$

Thus

$$d_N \leq C(1+2^{-p})N^{-p} \quad \forall N \geq N_0 \tag{4.1}$$

The computed ε-uniform convergence rates $p(N_0, k)$ are then defined by

$$p(N_0, k) = \min_{0 \leq l \leq k+1} p_{2^l N_0} \tag{4.2}$$

where

$$p_N = \log_2\left(\frac{d_N}{d_{2N}}\right). \tag{4.3}$$

The numerical determination of the approximate ε-uniform error constants and the approximate ε-uniform convergence rate of the solutions of an ε-uniform numerical method applied to a singularly perturbed problem is based on solving problems, corresponding to a sufficiently wide range of values of ε. The range of ε used depends, in general, on the problem being investigated and on the numerical method being used. In practice s is chosen such that

$$\max_{0 \leq l \leq k+1} |d_{2^l N_0, 2^{-s+1}} - d_{2^l N_0, 2^{-s}}| \leq \tau$$

where τ is a preassigned tolerance. The choice of the tolerance is $\tau = 10^{-6}$.

Having established the estimates $p(N_0, k)$ of the ε-uniform convergence rates $p(N_0)$ for various values of N_0 and k, it is now shown how estimates $C_q(N_0, k)$ of the corresponding ε-uniform error constant $C(N_0)$ may be obtained. Assuming that the numerical method is ε-uniform of order p for a sequence of meshes with ε-uniform error constant C, it follows from the definition of ε-uniform convergence that, for each q satisfying $0 \leq q \leq p$, there exists a C_q, independent of ε and N, such that for all $N \geq N_0$,

$$\sup_{0 < \varepsilon \leq 1} \|u - u_N\|_{\Omega_N} \leq C_q N^{-q} \tag{4.4}$$

where, clearly, C_q can always be chosen to be

$$C_q = C N_0^{-(p-q)} \leq C.$$

We need to use q because only the estimate $p(N_0, k)$ and not the true value of p is known. Furthermore, a small value of the error constant C may be preferable to the largest possible value of the convergence rate p.

From (4.1),
$$d_N \leq C_q(1 + 2^{-q})N^{-q} \quad \forall N \geq N_0 .$$

For each N let
$$D_{q,N} = N^q d_N \tag{4.5}$$

and let D_q be defined as
$$D_q(N_0, k) = \max_{0 \leq l \leq k+1} D_{q, 2^l N_0} .$$

Then
$$D_q \leq C_q(1 + 2^{-q})$$

The computed ε-uniform convergence constants $C_q(N_0, k)$ are then defined by
$$C_q(N_0, k) = \frac{D_q}{1 + 2^{-q}} . \tag{4.6}$$

In Table 4.1 values of $d_{2^l N_0, 2^{-r}}$ for $\tau = 10^{-6}, N_0 = 8, 0 \leq l \leq 4$ and $0 \leq r \leq s, s = 23$ are given for Problem 1. From the values of d_N given in Table 4.1 the quantities p_N for $N = 8, 16, 32, 64$ are then computed using (4.3).

Table 4.1 Double-mesh differences $d_{N,\varepsilon}$ and d_N for Problem 1.

ε	$N=8$	$N=16$	$N=32$	$N=64$	$N=128$
1	.950D-02	.327D-02	.192D-02	.103D-02	.529D-03
2^{-1}	.177D-01	.113D-01	.631D-02	.335D-02	.172D-02
2^{-2}	.417D-01	.272D-01	.159D-01	.857D-02	.444D-02
2^{-3}	.732D-01	.476D-01	.299D-01	.171D-01	.915D-02
2^{-4}	.550D-01	.424D-01	.385D-01	.308D-01	.176D-01
2^{-5}	.934D-01	.636D-01	.448D-01	.312D-01	.206D-01
2^{-6}	.117D+00	.825D-01	.571D-01	.384D-01	.244D-01
2^{-7}	.140D+00	.931D-01	.657D-01	.438D-01	.276D-01
2^{-8}	.153D+00	.103D+00	.703D-01	.472D-01	.297D-01
2^{-9}	.159D+00	.111D+00	.733D-01	.491D-01	.310D-01
2^{-10}	.162D+00	.114D+00	.754D-01	.501D-01	.317D-01
2^{-11}	.164D+00	.116D+00	.772D-01	.506D-01	.321D-01
2^{-12}	.165D+00	.117D+00	.781D-01	.509D-01	.323D-01
2^{-13}	.165D+00	.118D+00	.786D-01	.510D-01	.324D-01
2^{-14}	.165D+00	.118D+00	.788D-01	.511D-01	.325D-01
2^{-15}	.165D+00	.118D+00	.789D-01	.511D-01	.325D-01
2^{-16}	.165D+00	.118D+00	.790D-01	.512D-01	.325D-01
2^{-17}	.165D+00	.118D+00	.790D-01	.512D-01	.325D-01
2^{-18}	.165D+00	.118D+00	.790D-01	.512D-01	.325D-01
2^{-19}	.165D+00	.118D+00	.790D-01	.512D-01	.325D-01
2^{-20}	.165D+00	.118D+00	.790D-01	.512D-01	.325D-01
2^{-21}	.165D+00	.118D+00	.790D-01	.512D-01	.325D-01
2^{-22}	.165D+00	.118D+00	.790D-01	.512D-01	.325D-01
2^{-23}	.165D+00	.118D+00	.790D-01	.512D-01	.325D-01
d_N	.165D+00	.118D+00	.790D-01	.512D-01	.325D-01

The computed values of p_N are given in Table 4.2 for Problem 1 and in Table 4.3 values of $D_{q,N}$ are given for various values of q and N. The computed ε-uniform error constants $C_q(N_0, k)$ may then be obtained by taking the maximum D_q in each row of each table and then using (4.6) to compute C_q. The results are listed in Table 4.4 for a selection of values of q for each problem.

Table 4.2. Computed ε-uniform convergence rates p_N for Problem 1.

ε	$N=8$	$N=16$	$N=32$	$N=64$
p_N	0.48	0.58	0.63	0.65

Table 4.3. $D_{q,N}$ and D_q for Problem 1.

q	$N=8$	$N=16$	$N=32$	$N=64$	$N=128$	D_q
0.1	0.2031	0.1557	0.1117	0.0776	0.0528	0.2031
0.2	0.2501	0.2054	0.1580	0.1176	0.0858	0.2501
0.3	0.3079	0.2711	0.2234	0.1783	0.1393	0.3079
0.4	0.3791	0.3577	0.3160	0.2702	0.2263	0.3791
0.5	0.4667	0.4720	0.4469	0.4096	0.3677	0.4720
0.6	0.5746	0.6228	0.6320	0.6208	0.5973	0.6320
0.7	0.7074	0.8218	0.8938	0.9410	0.9704	0.9704
0.8	0.8709	1.0844	1.2640	1.4263	1.5763	1.5763
0.9	1.0722	1.4308	1.7876	2.1619	2.5608	2.5608
1.0	1.3200	1.8880	2.5280	3.2768	4.1600	4.1600

Table 4.4. Computed ε-uniform error constants C_q for Problem 1.

q	0.4	0.5	0.6	0.7	0.8	0.9	1.0
C_q	0.2156	0.2765	0.3808	0.6006	1.0013	1.6673	2.7733

In Table 4.5 values of $d_{2^l N_0, 2-2r}$ for $\tau = 10^{-6}$, $N_0 = 8$, $0 \le l \le 4$ and $0 \le 2r \le s$, $s = 30$ are given for Problem 2. From the values of d_N given in Table 4.5 the quantities p_N for $N = 8, 16, 32, 64$ are, as for Problem 1, computed using (4.3).

Table 4.5 Double-mesh differences $d_{N,\varepsilon}$ and d_N for Problem 2.

ε	$N=8$	$N=16$	$N=32$	$N=64$	$N=128$
1	.326D-02	.918D-03	.405D-03	.234D-03	.126D-03
2^{-2}	.206D-01	.104D-01	.536D-02	.273D-02	.138D-02
2^{-4}	.630D-01	.333D-01	.193D-01	.120D-01	.715D-02
2^{-6}	.784D-01	.468D-01	.253D-01	.148D-01	.907D-02
2^{-8}	.780D-01	.493D-01	.295D-01	.167D-01	.103D-01
2^{-10}	.778D-01	.497D-01	.311D-01	.172D-01	.109D-01
2^{-12}	.777D-01	.498D-01	.312D-01	.176D-01	.111D-01
2^{-14}	.777D-01	.499D-01	.312D-01	.178D-01	.112D-01
2^{-16}	.776D-01	.499D-01	.311D-01	.179D-01	.113D-01
2^{-18}	.776D-01	.499D-01	.311D-01	.179D-01	.113D-01
2^{-20}	.776D-01	.499D-01	.311D-01	.180D-01	.113D-01
2^{-22}	.776D-01	.499D-01	.311D-01	.180D-01	.113D-01
2^{-24}	.776D-01	.499D-01	.311D-01	.180D-01	.113D-01
2^{-26}	.776D-01	.499D-01	.311D-01	.180D-01	.113D-01
2^{-28}	.776D-01	.499D-01	.311D-01	.180D-01	.113D-01
2^{-30}	.776D-01	.499D-01	.311D-01	.180D-01	.113D-01
d_N	.784D-01	.499D-01	.312D-01	.180D-01	.113D-01

The computed values of p_N are given in Table 4.6 for problem 2. In Table 4.7 values of $D_{q,N}$ are given for various values of q and N. The computed ε-uniform error constants $C(N_0, k)$ may again be obtained by taking the maximum D_q in each row of each table and then using (4.6) to compute C_q. The results are listed in Table 4.8 for a selection of values of q for each problem.

Table 4.6. Computed ε-uniform convergence rates p_N for Problem 2.

ε	$N=8$	$N=16$	$N=32$	$N=64$
p_N	0.65	0.68	0.80	0.67

Table 4.7. $D_{q,N}$ and D_q for Problem 2.

q	$N=8$	$N=16$	$N=32$	$N=64$	$N=128$	D_q
0.1	0.0965	0.0658	0.0441	0.0273	0.0184	0.0965
0.2	0.1188	0.0869	0.0624	0.0414	0.0298	0.1188
0.3	0.1463	0.1146	0.0882	0.0627	0.0484	0.1463
0.4	0.1801	0.1513	0.1248	0.0950	0.0787	0.1801
0.5	0.2217	0.1996	0.1765	0.1440	0.1278	0.2217
0.6	0.2730	0.2634	0.2496	0.2183	0.2077	0.2730
0.7	0.3361	0.3475	0.3530	0.3308	0.3374	0.3530
0.8	0.4138	0.4586	0.4992	0.5014	0.5481	0.5481
0.9	0.5094	0.6051	0.7060	0.7600	0.8904	0.8904
1.0	0.6272	0.7984	0.9984	1.1520	1.4464	1.4464

Table 4.8. Computed ε-uniform error constants C_q for Problem 2.

q	0.4	0.5	0.6	0.7	0.8	0.9	1.0
C_q	0.1025	0.1299	0.1645	0.2185	0.3481	0.5797	0.9643

Applications of these approximate ε-uniform convergence rates and ε- uniform constants are discussed in the next section.

5. A priori and a posteriori estimates of the maximum error

In this section two important applications of ε-uniform error estimates are described. It is assumed throughout that the numerical method is ε-uniform on some sequence of meshes $\{\Omega_N^*\}_{N \geq N_0}$. The first application solves the following problem: given that the number of mesh points available is N, what precision δ can be guaranteed? Since, for all $N \geq N_0$

$$\delta = \sup_{0 < \varepsilon \leq 1} \|u - u_N\|_{\Omega_N^*} \leq C_q N^{-q}, \quad 0 < q \leq p \tag{5.1}$$

it follows that the pointwise error at each point is bounded above by $C_q N^{-q}$ for all N and ε, so

$$\delta \leq C_q N^{-q}, \quad 0 < q \leq p. \tag{5.2}$$

Applying this in the specific case of Problem 1 with $N = 64$, $q = 0.6$ ($C_{0.6} = 0.3808$) it is seen that the precision δ is bounded above by 0.0314.

The second application answers the following question: given a required guaranteed precision δ, what is the number N of mesh points that must be used? From (5.1), this accuracy is attained if

$$N \geq \max\{N_0, (\frac{C_q}{\delta})^{1/q}\}, \text{ for any } q \text{ where } 0 < q < p. \tag{5.3}$$

For example, in Problem 2, if $\delta = 0.01$ and $q = 0.6$ then taking the corresponding value of $C_{0.6}$ it follows that

$$N \geq \left(\frac{0.1645}{0.01}\right)^{1/0.6} = 107.$$

Then the use of $N = 107$ mesh points *guarantees* an error of less than 0.01, for *arbitrary* values of the singular perturbation parameter ε.

6. A practical approach

In this section a practical algorithm is given which allows the user to estimate quickly the ε-uniform error constants, without generating an extensive table (as in Table 4.1), for any particular problem of the form (2.1).

Assume that the numerical method is ε-uniform. The user must first guess (as high as they dare) the rate of ε-uniform convergence. Suppose that the user guesses that the numerical method is ε-uniform of at least 0.9 for all $N \geq N_0$. That is

$$\sup_{0<\varepsilon\leq 1} \|u - u_N\|_{\Omega_N} \leq C_{0.9} N^{-0.9}$$

where $C_{0.9} N^{-0.9}$ is to be estimated. For given values of ε and N choose convenient values N_1 and ε_1 close to the given values. Then use the numerical method to compute u_{N_1}, u_{4N_1} on the piecewise uniform meshes $\Omega^*_{N_1}, \Omega^*_{4N_1}$ respectively and define

$$\tilde{\delta} = \|u_{N_1} - u^I_{4N_1}\|_{\Omega^*_{N_1}} \ . \tag{6.1}$$

Then
$$\tilde{C}_{0.9} = N_1^{0.9} \tilde{\delta} \tag{6.2}$$

is taken as the approximate value of $C_{0.9}$.

The approximate value $\tilde{C}_{0.9}$ is then used in place of $C_{0.9}$, to estimate both the ε-uniform precision of a given number of mesh points and to determine the number of mesh points required to achieve a certain prescribed precision, using the methods in the last section. It should be noted that $\tilde{C}_{0.9}$ was obtained using two meshes and one value of ε. Clearly better approximations to $C_{0.9}$ can be obtained using more than two meshes and more values of ε.

References

1. A.F. Hegarty, J.J.H. Miller, E. O'Riordan, G.I. Shishkin, On a novel mesh for the regular boundary layers arising in advection-dominated transport in two dimensions. (submitted to Int. J. Numer. Meth. in Eng.).
2. A.F. Hegarty, J.J.H. Miller, E. O'Riordan, G.I. Shishkin, Special meshes for finite difference approximations to an advection-diffusion equation with parabolic layers (submitted to J.Comp.Phys.).

Numerical Simulation of Three–dimensional Non–stationary Compressible Flow in Complex Geometries

Monika Geiben
Institut für Angewandte Mathematik, Universität Freiburg
Hermann–Herder–Str. 10, D-79104 Freiburg, Germany

Summary

In this paper we shall present a numerical algorithm to solve the compressible Euler equations in three dimensional geometries. The numerical algorithm consists of an cell-centered upwind finite volume scheme of higher order on a grid of tetrahedra and the possibility to refine and coarse locally the grid appropriate to the approximated solution. In the first section we shall describe the flow problem we want to solve, in the second a new higher order upwind finite volume scheme, in the third the way how we do the local adaption of the grid and in the last we present some numerical solutions we got for different flow problems in 3D.

1. Flow Problem

We want to solve the compressible Euler equations in geometries with inlets and outlets (cylinder model of a two–stroke–engine).

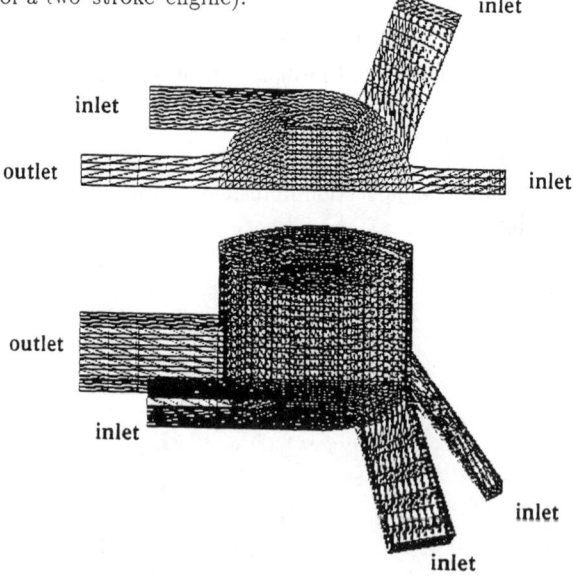

In this geometry a piston should move up and down. The piston motion will cause very high velocities, if an inlet or outlet is nearly closed. This motivates the use of an upwind scheme. To get high resolution of this flow a higher order scheme in space is developed and an algorithm is implemented to adapt locally the fineness of the grid. The affected regions for local adaption are detected by a residuum expression.

The description to the modelling of the piston motion can be found in [13]. With this new approach on a grid of tetrahedra a way to model the piston motion is found with no lack of conservation of the flux quantities.

The conservation of mass, momentum and energy in a non-stationary compressible inviscid fluid is described through the Euler equations of gas dynamics. These equations are given in conservation form:

$$\partial_t u + \partial_x f_1(u) + \partial_y f_2(u) + \partial_z f_3(u) = 0, \tag{1.2}$$

where

$$u = \begin{pmatrix} \rho \\ \rho u_1 \\ \rho u_2 \\ \rho u_3 \\ e \end{pmatrix}, \quad f_i(u) = \begin{pmatrix} \rho u_i \\ \rho u_i u_1 + \delta_{i1} p \\ \rho u_i u_2 + \delta_{i2} p \\ \rho u_i u_3 + \delta_{i3} p \\ (e+p) u_i \end{pmatrix} \tag{1.3}$$

and with the equation of state

$$p = (\gamma - 1)\left(e - \frac{\rho}{2}(u_1^2 + u_2^2 + u_3^2)\right). \tag{1.4}$$

The size ρ describes the density, $\mathbf{u} = (u_1, u_2, u_3)^t$ the velocity, e the total energy and p the pressure in the fluid.

The numerical flux we used for the calculations of the Burgers equation is the Engquist–Osher flux [9] and that for the system of Euler equations the Steger–Warming flux [23].

2. Definition of a New Higher Order Scheme in Space

In this section we will present a new upwind finite volume scheme of higher order on unstructured grid of triangles. In [12] we discussed some approaches to get such a scheme based on the MUSCL–approach ((Monotone Upstream Centered Schemes for Conservation Laws), initially developed on one–dimensional finite difference grids [25, 26].

The motivation for the definition of the three schemes A, B and C in [12] is, that these schemes fulfill with a suitable limiter function for the reconstruction function the constraints which are necessary to prove convergence of the approximated solution to the entropy solution in the case of scalar conservation laws [18].

There are several approaches to get methods of higher order in multidimensions on unstructured grids (citations see in [12]):

> essentially non–oscillatory method; streamline–diffusion–method; streamline–diffussion–shock–capturing–method; finite element methods; discontinuous–Galerkin–method; upwind finite volume method of higher order; maximum and minimum bound (Mmb) scheme

So far only for the streamline–diffusion–shock–capturing–method [17], the discontinuous–Galerkin–method [16, 18] and the cell–centered upwind finite volume method of higher order in [4], [6], [18] and [20] theoretical convergence in the case of scalar conservations laws has been proved. The upwind finite volume scheme with the MUSCL approach, as opposed to the streamline–diffusion–shock–capturing–method and the discontinuous–Galerkin–method, is very easy to implement, also for systems of conservation laws in three dimensions.

After some basic notations and assumptions we shall present the result of convergence from [18]. Then we shall define ansatz C and compare this approach with other ones on numerical results for the Burgers equation and the system of the Euler equations.

2.1 Definitions, Notations and Assumptions

DEFINITION 2.2 *(of notations)*
Let I be an index set and for $h > 0$ let $\mathcal{T}_h := \{T_i | i \in I\}$ denote an unstructured triangular grid of $\mathbb{R}^2$.
$h := \sup_i diam(T_i)$.
Δt^j : time step of the jth iteration step of the numerical scheme.
t^n: the time after n time steps.
S_{ij}: j^{th} edge of T_i; n_{ij}: j^{th} outward unit normal on edge S_{ij}; ν_{ij}: j^{th} outward normal with length $|S_{ij}|$ on edge S_{ij} .
x_i: center of gravity of T_i; x_{ij}: center of gravity of the jth neighbour T_{ij} of T_i; z_{ij}: edge mid point on S_{ij}.
u_i^n: constant approximation of the solution on triangle T_i at time $n \cdot \Delta t$; u_{ij}^n: the value on T_{ij}.
$u_h(x,t) := u_i^n$ for $x \in T_i$ and $t^n \leq t < t^{n+1}$.
L_i^n: reconstruction function for the triangle T_i depending linearly on the constant values $\{u_i^n, i \in I\}$.
L_{ij}^n: reconstruction function for the triangle T_{ij} depending linearly on the constant values $\{u_i^n, i \in I\}$.

We consider in this section the following general scalar conservation law

$$\frac{\partial}{\partial t}u(x,t) + \partial_x f_1(u(x,t)) + \partial_y f_2(u(x,t)) = 0, \tag{2.5}$$

with initial values
$$u(x,0) = u_0(x), \tag{2.6}$$

where $x \in \mathbb{R}^2, t \in \mathbb{R}_+, u : \mathbb{R}^2 \times \mathbb{R}^+ \to \mathbb{R}^1$, $f_1, f_2 \in C^1(\mathbb{R}^1, \mathbb{R}^1)$ and $u_0 \in L^1 \cap L^\infty(\mathbb{R}^2)$.

DEFINITION 2.3 *(of the triangulation)*
The set
$$\mathcal{T} := \{T_i | T_i \text{ is a triangle for } i \in I \subseteq \mathbb{N}\}$$

where $I \subseteq \mathbb{N}$ is an index set, is called an admissible triangulation of $\Omega \subset \mathbb{R}^2$ if the following two properties are satisfied:

1) $\Omega = \bigcup_{i \in I} T_i$,

2) For two different T_i, T_j we have $T_j \cap T_i = \emptyset$ or

$T_j \cap T_i = $ a common vertex of T_i, T_j or

$T_j \cap T_i = $ a common edge of T_i, T_j.

ASSUMPTION 2.4 *(for the triangulation and the quotient of Δt^n and h)*

a) The triangulations $\mathcal{T}_h$ are regular.

b) We assume that there is a constant c_1 such that as $\Delta t, h \to 0$

$$0 < c_1 \leq \frac{\Delta t}{h} \ . \tag{2.7}$$

ASSUMPTION 2.5 *(for the numerical flux)*
We assume that the numerical flux $g_{ij}(u,v)$ is consistent with $f(u) \cdot \nu_{ij}$, i.e.

$$g_{ij}(u,u) = f(u) \cdot \nu_{ij}, \text{ where } f(u) := \begin{pmatrix} f_1(u) \\ f_2(u) \end{pmatrix}. \tag{2.8}$$

We assume that g_{ij} is Lipschitz-continuous. In particular, suppose that for all $M > 0$ there is a constant $C_g = C_g(M)$ such that for all $u, u', v, v' \in B_M \subset \mathbb{R}$

$$|g_{ij}(u,v) - g_{ij}(u',v')| \leq C_g(M)|S_{ij}|(|u - u'| + |v - v'|) \tag{2.9}$$

and that g_{ij} is conservative, i.e. if $T_{ij} = T_k$ and $T_{kl} = T_i$ then

$$g_{ij}(u,v) = -g_{kl}(v,u). \tag{2.10}$$

Moreover, assume that g_{ij} is monotone:

$$\frac{\partial}{\partial u} g_{ij}(u,v) \geq 0 \geq \frac{\partial}{\partial v} g_{ij}(u,v) \tag{2.11}$$

in the sense of distributions.

NOTE 2.6 The Engquist-Osher numerical flux (see [9]) and the Lax-Friedrichs numerical flux (see [19]) (and all their convex combinations) satisfy the conditions (2.8) – (2.11).

DEFINITION 2.7 *(explicit cell–centered upwind finite volume scheme in 2D)*
For given initial values $u_0 \in L^\infty(\mathbb{R}^2)$ define for instance

$$u_i^0 := u_0(x_i), \qquad x_i \in T_i \qquad \text{(compare also with Theorem 2.9)}. \tag{2.12}$$

Let

- g_{ij} be the numerical flux, which should fulfill Assumption 2.5.

- L_i^n and L_{ij}^n be the linear reconstruction functions on the triangles T_i and T_{ij} respectively depending on $\{u_i^n, i \in I\}$. Examples for the values $L_i^n(z_{ij})$ and $L_{ij}^n(z_{ij})$ will be presented in the following section.

The First Order Scheme:
$$u_i^{n+1} := u_i^n - \frac{\Delta t}{|T_i|} \sum_{j=1}^{3} g_{ij}(u_i^n, u_{ij}^n). \tag{2.13}$$

The Higher Order Scheme in Space:
$$u_i^{n+1} := u_i^n - \frac{\Delta t}{|T_i|} \sum_{j=1}^{3} g_{ij}(L_i^n(z_{ij}), L_{ij}^n(z_{ij})). \tag{2.14}$$

The Higher Order Scheme in Space and Time:
(*TVD – Runge – Kutta method of second order concerning the time discretization [21]*)

$$u_i^{n+1/2} := u_i^n - \frac{\Delta t}{|T_i|} \sum_{j=1}^{3} g_{ij}(L_i^n(z_{ij}), L_{ij}^n(z_{ij}))$$

$$u_i^{n+1} := \frac{1}{2}\left(u_i^n + u_i^{n+1/2} - \frac{\Delta t}{|T_i|} \sum_{j=1}^{3} g_{ij}(L_i^{n+1/2}(z_{ij}), L_{ij}^{n+1/2}(z_{ij}))\right) \tag{2.15}$$

with the linear functions $L_i^{n+1/2}$ and $L_{ij}^{n+1/2}$ defined with the intermediate values $u_i^{n+1/2}$.

DEFINITION 2.8 *(finite stencil of a scheme)*
We say that the scheme defined in Definition 2.7 has a finite stencil if there is a constant $K \in \mathbb{N}$ such that the update u_i^{n+1} is only affected by the values $\{u_k|, |x_k - x_i| < Kh\}$.

THEOREM 2.9 *(Convergence result for the scheme for scalar conservation laws, see [18])*
Assume that the Assumption 2.4 holds for the triangulation, 2.5 for the numerical flux. Let $\alpha \in (\frac{1}{2}, 1]$ and $\gamma \in (0, \frac{2\alpha - 1}{4})$. Given initial data $u_0 \in L^1 \cap L^\infty(\mathbb{R}^2)$, define

$$u_i^0 := \begin{cases} \frac{1}{|T_i|} \int_{T_i} u_0(x) dx & \text{for } |x_i| \leq C_0 h^{-\gamma} \\ 0 & \text{otherwise} . \end{cases}$$

Let u_h be the sequence of approximate solutions of (2.5) and (2.6) defined by

$$u_i^{n+1} := u_i^n - \frac{\Delta t}{|T_i|} \sum_{j=1}^{3} g_{ij}(u_i^n + u_{ij}^{n,int}, u_{ij}^n + u_{ij}^{n,ext}). \tag{2.16}$$

Suppose that this scheme has a finite stencil in the sense of Definition 2.8 and that for all h

i) there is a constant $M < \infty$ independent of h such that

$$\|u_h\|_{L^\infty(\mathbb{R}^2 \times [0,T])} \leq M, \tag{2.17}$$

ii)

$$\sup_i \frac{\Delta t}{|T_i|} C_g(M) \sum_{j=1}^{3} |S_{ij}| \leq 1 \tag{2.18}$$

iii) if $T_{ij} = T_k$ and $T_{kl} = T_i$ then

$$u_{ij}^{n,int} = u_{kl}^{n,ext}, \qquad (2.19)$$

iv) there is a constant $C_1 > 0$ independent of h such that for all i,j

$$|u_{ij}^{n,int}| \leq C_1 h^\alpha, \qquad (2.20)$$

v) for all i,j

$$u_{ij}^{n,int} \cdot (u_i^n - u_{ij}^n) \leq C_2 h^{2\alpha} \qquad (2.21)$$

with a constant C_2 independent of h.

Then for any given $T > 0$, $u_h \in L^1 \cap L^\infty(\mathbb{R}^2 \times [0,T])$, and as $h \to 0$ there is a subsequence of $(u_h)_h$ which converges to the Kruzkov solution of (2.5) and (2.6) strongly in $L^1_{loc}(\mathbb{R}^2 \times [0,T])$.

2.2 Ansatz C

DEFINITION 2.10 *(the limiter function of ansatz C in 2D)*
Determine

$$u_i^{max} := max\{u_i^n, u_{ij}^n, u_{ij+1}^n, u_{ij-1}^n\} \text{ and } u_i^{min} := min\{u_i^n, u_{ij}^n, u_{ij+1}^n, u_{ij-1}^n\}.$$

*Construct for each triangle a linear reconstruction function of the form $F_i^n(x) = u_i^n + s_i^n \cdot (x - x_i)$ (for example with the approach in [1], [8], [10], [11] or [14]).
Define then*

$$term(j) := s_i^n \cdot (z_{ij} - x_i), \qquad crit(j) := \begin{cases} 1, & \text{if } sign(term(j)) = sign(u_{ij}^n - u_i^n) \\ 0, & \text{otherwise} . \end{cases}$$

With this terminology we define then $u_{ij}^{n,int}$ with the following algorithm:
if $crit(j) = 1$ for all j:
  $u_{ij}^{n,int} := \beta \cdot s_i^n \cdot (z_{ij} - x_i)$ with $\beta \in [0,1]$ such that :
    a) $\beta \cdot |s_i^n \cdot (z_{ij} - x_i)| \leq C_1 h^\alpha, \ \forall j$
    b) $u_i^{min} \leq u_i^n + \beta \cdot s_i^n \cdot (z_{ij} - x_i) \leq u_i^{max}, \ \forall j$
If $crit(j) = 0$ only for one j and $sign(term(j-1)) \neq sign(term(j+1))$:
  $u_{ij}^{n,int} = 0$
  $u_{ij+1}^{n,int} = \beta \cdot sign(s_i^n \cdot (z_{ij+1} - x_i)) \cdot minvalue$
  $u_{ij-1}^{n,int} = -u_{ij+1}^{n,int}$
  with $minvalue = min(|s_i^n \cdot (z_{ij+1} - x_i)|, |s_i^n \cdot (z_{ij-1} - x_i)|)$
  and $\beta \in [0,1]$ such that
    a) $\beta \cdot minvalue \leq C_1 h^\alpha, \ \forall j$
    b) $u_i^{min} \leq u_i^n + \beta \cdot minvalue \leq u_i^{max}, \ \forall j$
for all other cases set:
  $u_{ij}^{n,int} := 0$ for all j
To define the value in the neighbouring cell we define the indices k and l by $T_k = T_{ij}$ with $T_{kl} = T_i$:

$$u_{ij}^{n,ext} := u_{kl}^{n,int}$$

THEOREM 2.11 *(for the proof see [12]) Suppose additionally to the assumptions in Theorem 2.9 that the CFL–condition to determine Δt^n should be*

$$\sup_i \frac{\Delta t^n}{|T_i|} C_g(M) \sum_j |S_{ij}| \leq \frac{1}{3} \tag{2.22}$$

with $C_g(M)$ the Lipschitz constant of the numerical flux defined in Assumption 2.5 for $M = \|u_0\|_{L^\infty(\mathbf{R}^2)}$. Then the conditions (2.17) - (2.21) in Theorem 2.9 are fulfilled if we define $u_{ij}^{n,int}$ and $u_{ij}^{n,ext}$ as in Definition 2.10.

NOTE 2.12 i) *The explicitly inserted $C_1 h^\alpha$–term is necessary to force the convergence of the numerical solution to the entropy solution. But the constant C_1 can not be adjusted automatically by the scheme for every individual problem. In the case of an approximated solution with a discontinuity there would exist for a chosen constant C_1 an $h_0(C_1)$ such that $C_1 h_0(C_1)^\alpha$ would restrict $u_{ij}^{n,int}$ and therefore the resolution of this discontinuity. A C_1 that is too small would limit an increase of the resolution of the discontinuity below a certain h_0. In the case of a smooth solution ($u \in C^1$ and $|\nabla u| < C$) C_1 can be determined explicitly and the term $C_1 h^\alpha$ would not matter in the limiting process. Therefore the term $C_1 h^\alpha$ is dropped in the numerical realization of these limiters. The numerical results in section 2.3 show that nevertheless the numerical solution converges to the entropy solution. A similiar argument can be found in [7].*

ii) *(example for the implementation of ansatz C)*
Determine values $\tilde{u}_{ij}^n, j = 1, 2, 3$ at the vertices P_{ij} of the triangle T_i by averaging the piecewise constant values to the triangles. This can be done for example with

$$\tilde{u}_{ij}^n = \frac{\sum_{k \in I, P_{ij} \in T_k} \frac{u_k^n}{r_k}}{\sum_{k \in I, P_{ij} \in T_k} \frac{1}{r_k}}; \qquad r_k = |P_{ij} - x_k|\,.$$

Let $\tilde{u}_{ij+1}^n$ and $\tilde{u}_{ij+2}^n$ the values at the vertices of the jth edge of T_i. Define then:

$$s_i^n \cdot (z_{ij} - x_i) := \frac{\frac{1}{2}(\tilde{u}_{ij+1}^n + \tilde{u}_{ij+2}^n) - \tilde{u}_{ij}^n}{3}. \tag{2.23}$$

This is the reconstruction suggested in [11]. The essential property of this ansatz

$$\sum_j s_i^n \cdot (z_{ij} - x_i) = 0$$

follows immediatly.

iii) *The main advantage of the approach C is that no further constraints onto the triangulation are necessary to get the L^∞–bound. Ansatz $A(\varphi)$ need an underlying B–triangulation.*

iv) *The condition $u_i^{min} \leq u_i^n + u_{ij}^{n,int} \leq u_i^{max}$ can also be replaced by a more global one, for example $\min_{i \in I} u_i^0 \leq u_i^n + u_{ij}^{n,int} \leq \max_{i \in I} u_i^0$. In the case of the system of the Euler equations it is more useful to work with the local minimum and maximum values.*

v) *The definition of ansatz C in 3D is similiar to the 2D case, see [13].*

2.3 Numerical Order of Convergence

To value the new ansatz C with other higher order approaches we compared it in [12] with other ones. Denote by u_h the approximated solution of the first order scheme, by u_h^C that with the ansatz C, by u_h^S that we get with the superbee approach in [8] and by u_h^F the unlimited interpolation approach described in ii) of Note 2.12. We made calculations on different kinds of triangulations and problems with smooth solutions and those containing disontinuities. We determined the EOC ("experimental order of convergence"):

$$\text{EOC} = \ln\left[\frac{\|u - u_{h_j}\|_{L^1}}{\|u - u_{h_{j+1}}\|_{L^1}}\right] \bigg/ \ln\left[\frac{h_j}{h_{j+1}}\right]$$

where u is the exact solution, u_{h_j} the discrete one on a grid where the maximal length of an edge is h_j and $\|u - u_{h_j}\|_{L^1} = \sum_{i \in I} |T_i| |u(x_i, T) - u_{h_j}(x_i, T)|$. To consider the correct development of the L^1-error in time and space we take $\Delta t^n = \Delta t^0$ for all numerical calculations, where Δt^0 is determined with the indicated CFL-number.

a) Approximation of a smooth solution of the Burgers equation
Solve $\partial_t u + \partial_x u^2 + \partial_y u^2 = f$ in $[0,1]^2$ with $u(x,y,t) = sin(2\pi(x-t)) \cdot sin(2\pi(y-t))$
Grid:

h	$\|u - u_h\|_{L^1}$	EOC	$\|u - u_h^C\|_{L^1}$	EOC	$\|u - u_h^F\|_{L^1}$	EOC
0.3536	0.063526		0.028382		0.028139	
0.1768	0.029926	1.09	0.008230	1.79	0.006953	2.02
0.0884	0.014833	1.01	0.001963	2.07	0.001731	2.01
0.0442	0.007503	0.98	0.000478	2.04	0.000450	1.94
0.0221	0.003790	0.99	0.000118	2.01	0.000117	1.94

(Evaluation at T = 0.3, determination of Δt^0 with CFL = 0.5, we used the scheme (2.15) as higher order scheme.)

b) Solution of the quasi two–dimensional shock tube problem of the system of the Euler equations (1.2)

Grid:

Initial values

$$\rho(x,y,0) = \begin{cases} 4, & x < 0 \\ 1, & x > 0 \end{cases} ; \; p(x,y,0) = \begin{cases} 1.6, & x < 0 \\ 0.4, & x > 0 \end{cases} ;$$

$$u_1(x,y,0) = u_2(x,y,0) = 0$$

(Evaluation at T = 0.4, determination of Δt^0 with CFL = 0.7, we used the scheme (2.14) as higher order scheme.)

h	$\|u - u_h\|_{L^1}$	EOC	$\|u - u_h^C\|_{L^1}$	EOC	$\|u - u_h^S\|_{L^1}$	EOC
0.2500	0.391090		0.223685		0.279411	
0.1250	0.305952	0.35	0.111804	1.00	0.184999	0.59
0.0625	0.226816	0.43	0.064125	0.80	0.137332	0.43
0.0313	0.165686	0.45	0.042131	0.61	0.111588	0.30
0.0156	0.124704	0.41	0.031769	0.41	0.100700	0.15

3. Local Adaption

We are using the local adaption algorithm on a grid of tetrahedra proposed in [2]. With this technique of bisection of tetrahedra it is very easy to project the data between two grids under the constraint of conservation. Details to the implementation can also be found in [5].

3.1 The Used Local Adaption Criterion

The aim of local increasing and decreasing of the grid fineness should be the reducing of the numerical error of the scheme. Therefore no criterion to adapt the grid to the solution but to adapt the grid to the numerical error should be used.

There exists no a–posteriori estimator for the upwind finite volume schemes for the system of Euler equations. With theoretical results in [15] and [24] for scalar conservation laws the considering of the residuum is the most promising approach.

Therefore our approach is based on the residuum, where we define a local residuum for each element. Since the cell–centered upwind finite volume scheme approximates the solution piecewise constant on each tetrahedron we have to define again a linear reconstruction function K_i^n of the piecewise constant values.

DEFINITION 2.13 (*of the used adaption criterion*)
Denote by K_i^n a linear function of the tetrahedron T_i determined with the piecewise constant values $\{u_i^n, i \in I\}$.

$$S_i^n := \max_j \{|S_{ij}|, S_{ij} \text{ face of } T_i\}$$

$$r^n(T_i) := \frac{u_i^{n+1} - u_i^n}{\Delta t^n} + f_1'(u_i^n)\partial_x K_i^n(x_i) + f_2'(u_i^n)\partial_y K_i^n(x_i) + f_3'(u_i^n)\partial_z K_i^n(x_i)$$

with

$$\partial_\xi K_i^n(x_i) = \frac{1}{|T_i|}\int_{\partial T_i} K_i^n(x) n_i^\xi = \frac{1}{|T_i|}\sum_{j=1}^4 |S_{ij}| K_i^n(z_{ij}) n_{ij}^\xi$$

Denote with $(r^n(T_i))_l$ the lth component of the local residuum $(r^n(T_i))$.

$$|r^n(T_i)|_r := S_i^n \sqrt{|T_i|} \sum_{l=1}^5 |(r^n(T_i))_l|$$

(For the motivation of this expression see the next note.)

$$mean\_value\_of\_res := \frac{1}{NT}\sum_{i=1}^{NT}|r^n(T_i)|_r \qquad (NT: \text{number of tetrahedra})$$

We mark then the tetrahedra to refine or to coarse as follows:

$$if\,(|r^n(T_i)|_r > ref \cdot mean\_value\_of\_res)\,and\,(number\,of\,refinements\,for\,T_i < 8)$$
$$refine\ T_i$$
$$if\,(|r^n(T_i)|_r < 0.1 \cdot mean\_value\_of\_res)$$
$$coarse\ T_i$$

ref is initially chosen equal to 0.9, but it is adjusted if the quotient of NT and number of marked tetrahedra to refine is not in the intervall $[10,50]$.

NOTE 2.14   i) In this definition we used the result in [22] for the stationary compressible Euler equations in 2D:

> For a cell–vertex upwind finite volume method the solution is approximated at the vertices. With ω_h^n denote then the continuous piecewise linear (on each triangle) solution at time t^n. So in [22] the local residuum for each triangle is defined as
>
> $$r^n(T_i) := f_1'(\omega_h^n)\partial_x\omega_h^n + f_2'(\omega_h^n)\partial_y\omega_h^n + f_3'(\omega_h^n)\partial_z\omega_h^n\ .$$
>
> The additional power of h in the final expression to locate the refinement and coarsening zones is motivated by an very simple consideration in 1D: The expression without the power of h blows up at discontinuities. Another way to motivate the final expression is the consideration of another norm in [22].

We modified this result then for the unsteady case by adding the term $\frac{u_i^{n+1}-u_i^n}{\Delta t^n}$ as approximation of $\partial_t u_h$ on T_i at time t^n in the local residuum expression.

ii) (choice of K_i^n) We get the numerical results in section 4 with $K_i^n(x) = u_i^n + \nabla K_i^n(x - x_i)$ and $\nabla K_i^n(x - x_i) = s_i^n \cdot (x - x_i)$ as defined in 2.23.

4. Numerical Results

a) Forward facing step problem in 3D: The 2D test example described in [27] is used to test the higher order scheme and the adaption criterion. We see the isolines of the density and the colored fineness of the grid on three cutting planes. Furthermore the grid on the outer boundary of the step is displayed to get an 3D impression.

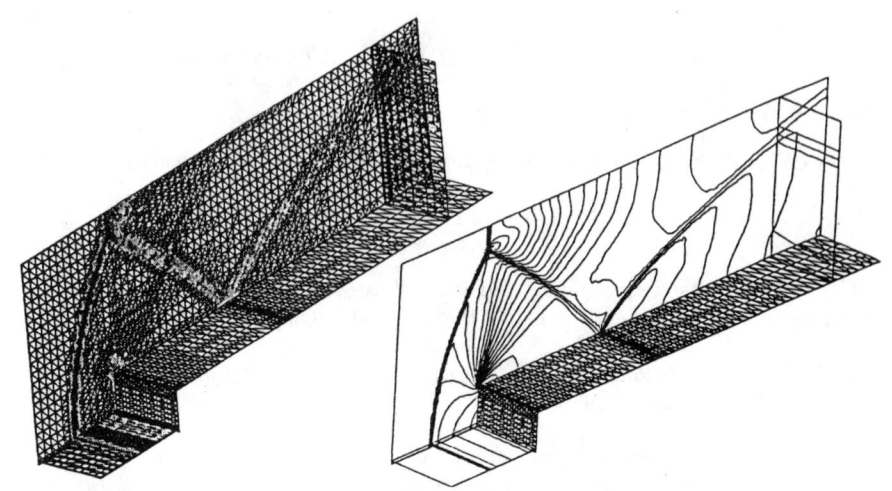

b) Visualization of the flow in the cylinder model of a two–stroke–engine: We see the velocity field on three cutting planes through the geometry and some particle traces.

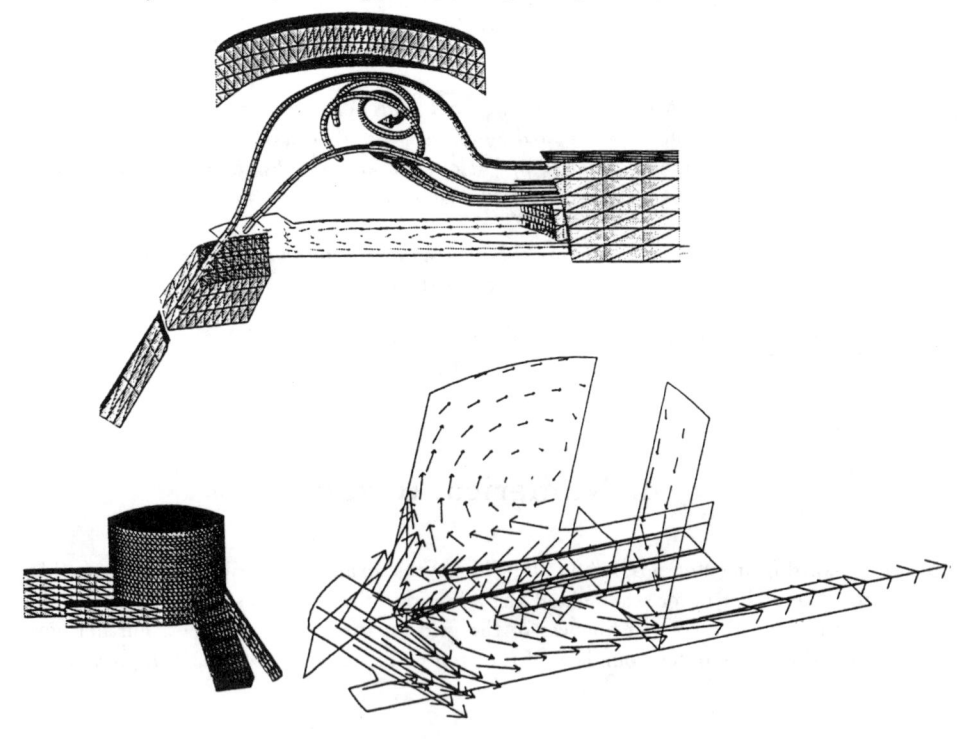

References

[1] Barth, T. J.; Jesperson, D. C.: *The Design and Application of Upwind Schemes on Unstructured Meshes.* AIAA-98-0366.

[2] Bänsch, E.: *Local Mesh Refinement in 2 and 3 Dimensions.* IMPACT of Computing in Science and Engineering. No. 3, 181 – 191, 1991.

[3] Barth, T. J.: *Numerical Aspects of Computing Viscous High Reynolds Number Flows on Unstructured Meshes.* AIAA 91-0721.

[4] Benharbit, S.; Chalabi, A.; Vila, J. P.: *Numerical Viscosity, Entropy Condition and Convergence of Finite Volume Schemes for General Multidimensional Conservation Laws.* in: Fourth International Conference of Hyperbolic Problems: Theory, Numerical Methods and Applications. Taormina, 1992.

[5] Biesenbach, H.; Geiben, M.: *Implementation of Adaptive Tools for Tetrahedral Meshes* Report to appear, SFB 256, 1994.

[6] Cockburn, B.; Coquel, F.; LeFloch, P.: *An Error Estimate for High–Order Accurate Finite Volume Methods for Scalar Conservation Laws.* Preprint of AHPCRC Institute, Minneapolis, Minnesota (1991). (to appear in Math. Comp. July 94)

[7] Cockburn, B.; Shu, C.W.: *TVB Runge-Kutta Local Projection Discontinuous Galerkin Finite Element Method for Conservation Laws II: General Framework.* Math. of Comp. 52 (1989), 411–435.

[8] Durlofsky, L. J.; Engquist, B.; Osher, S.: *Triangle Based Adaptive Stencils for the Solution of Hyperbolic Conservation Laws.* J. of Comp. Physics 98 (1992), 64–73.

[9] Engquist, B.; Osher, S.: *One Sided Difference Approximations for Nonlinear Conservation Laws.* Math. of Comp. 36 (1981), 321–351.

[10] Fukuda, H.; Yamamoto, K.: *An Upwind Method Using Unstructured Triangular Meshes.* Volume 1 of Fourth International Symposium on Computational Fluid Dynamics, 1991.

[11] Frink, N. T.; Parikh, P.; Pirzadeh, S.: *A Fast Upwind Solver for the Euler Equations on Three-Dimensional Unstructured Meshes.* AIAA-91-0102.

[12] Geiben, M.: *Convergence of MUSCL–Type Upwind Finite Volume Schemes on Unstructured Triangular Grids.* Preprint 318, SFB 256, Bonn 1994. (submitted to Math. of Comp.).

[13] Geiben, M.: *Higher Order Upwind Schemes for the Compressible Euler Equations in Timedependent Geometries in 3D.* Dissertation, in preparation, Freiburg, 1994.

[14] Glinsky, N.; Fézoui, L.; Ciccoli, M. C.; Desidéri, J.-A.: *Non–Equilibrium Hypersonic Flow Computations by Implicit Second–Order Upwind Finite–Elements.* In: Proceedings of the Eight GAMM- Conference on Numerical Methods in Fluid Mechanics, Notes on Numerical Fluid Mechanics, Vol. 29, Vieweg, Braunschweig, 1990.

[15] Johnson, C.; Szepessy, A.: *Adaptive Finite Element Methods for Conservation Laws Based on A-posteriori Estimates.* Preprint 1992-31 — ISSN 0347-2809, Dep. of Math. Chalmers Univ. of Techenology, The University of G"oteborg.

[16] Jaffre, J.; Johnson, C.; Szepessy, A.: *Convergence of the Discontinuous Galerkin Finite Element Method for Hyperbolic Conservation Laws.* to appear.

[17] Johnson, C.; Szepessy, A.: *Convergence of a Finite Element Method for a Nonlinear Hyperbolic Conservation Law.* Math. of Comp. 49 (1987), 427–444.

[18] Kröner, D.; Noelle, S.; Rokyta, M.: *Convergence of Higher Order Finite Volume Schemes on Unstructured Grids for Scalar Conservation Laws in Two Space Dimensions.* Preprint 268, 1993, SFB256, Bonn, Germany.

[19] Lax, P.: *Weak Solutions of Nonlinear Hyperbolic Equations and Their Numerical Computation.* Comm. Pure Appl. Math. VII (1954), 159–193.

[20] Noelle, S.: *Convergence of Higher Order Finite Volume Schemes on irregular grids.* SFB256 Preprint 295, 1993, Bonn. (submitted to Mathematics of Computations).

[21] Shu, C. W.; Osher, S.: *Efficient Implementation of Essentially Non-Oscillatory Shock-capturing Schemes.* Journal of Computationl Physics, 77 (1988), 439-471.

[22] Sonar, T.: *Strong and Weak Norm Error Indicators Based on the Finite Element Residual for Compressible Flow Computations.* Preprint 92-07, DLR G"ottingen, 1992. (submitted to Impact of Computing in Science and Engineering).

[23] Steger, J. L.; Warming R. F.: *Flux Vector Splitting of the Inviscid Gasdynamic Equations with Application to Finite-Difference Methods.* Journal of Computational Physics 40 (1981), 263-293.

[24] Nessyahu, H.; Tassa, T.; Tadmor, E.: *The Convergence Rate of Godunov Type Schemes.* (submitted to SIAM J. Numer. Anal.).

[25] Van Leer, B.: *Towards the Ultimate Conservative Difference Scheme. IV: A New Approach io Numerical Convection.* JCP 23, 276-299, 1977.

[26] Van Leer, B.: *Towards the Ultimate Conservative Difference Scheme. V: A Second Order Sequel to Godunov's Method.* JCP 32, 101-136, 1979.

[27] Woodward, P. R.; Colella, P.: *The Numerical Simulation of Two–Dimensional Flow with Strong Shocks.* Journal of Computational Physics 54, 115-173, 1984.

On the Numerical Simulation of Coupled Transient Problems on MIMD Parallel Systems

Ulrich Groh, Stefan Meinel, Arnd Meyer
Technische Universität Chemnitz-Zwickau, Fakultät für Mathematik,
Postfach 964, D-09009 Chemnitz

Abstract

The proposed algorithms for coupled transient flow and transport problems are based on stabilized FE-discretizations of the governing equations, semiimplicit time-stepping and a pressure projection method for the flow problem. Preconditioned conjugate gradient solvers for symmetrical and non-symmetrical linear systems are applied.

The authors use a domain decomposition concept for the parallelization of data and computational work. The algorithms are designed for MIMD parallel computers and implemented for this time on a transputer system. Numerical results are presented for 2D combustion simulation.

1 Introduction

Coupled transient flow and transport problems play a major role in technical and environmental applications. They are characterized by a nonuniform problem dependent complexity, frequently strong nonlinearities resulting from various source terms and more or less strong coupling between the quantities. The numerical simulation of such processes is of practical importance and it can be a very challenging task even for powerful computer systems. Therefore, this class of problems is a potential application field for parallel computing, and it is necessary to investigate and implement algorithms that are stable and accurate enough and run effectively on the parallel computers.

The methods based on domain decomposition techniques are well developed for elliptic problems, see e.g. [4], [7]. For the more involved coupled nonlinear convective-diffusive problems an adaption of the results is possible, but there are new aspects having some effects on the theoretical considerations. Because of the variety of occuring physical situations we have concentrated upon a typical model problem, the simulation of a very simplified combustion process. Therefore, we cannot expect general results, but we can identify some features representative for this class of problems.

2 Governing equations, discretization

In the 2D domain Ω we consider the following initial boundary value problem for an incompressible flow described by the Navier-Stokes equations

$$\varrho \left\{ \frac{\partial \vec{u}}{\partial t} + (\vec{u} \cdot \nabla) \vec{u} \right\} = -\nabla p + \nabla \cdot \left\{ \eta \left(\nabla \vec{u} + (\nabla \vec{u})^{\mathsf{T}} \right) \right\} + \varrho \vec{f}, \qquad (1)$$

$$\nabla \cdot \vec{u} = 0, \qquad (2)$$

and the convective-diffusive transport of N_{tr} scalar quantities Φ_k described by

$$\varrho \left\{ \frac{\partial \Phi_k}{\partial t} + (\vec{u} \cdot \nabla) \Phi_k \right\} = \nabla \cdot \{\kappa_k \nabla \Phi_k\} + \varrho\, q_k, \qquad k = 1, 2, \ldots, N_{tr}. \tag{3}$$

The number N_{tr} of transport processes depends on the problem considered and can reach from one or two in simple thermal or turbulent models up to more then hundred for realistic combustion processes.

In (1) ϱ denotes the constant density, $\vec{u}$ the velocity vector, p the pressure, η the dynamical viscosity and $\vec{f}$ an external force vector. In (3) κ_k stands for a diffusivity and q_k for a specific source term of the quantity Φ_k.

The boundary $\partial \Omega$ is divided into parts Γ_1 and Γ_2 corresponding to the boundary conditions of Dirichlet or von Neumann type, but this subdivision is not nessecarily the same for the velocity components and the transported scalars. Typical homogeneous von Neumann ("do nothing") conditions for the flow problem are the outflow conditions. For the transport problems the diffusive fluxes can be prescribed.

For the spatial discretization we use a finite element method with continuous, piecewise linear functions on triangular meshes or continuous, piecewise bilinear functions on quadrilateral meshes. The pressure is discretized in the same manner like the cartesian components of the velocity, but in a one level coarser mesh (fig. 1), fulfilling the LBB-condition.

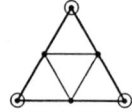

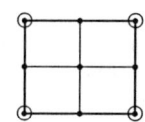

○ pressure nodes

• velocity nodes, transport quantities

Figure 1: The finite elements used

The time-stepping for the flow problem (1), (2) is carried out by the semiimplicit projection method 2 of GRESHO [5], using a two-level weighted scheme. In matrix form this leads to the following discrete system to compute the vectors $\underline{v}^{n+1}, \underline{p}^{n+1}$ of the velocity and pressure nodal values at time level $n+1$

$$[\bar{\mathbf{M}} + \Delta t\, \theta\, \nu_0\, \bar{\mathbf{A}} + \Delta t\, \bar{\mathbf{A}}_{BTD}(\underline{v}^n)]\, \tilde{\underline{v}}^{n+1} = \bar{\mathbf{M}} \underline{v}^n + \Delta t \left\{ \theta\, \underline{\bar{F}}^{n+1} + (1-\theta)(\underline{\bar{F}}^n - \nu_0\, \bar{\mathbf{A}}\, \underline{v}^n) - \bar{\mathbf{C}}(\underline{v}^n)\, \underline{v}^n - \bar{\mathbf{M}} \bar{\mathbf{M}}_L^{-1} \mathbf{B}\, \underline{p}^n \right\}, \tag{4}$$

$$\mathbf{B}^{\mathsf{T}} \bar{\mathbf{M}}_L^{-1} \mathbf{B}\, \underline{\varphi}^{n+1} = -\underline{G}^{n+1} + \mathbf{B}^{\mathsf{T}} \tilde{\underline{v}}^{n+1}, \tag{5}$$

$$\underline{v}^{n+1} = \tilde{\underline{v}}^{n+1} - \bar{\mathbf{M}}_L^{-1} \mathbf{B}\, \underline{\varphi}^{n+1},$$

$$\underline{p}^{n+1} = \underline{p}^n + \frac{\gamma}{\Delta t} \underline{\varphi}^{n+1} \qquad \text{with} \quad 0 < \gamma \leq 2.$$

In (4), (5) $\bar{\mathbf{M}}$ and $\bar{\mathbf{M}}_L$ stand for the mass matrix of the velocity basic functions and the corresponding lumped mass matrix, respectively. $\bar{\mathbf{A}}$ is the viscous matrix, normalized by an intermediate viscosity ν_0, $\mathbf{B}$ and $\mathbf{B}^{\mathsf{T}}$ represent the discrete gradient and divergence operators, $\bar{\mathbf{C}}(\underline{v}^n)$ the convective part operator. $\underline{F}^n$ and $\underline{G}^n$ are right-hand sides produced by source terms and inhomogenous Dirichlet boundary conditions referred to time level n.

The so-called balancing tensor diffusivity matrix $\bar{\mathbf{A}}_{BTD}(\underline{v}^n)$ comes from the discretization of

$$\frac{\Delta t}{2} \nabla \cdot ((\vec{u}_n \circ \vec{u}_n) \cdot \nabla \vec{u}_{n+1})$$

and is intended to stabilize the convective part in the momentum equation (1). In many cases this simple method works well and allows Courant numbers up to the order of 10, but it has unsatisfactory damping properties for higher Reynolds numbers and strongly curved streamlines. For the further developement we think about more sophisticated methods of streamline upwind Petrov-Galerkin (SUPG) or streamline diffusion type.

For the transport equations we use a weighted implicit time-stepping scheme with a SUPG-stabilization (BROOKS, HUGHES [2]). This leads to the following discrete systems in matrix formulation for the vectors of nodal values $\underline{\Phi}_k^{n+1}$, $k = 1, 2, \ldots, N_{tr}$, of the transport quantities at time level $n + 1$

$$[\mathbf{M} + \mathbf{M}_{k,\tau} + \Delta t\,\theta \{\kappa_{k,0}\,\mathbf{A}_k(\hat{\varepsilon}_k) + \mathbf{C}(\underline{v}^n) + \mathbf{C}_{k,\tau}(\underline{v}^n)\}]\,\underline{\Phi}_k^{n+1} = [\mathbf{M} + \mathbf{M}_{k,\tau}]\,\underline{\Phi}_k^n$$
$$+ \Delta t \left\{ \theta \left(\underline{F}_k^{n+1} + \underline{F}_{k,\tau}^{n+1} \right) + (1-\theta)\left(\underline{F}_k^n + \underline{F}_{k,\tau}^n - \kappa_{k,0}\,\mathbf{A}_k(\hat{\varepsilon}_k) - \mathbf{C}(\underline{v}^n) - \mathbf{C}_{k,\tau}(\underline{v}^n) \right) \right\}. \quad (6)$$

The boldfaces in (6) have almost the same meaning as before. The subscript τ denotes those matrix and vector parts generated by the discretization of the Petrov-Galerkin extension $\tau\vec{u}_n \cdot \nabla w$ of the set of test functions w. The $\kappa_{k,0}$ is an average diffusivity and $\mathbf{A}_k(\hat{\varepsilon}_k)$ is a diffusivity matrix calculated with a streamline diffusion coefficient (JOHNSON [8])

$$\hat{\varepsilon}_k = \frac{1}{\kappa_{k,0}} \max\left\{ \kappa_k,\ c_1 h\,\frac{R_k(\underline{\Phi}_k^n)}{|\nabla_h \underline{\Phi}_k^n| + h},\ c_2 h^{\frac{3}{2}} \right\},$$

where $R_k(.)$ is the residuum of the transport equation and h is a mesh parameter.

The nonlinearities and the coupling between the equations are handled with a simple successive substitution process. That is, after updating the coupling and nonlinear terms the new iterates of *all* quantities are calculated in a decoupled manner. It turned out that this procedure is very efficient in many situations and it is enough to perform 2 – 4 iterations. Only if strong nonlinearities occur, e.g. as a result of very fast chemical reactions, the convergence of this procedure slows down and 8, 10 or more iterations are necessary in some cases.

3 Parallelization, solution of linear systems, preconditioning

The question of parallelization is tightly connected with the architecture of the parallel systems. In our opinion the *message passing concept* on the basis of powerful processor units and high speed communication networks is the most promising in high performance scientific computing. From this our efforts are directed to methods and algorithms well suited for MIMD systems. The present investigations are realized on T800 transputer systems. The programs are readily portable to other MIMD computers with little modifications concerning the fundamental communication procedures and the I/O part.

As the basic parallelization principle with respect to the geometry and data we use a partitioning of the solution domain Ω into N_{proc} subdomains Ω_s, $s = 1(1)N_{proc}$,

$$\overline{\Omega} = \bigcup_{s=1}^{N_{proc}} \overline{\Omega}_s, \qquad \Omega_s \cap \Omega_t = \emptyset \quad \text{for} \quad s \neq t,$$

and its assignment to N_{proc} processors of the parallel computer, see fig. 2.

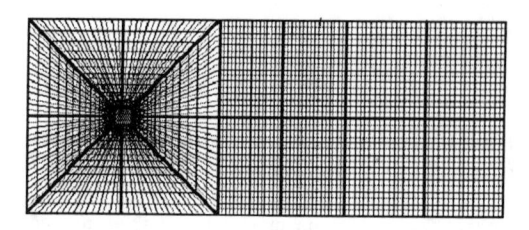

Figure 2: Partitioning of the mesh, $N_{proc} = 16$

Under the assumption, that each subdomain contains an equal number of elements, the generation of the local parts of the FEM matrices and right-hand sides on each processor requires no communication and the load of this process is balanced. In particular for time dependent nonlinear problems this can be a substantial part of the work. To get an efficient parallelization the subdomains should not be smaller than a certain minimal size, depending on processor performance, communication speed and the solution algorithm. For present transputer systems a minimal size of about 100 – 200 nodal points per processor can still be accepted, whereas for powerful processors 1000 or more nodal points should be the minimum. If a larger number of processors is available, additional parallelization principles like distribution of the differential equations (if there is a large number of transport equations) or "time step parallelization" [1] become important.

The crucial point in parallelization is the solution of the linear systems (4) – (6). We use a preconditioned conjugate gradient method (PCG) for the symmetric systems (4) and (5) and a variant of the preconditioned conjugate gradient squared algorithm (PCGS) for the non-symmetric systems (6).

Neglecting the preconditioner a CG-iteration step needs only matrix by vector and dot products, which allow a high degree of parallelization on the given DD-data distribution, see [10], [7]. The conventional CG method implies two scalar products requiring global communication and one exchange of data over the nodes of the coupling boundaries per iteration. The CGS method needs two such exchanges over the coupling boundaries.

It remains the important problem to have a good preconditioner requiring less additional communication. As a preconditioner for the discrete momentum equation (4) we use a block structured matrix $\mathcal{C}$ with

$$\mathcal{C}^{-1} = \begin{pmatrix} \mathbf{C}_1^{-1} & \mathbf{0} \\ \mathbf{0} & \mathbf{C}_2^{-1} \end{pmatrix}, \qquad (7)$$

$$\mathbf{C}_i^{-1} = \begin{cases} \mathbf{J}_i^{-\frac{1}{2}} \mathbf{S} \mathbf{K}_0^{-1} \mathbf{S}^\mathsf{T} \mathbf{J}_i^{-\frac{1}{2}} & \text{if } \Delta t\, \theta\, \nu_0 \geq c_\nu > 0, \\ \mathbf{J}_i^{-1} & \text{otherwise}. \end{cases} \qquad (8)$$

The diagonal matrices $\mathbf{J}_i$ ($i = 1, 2$) contain the main diagonals of the two diagonal blocks of the matrix $\bar{\mathbf{K}} = \bar{\mathbf{M}} + \Delta t\, \theta\, \nu_0\, \bar{\mathbf{A}} + \Delta t\, \bar{\mathbf{A}}_{BTD}$. The matrix $\mathbf{C}_H^{-1} = \mathbf{S}\, \mathbf{K}_0^{-1}\, \mathbf{S}^\mathsf{T}$ refers to the hierarchical preconditioner of YSERENTANT [12] being spectrally equivalent to the discrete Laplacian. It encloses two linear operations with $\mathbf{S}^\mathsf{T}$ and $\mathbf{S}$, each of them having a computational cost of $3N$ (N - number of nodes) floating point operations. The application of the hierarchical preconditioner requires a hierarchical grid structure. Under certain assumptions on the distribution of the grid the multiplication by $\mathbf{S}^\mathsf{T}$ and $\mathbf{S}$ can be realized without any communication, see section 4. $\mathbf{K}_0^{-1}$ is mainly the inverse of the discrete operator on the coarsest grid of the hierarchy and the identity operator for all nodal variables not belonging to the coarsest grid. In our computations so far we simply use $\mathbf{K}_0 = \mathbf{I}$.

For the matrix $\mathbf{B}^T \bar{\mathbf{M}}_L^{-1} \mathbf{B}$ in the pressure correction equation (5) a spectral equivalence to the discrete Laplacian $\mathbf{A}_p$ in the pressure space can be shown [11]. As a preconditioner for $\mathbf{B}^T \bar{\mathbf{M}}_L^{-1} \mathbf{B}$ we use

$$\mathcal{C}_p^{-1} = \mathbf{J}_p^{-\frac{1}{2}} \mathbf{S}_p \mathbf{A}_{p,0}^{-1} \mathbf{S}_p^T \mathbf{J}_p^{-\frac{1}{2}}, \tag{9}$$

with $\mathbf{J}_p$ the main diagonal matrix of the pressure Laplacian and $\mathbf{S}_p \mathbf{A}_{p,0}^{-1} \mathbf{S}_p^T$ the hierarchical preconditioner in the pressure space. Here we have also applied $\mathbf{A}_{p,0} = \mathbf{I}$. In recent tests using a kind of coarse grid solver we obtained reductions in the number of iterations of 10 – 50 % for solving (5).

Within the parallel implementation the matrix product $\mathbf{Q} = \mathbf{B}^T \bar{\mathbf{M}}_L^{-1} \mathbf{B}$ requires special attention. The computation of $\mathbf{Q}$ and its local multiplication by vectors demands additional nonstandard communications [6] and, therefore, the matrix $\mathbf{Q}$ is never built. The product $\mathbf{B}^T \bar{\mathbf{M}}_L^{-1} \mathbf{B}\,\varphi$ is calculated in each iteration by three multiplications matrix by vector, with only one communication over the coupling boundaries after the multiplication by $\mathbf{B}$.

Structurally the same preconditioner as for the discrete momentum equation is used for the discrete transport equations

$$\mathcal{C}_k^{-1} = \begin{cases} \tilde{\mathbf{J}}_k^{-\frac{1}{2}} \mathbf{S} \mathbf{K}_{k,0}^{-1} \mathbf{S}^T \tilde{\mathbf{J}}_k^{-\frac{1}{2}} & \text{if } \Delta t\,\theta\,\kappa_{k,0} \geq c_{\kappa,k} > 0, \\ \tilde{\mathbf{J}}_k^{-1} & \text{otherwise}. \end{cases} \tag{10}$$

Obviously there is no special treatment for preconditioning the non-symmetric parts of the transport operators. Thus, in the case of high CFL- *and* Peclet numbers the CGS-convergence rate deteriorates, leading to an effective time step constraint. However, it turns out, that the time step limitations due to the convergence of the nonlinear iterations and due to stability are more severe *for the considered class of problems.*

4 Grid generation

For the hierarchical preconditioning it is inevitable to have a hierarchical grid structure. From theory this structure should be built up on a very coarse basic mesh. In this case the coarse grid solvers $\mathbf{K}_0^{-1}, \ldots$ are not expensive or even can be swept away.

Up to now we used simple coarse grids (more or less generated by hand) with $O(10^1)$ to $O(10^2)$ nodal points. The hierarchical refinement is carried out by subdividing *all* edges and produces four times the amount of elements in the next higher level. If the coarse mesh elements are uniformly distributed over the available processors (in the best case each processor has the same number of coarse mesh elements), the procedure automatically guarantees a good load balance. Furthermore no communication within the operations $\mathbf{S}^T$ and $\mathbf{S}$ is necessary since there is a complete hierarchical substructure in the subdomain *on each processor.*

As a problem of this procedure appears that the resulting fine grid becomes uniform in some sense and grid adaption is mainly the result of an adaptive coarse grid. Therefore, for practical purposes it seems to be necessary to construct meshes with local hierarchical refinement depending on the transient solution. From this some new questions of load balancing, communication and redistribution strategies of subdomains arise.

5 Combustion simulation

5.1 Model problem

The simulation of realistic combustion processes is a very complex task and, therefore, a subject of permanent extensive research. In this paper only a rough modeling is used to obtain an insight into some typical features of the problem. The following assumptions hold:

- 2D laminar incompressible flow
- forced or buoyancy driven flow around a solid body emitting a fuel gas
- Boussinesq type approximation for buoyancy effects
- pyrolysis on the solid surface, Arrhenius form, not exhausting
- one-step second order combustion reaction, Arrhenius rate
- radiation neglected
- constant material parameters and thermal properties.

The solution domain of the flow problem (1), (2) and for the species concentrations y_k is Ω_g, whereas the temperature T must be calculated both in the solid Ω_s and in the gas Ω_g. We obtain the following scalar transport equations:

Energy balance
gas phase:
$$\varrho\, c_p \left\{ \frac{\partial T}{\partial t} + (\vec{u} \cdot \nabla) T \right\} = \nabla \cdot \{\lambda \nabla T\} + \Delta H\, \varrho^2 A_g\, e^{-\frac{E_g}{RT}}\, y_o\, y_f ,$$

solid phase:
$$\varrho_s\, c_s\, \frac{\partial T}{\partial t} = \nabla \cdot \{\lambda_s \nabla T\} - \varrho_s\, L\, A_s\, e^{-\frac{E_s}{RT}} ,$$

Species mass balances
for species concentrations y_k, $k \in \{f, o, p, n\}$, where f – fuel, p – product, o – oxygen, n – nitrogen.

$$\varrho \left\{ \frac{\partial y_k}{\partial t} + (\vec{u} \cdot \nabla) y_k \right\} = \nabla \cdot \{\varrho D \nabla y_k\} + q_k , \qquad k = f, o, p , \qquad (11)$$

$$\sum_{k \in \{f,o,p,n\}} y_k = 1 ,$$

with

$$q_f = -\varrho\, \dot{m} ,$$

$$q_o = -\varrho\, s_o\, \frac{M_o}{M_f}\, \dot{m} ,$$

$$q_p = +\varrho\, s_p\, \frac{M_p}{M_f}\, \dot{m} ,$$

$$\dot{m} = \varrho\, A_g\, e^{-\frac{E_g}{RT}}\, y_o\, y_f .$$

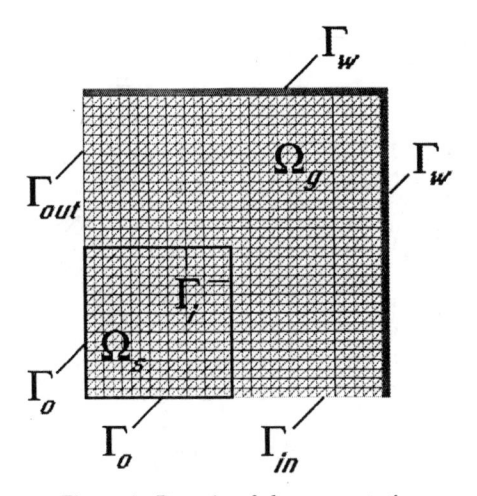

Figure 3: Domain of the computation

In (11) an one step chemical reaction

$$F + s_o\, O \longrightarrow s_p\, P$$

is assumed. The material constants are partially adopted from DI BLASI [3] and correspond to the reaction of a hydrocarbon with air.

The following boundary conditions hold, see fig. 3:

Γ_{in}	inflow	Dirichlet b.c. for	$\vec{u}$, T, y_f, y_o, y_p
Γ_{out}	outflow	$\frac{\partial \vec{u}}{\partial n} = \vec{o},\quad \frac{\partial T}{\partial n} = \frac{\partial y_f}{\partial n} = \frac{\partial y_o}{\partial n} = \frac{\partial y_p}{\partial n} = 0$	
Γ_w	wall	$\vec{u} = \vec{o},\quad \frac{\partial T}{\partial n} = \frac{\partial y_f}{\partial n} = \frac{\partial y_o}{\partial n} = \frac{\partial y_p}{\partial n} = 0$	
Γ_i	interface	$\vec{u} = \vec{u}_f(T),\quad y_f = 1,\quad y_o = y_p = 0$	
Γ_o	outer boundary of the solid	$\frac{\partial T}{\partial n} = 0$.	

The von Neumann conditions are fulfilled as natural conditions in the FEM context. The ignition of the reaction is achieved by a preheated inflow acting only a very short time. As initial conditions constant values of the transport quantities and a velocity field fulfilling the boundary conditions are used.

5.2 Discussion of the numerical results

Previous investigations of the authors [9] have shown that the flow solver is very robust and suitable for the simulation of complex flow problems. The calculated distributions of the temperature and the concentrations (fig. 5–8) seem reasonable.

There is no exact solution available, but we have made some tests to check the results indirectly. So, if we compare the solution on three different meshes with increasing refinement a convergent

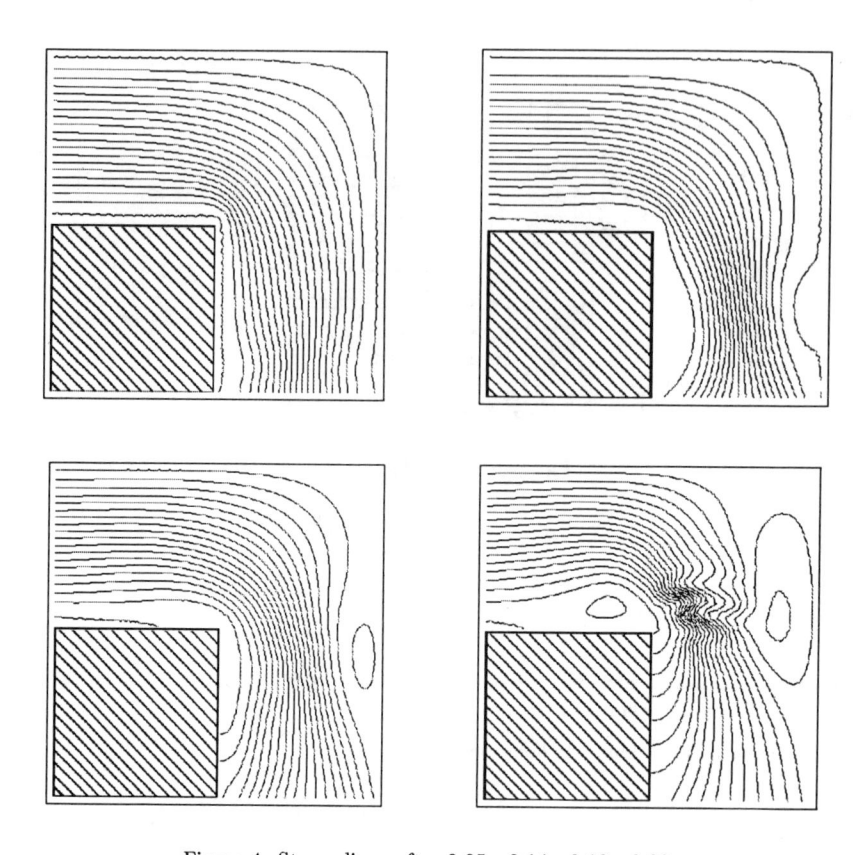

Figure 4: Streamlines after 0.05s, 0.14s, 0.18s, 0.22s

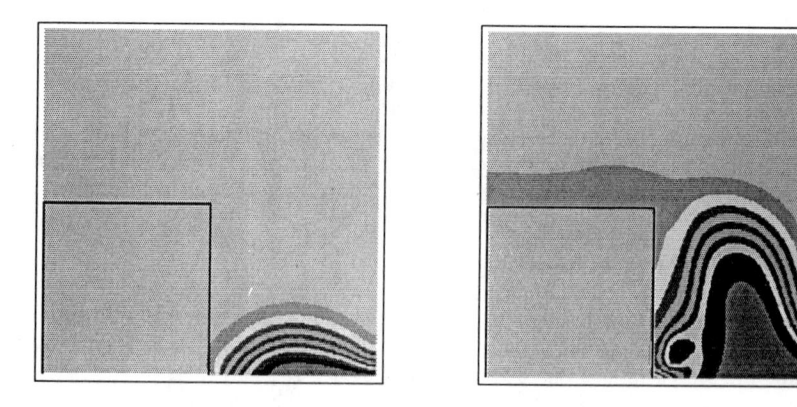

Figure 5: Temperature after 0.05s, 0.14s

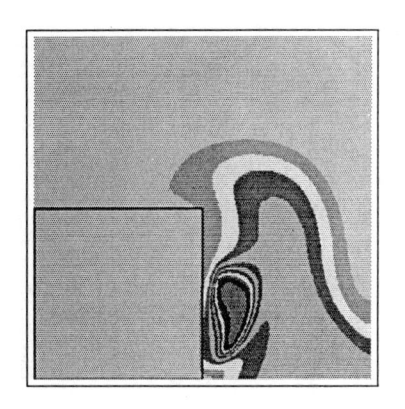

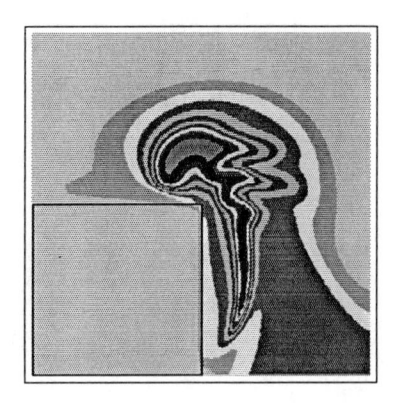

Figure 6: Temperature after 0.18s, 0.22s

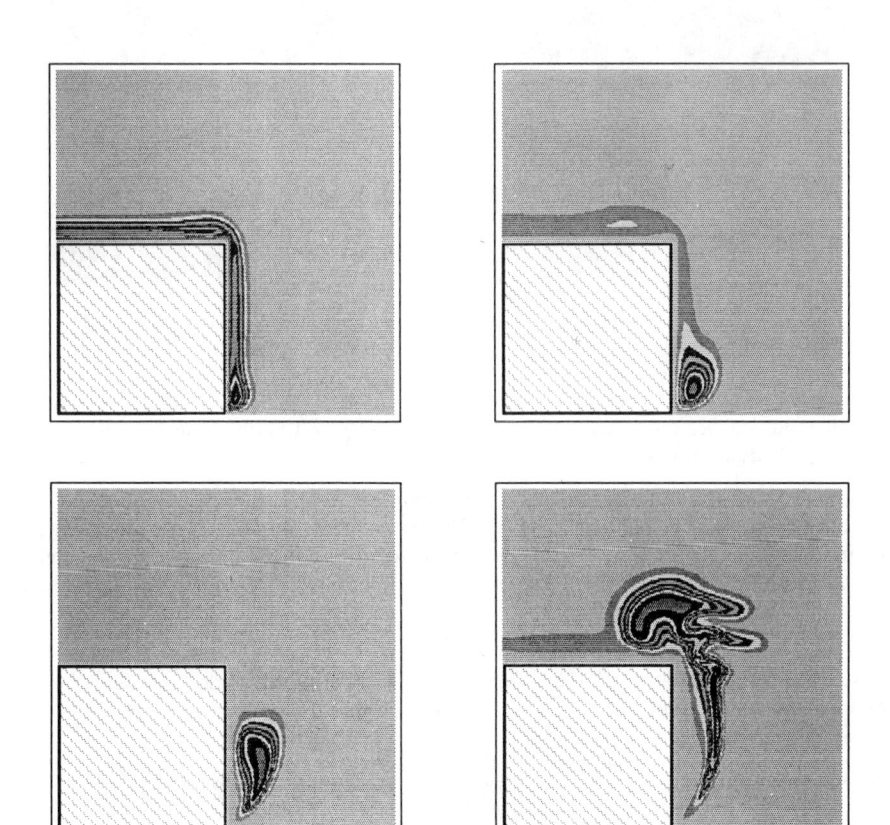

Figure 7: Concentration of product after 0.05s, 0.14s, 0.18s, 0.22s

117

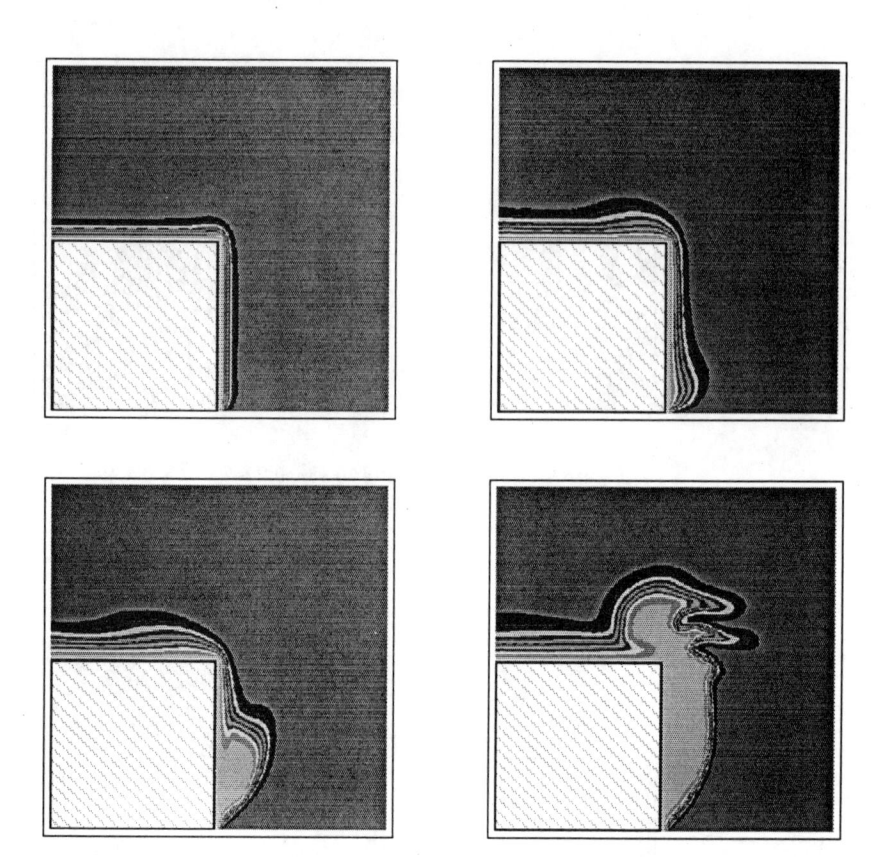

Figure 8: Concentration of oxygen after 0.05s, 0.14s, 0.18s, 0.22s

behaviour of all quantities can be noticed. A stabilization of the final results is also observable, if smaller and smaller time steps are used and the number of CG-iterations and the number of nonlinear iterations are increased.

A conspicuous phenomenon in this and other calculations is the appearance of "wiggles" in the streamlines near the "flame" after a specified period of time, see the last streamline figure. If the grid is fine enough, these wiggles are stable and can be observed over a longer time interval. It is not clear yet whether this is a physical effect or a result of Boussinesq type approximation of the buoyancy force.

The combustion process itself is characterized by extremly sharp internal layers moving through the solution domain and requiring a locally fine grid for the approximation. On the other hand, the rapid temporal change of the quantities at the flame front results in a severe time step limitation ($\Delta t \sim 10^{-3}$).

The snapshots (fig. 4–8) give only a rough impression about the results, because the "colour" scale of each picture depends on the range of the solution itself.

Table 1 compares the CPU times of a test example. The problem size scales with the number of processors, giving a measure of efficiency both for the algorithm and for the parallelization. As expected the generation of matrices and right-hand sides is fully parallelized. The increase in CPU time for the solvers is essentially caused by the computation of the pressure correction in (5). However, this part can be reduced substantially by using a coarse grid solver in the preconditioner for the matrix $\mathbf{Q}$ not used in this test.

Table 1: Computing time on a transputer T800 system, $\varepsilon = 10^{-4}$, $\varepsilon_p = 10^{-3}$

		# of iterations			CPU time [s]		
# of processors		1	4	16	1	4	16
# of nodes		2145	8385	33153	2145	8385	33153
nodes per processor		2145	2145	2145	2145	2145	2145
1. outer iteration	assembly	–	–	–	95	95	95
	solver $\Rightarrow \underline{v}$	8	7	9	8	7	9
	solver $\Rightarrow \underline{p}$	42	63	94	22	40	71
	solver $\Rightarrow \underline{T}, \underline{y_f}, \underline{y_o}, \underline{y_p}$	8,6,6,6	19,6,6,5	41,6,6,6	12	17	30
2. outer iteration	assembly	–	–	–	67	66	72
	solver $\Rightarrow \underline{v}$	8	7	8	5	6	7
	solver $\Rightarrow \underline{p}$	43	59	96	24	38	73
	solver $\Rightarrow \underline{T}, \underline{y_f}, \underline{y_o}, \underline{y_p}$	9,5,5,6	15,5,5,5	35,5,5,7	10	14	25
3. outer iteration	assembly	–	–	–	68	66	72
	solver $\Rightarrow \underline{v}$	6	6	7	5	6	9
	solver $\Rightarrow \underline{p}$	34	62	100	19	39	72
	solver $\Rightarrow \underline{T}, \underline{y_f}, \underline{y_o}, \underline{y_p}$	8,4,4,4	14,5,5,4	9,4,4,4	9	13	14
1.+2.+3. outer iterations	assembly	–	–	–	230	227	239
	solver $\Rightarrow \underline{v}$	22	20	24	18	19	25
	solver $\Rightarrow \underline{p}$	119	184	290	65	117	216
	solver $\Rightarrow \underline{T}, \underline{y_f}, \underline{y_o}, \underline{y_p}$	71	94	132	31	44	69
TOTAL for 1 time step					344	407	549

6 Conclusions

We have implemented an efficient, robust numerical algorithm for the solution of coupled transient flow and transport problems on a MIMD parallel computer. The parallelization based on domain decomposition leads to a good balanced distribution of data and computational work.

The linear systems resulting from a semiimplicit time-stepping are solved with parallelized conjugate gradient type algorithms. The preconditioners representing combinations of hierarchical bases methods and diagonal scaling allow a nearly optimal parallel implementation.

The solution of more complex problems requires further development of the algorithms and of the parallel implementation. The inclusion of adaptive grids in connection with questions of load balancing and redistribution of data between the processors is of a special importance. For strong nonlinearities a Newton linearization should be implemented.

In future considerations we will include effects of variable density and radiation and also the calculation of 3D problems.

References

[1] Bastian, P., Burmeister, J., Horton, G.: Implementation of a parallel multigrid method for parabolic partial differential equations. In: Hackbusch, W. (ed.), Parallel Algorithms for Partial Differential Equations, Proceedings of the Sixth GAMM-Seminar, Kiel 1990, Vieweg, 1991.

[2] Brooks, A.N., Hughes, T.J.R.: Streamline upwind / Petrov-Galerkin formulations for convection dominated flow with particular emphasis on the incompressible Navier-Stokes equations. Comp. Meths. Appl. Mech. Engng., *32*, 199–259 (1982).

[3] Di Blasi, C., Continillo, G., Crescitelli, S., Russo, G.: Numerical Simulation of Opposed Flow Flame Spread over a Thermally Thick Solid Fuel. Combust. Sci. and Tech., *54*, 25–36 (1987).

[4] Glowinski, R., Kuznetsov, Y.A., Meurant, G., Périaux, J., Widlund, O.B. (eds.): Domain Decomposition Methods for Partial Differential Equations. SIAM, 1991. Proceedings of the Fourth International Symposium on Domain Decomposition Methods, Moscow 1990.

[5] Gresho, P.M., Chan, S.T.: On the theory of semi-implicit projection methods for viscous incompressible flow and its implementation via a finite element method that also introduces a nearly consistent mass matrix, part 1,2. Int. J. Num. Meths. Fluids, *11*, 587–659 (1990).

[6] Groh, U.: Lokale Realisierung von Vektoroperationen auf Parallelrechnern. Preprint-Reihe der Chemnitzer DFG-Forschergruppe SPC, Preprint SPC 94-2, 1994.

[7] Haase, G., Langer, U., Meyer, A.: Parallelisierung und Vorkonditionierung des CG-Verfahrens durch Gebietszerlegung. In: Bader G., Rannacher R., Wittum G. (Hrsg.), Numerische Algorithmen auf Transputersystemen, B. G. Teubner Stuttgart 1993.

[8] Johnson, C.: The streamline diffusion finite element method for compressible and incompressible fluid flow. VKI Lecture Series 1990-03, 1990.

[9] Meinel, St., Meyer, A.: A parallel algorithm for simulation of incompressible transient flows. In: Bader, G., Rannacher R., Wittum G. (Hrsg.), Beiträge zum Workshop "Parallel Solution Methods for Differential Equations", 6.–7. Nov. 1992 an der Universität Heidelberg.

[10] Meyer, A.: A parallel preconditioned conjugate gradient method using domain decomposition and inexact solvers on each subdomain. Computing, *45*, 217–234 (1990).

[11] Meyer, A.: Preconditioning the Pseudo-Laplacian for Finite Element Simulation of Incompressible Flow. Preprint-Reihe der Chemnitzer DFG-Forschergruppe SPC, Preprint SPC 94-3, 1994.

[12] Yserentant, H.: On the multilevel-splitting of finite element spaces. Numer. Math., *49*, 379–412 (1986).

Fast solvers for coupled FEM-BEM equations I

M. Hahne, E.P. Stephan, W. Thies
Institut für Angewandte Mathematik, Universität Hannover
Welfengarten 1, 30167 Hannover, Germany

Summary

This note concerns the problem of solving efficiently systems of linear equations which arise from the symmetric coupling of finite elements (FEM) and boundary elements (BEM) when dealing with elliptic transmission problems. Within the framework of a Block Gauß Seidel iteration, two efficient preconditioners are discussed. A domain decomposition preconditioner based on Schur complement techniques deals with the degrees of freedom in the interior domain. Those unknowns which are associated with the coupling boundary, are treated by an inner-outer iteration method. The convergence rate of the Block Gauß Seidel algorithm is demonstrated experimentally.

1 Introduction

We consider transmission problems of the form

$$
\begin{aligned}
-\Delta u_1 &= 0 \quad \text{in } \Omega \\
-\Delta u_2 &= 0 \quad \text{in } \bar{\Omega} := \mathbb{R}^2 \setminus \Omega \\
u_1 &= u_2 + f \quad \text{on } \Gamma \\
\frac{\partial u_1}{\partial n} &= \frac{\partial u_2}{\partial n} + g \quad \text{on } \Gamma.
\end{aligned}
\quad (1.1)
$$

Here, $\Omega \subset \mathbb{R}^2$ is a bounded domain with Lipschitz boundary Γ. The symmetric coupling of FEM and BEM (see e.g. [3]) leads to a block matrix equation of the form

$$
\begin{pmatrix} A & B & 0 \\ B^T & C+W & K^T - I \\ 0 & K - I & -V \end{pmatrix} \begin{pmatrix} u_f \\ u_b \\ \psi \end{pmatrix} = \begin{pmatrix} o \\ r_{fc} \\ r_b \end{pmatrix}
\quad (1.2)
$$

briefly, $Mx = r$, with symmetric indefinite matrix M.

Here A, B and C denote different contributions to the part of the stiffness matrix which results from the FEM trial and test functions which are supposed to be piecewise bilinear with respect to a certain triangulation of Ω consisting of rectangles. First, A stems from the finite element functions which are centered in Ω, C belongs to those functions which are centered on Γ, and B describes the mixed terms. Further, V, K, K^T, W are the Galerkin discretizations of the single layer potential operator, the double layer potential operator and its adjoint, and the hypersingular operator. These operators are defined by

$$V\psi(x) := -\frac{1}{\pi}\int_\Gamma \psi(y)\ln|x-y|ds_y, \quad x \in \Gamma \tag{1.3}$$

$$K\psi(x) := -\frac{1}{\pi}\int_\Gamma \psi(y)\frac{\partial}{\partial n_y}\ln|x-y|ds_y, \quad x \in \Gamma \tag{1.4}$$

$$K'\psi(x) := -\frac{1}{\pi}\int_\Gamma \psi(y)\frac{\partial}{\partial n_x}\ln|x-y|ds_y, \quad x \in \Gamma \tag{1.5}$$

$$W\psi(x) := \frac{\partial}{\partial n_x}\frac{1}{\pi}\int_\Gamma \psi(y)\frac{\partial}{\partial n_y}\ln|x-y|ds_y, \quad x \in \Gamma. \tag{1.6}$$

The note is organized as follows: In Section 2, the Block Gauß Seidel (BGS) algorithm is set up. At each step of the iteration, two subproblems have to be solved. These are formed by the finite element discretization in Ω, that is

$$M_{11} := A,$$

and the Schur complement of the boundary parts of M, which is

$$M_{22} = \begin{pmatrix} C+W & K^T - I \\ K - I & -V \end{pmatrix}.$$

In Section 3, we describe an approach where the direct inversion of M_{11} is replaced by an iterative relaxation method (IRM) by Marini and Quarteroni [4] which is based on the decomposition of the domain into two subdomains and an efficient treatment of the Schur complement system.

In Section 4, an inner-outer iteration method (INO) designed by Axelsson and Vassilevski [1] is considered, which replaces the direct inversion of M_{22}. INO is a method of iterative refinement, with a gradient method being applied to the Schur complement system at each step of the iteration. Note that the inverse of the single layer potential operator is involved in forming the Schur complement matrix. Therefore its product times a vector should be worked out by another iteration which is carried out at each step of the gradient method. With INO algorithm, the cg-method takes this role.

Finally, in Section 5, numerical results are presented for the case of a Laplacian transmission problem with an L-shaped boundary. The total number of BGS iterations which is needed in order to achieve an accuracy up to the order of the error of the Galerkin method grows moderately with the number of degrees of freedom. The preconditioning iterations are both observed to converge optimally w.r.t. the meshsize h. A modification of the domain decomposition preconditioner is addressed, aiming at parallelization of the algorithm.

2 The Block-Gauß-Seidel algorithm

We write system (1.2) as

$$\begin{pmatrix} M_{11} & M_{12} \\ M_{21} & M_{22} \end{pmatrix} \begin{pmatrix} x_1 \\ x_2 \end{pmatrix} = \begin{pmatrix} o \\ r_2 \end{pmatrix}. \tag{2.1}$$

The BGS-algorithm is as follows: Let $x^\circ = \begin{pmatrix} x_1^\circ \\ x_2^\circ \end{pmatrix}$ be given. Then

FOR $n = 1, 2, \ldots$

$$\left. \begin{array}{rcl} M_{11}\delta_1^n & = & -M_{11}x_1^{n-1} - M_{12}x_2^{n-1} \\ x_1^{n-1/2} & = & x_1^{n-1} + \delta_1^n \\ x_2^{n-1/2} & = & x_2^{n-1} \end{array} \right\} \tag{2.2}$$

$$\left. \begin{array}{rcl} M_{22}\delta_2^n & = & r_2 - M_{12}^T x_1^{n-1/2} - M_{22}x_2^{n-1/2} \\ x_1^n & = & x_1^{n-1/2} \\ x_2^n & = & x_2^{n-1/2} + \delta_2^n \end{array} \right\} \tag{2.3}$$

IF $(e_\psi^n < e_\psi)$ .AND. $(e_u^n < e_u)$ STOP

NEXT .

Here

$$e_u := \left\{ \int_\Omega \left((u(x) - u_f(x))^2 + (\nabla(u(x) - u_f(x)))^T \nabla(u(x) - u_f(x)) \right) dx \right\}^{1/2}$$

and

$$\begin{array}{rcl} e_\psi & := & \left\{ < V(\frac{\partial u}{\partial n} - \psi), \frac{\partial u}{\partial n} - \psi > \right. \\ & = & \left. < V \frac{\partial u}{\partial n}, \frac{\partial u}{\partial n} > - 2 < V\psi, \frac{\partial u}{\partial n} > + < V\psi, \psi > \right\}^{1/2}, \end{array} \tag{2.4}$$

where u is the exact solution of (1.1), whose normal derivative on Γ is $\frac{\partial u}{\partial n}$, u_f is the piecewise linear function defined on $\bar{\Omega}$ by the values of the vector $u_f = \begin{pmatrix} u_{fi} \\ u_{fc} \end{pmatrix}$, $<\cdot,\cdot>$ denotes the duality pairing between $H^{1/2}(\Gamma)$ and $H^{-1/2}(\Gamma)$, and ψ is the piecewise constant function on Γ given by the values of the vector $\boldsymbol{\psi}$ (see (1.2)).

3 A domain decomposition preconditioner for the interior FEM part

First, we note that step (2.2) of the BGS-algorithm may be rewritten as

$$M_{11} x_1^{n-1/2} = -M_{12} x_2^{n-1}. \tag{3.1}$$

Let φ_{n-1} be the piecewise linear function on Γ whose values at the meshpoints are given by the elements of the vector x_2^{n-1}. Then (3.1) can be interpreted as a system of stiffness equations resulting from the finite element discretization of the boundary value problem

$$-\Delta \hat{u} = 0 \quad \text{in } \Omega \tag{3.2}$$
$$\hat{u} = \varphi_{n-1} \quad \text{on } \Gamma \tag{3.3}$$

with weak formulation

$$\text{Find } \hat{u} \in w + H_0^1(\Omega) : a(\hat{u}, v) = (f, v) \quad \forall v \in H_0^1(\Omega). \tag{3.4}$$

where $w \in H^1(\Omega)$ is an extension of the Dirichlet data φ_{n-1} to Ω. Here a is the bilinear form defined by

$$a(v_1, v_2) := \int_\Omega \nabla v_1 \nabla v_2 \, dx \quad \forall v_1, v_2 \in H^1(\Omega),$$

and $(\cdot, \cdot)$ is the usual scalar product on $L^2(\Omega)$.

Now the ideas of [4] can be applied. The domain of consideration Ω is split into two subdomains Ω_1 and Ω_2 with interface $\Gamma_{12} := \partial \Omega_1 \cap \partial \Omega_2$, and the weak problem (3.4) is replaced by an equivalent set of two coupled problems, each of them defined on one of the subdomains. Hence, a discrete iterative procedure is derived and proved to converge geometrically to the solution of (3.1). Here the error reduction factor at the n-th step of the iteration depends on the actual relaxation parameter θ_n which must be bounded from below and above by positive constants θ' and θ'' which are independent of the meshsize h. The following algorithm which is proposed in [4] involves the attempt to choose θ_n optimally so as to accelerate convergence.

We denote by S_h the space of functions which are continuous over Ω, bilinear on each element of a triangulation of Ω consisting of rectangles, and vanish on $\partial \Omega$. An extension of φ_{n-1} to Ω, which is piecewise bilinear and vanishes at all nodes except for those on Γ, is introduced and called w_h. Then the discrete analogon of the weak problem (3.4) is

$$\hat{u}_h \in w_h + S_h : a(\hat{u}_h, v_h) = (f, v_h) \quad \forall v_h \in S_h. \tag{3.5}$$

The formulation of the domain decomposition algorithm requires the definition of the discrete function spaces

$$S_0^{i,h} = \{v_h|_{\Omega_i} : v_h \in S_h\} \tag{3.6}$$
$$S_{00}^{i,h} = \left\{v_h \in S_0^{i,h} : v_h = 0 \text{ on } \Gamma\right\} \tag{3.7}$$

for $i = 1, 2$, and the space

$$\Phi_h = \{v_h|_{\Gamma_{12}} : v_h \in S_h\} \ . \tag{3.8}$$

Furthermore, for any $\phi_h \in \Phi_h$ we let $\rho_{i,h}\phi_h$ be the extension of ϕ_h to Ω_i which vanishes at all the nodes of the mesh which do not lie on Γ_{12}, and set $w_{i,h} := w_h|_{\Omega_i}$. The norms $\|\cdot\|_1^2$ and $\|\cdot\|_2^2$ denote the H^1-norms over Ω_1 and Ω_2, $a_i(\cdot,\cdot)$ is the bilinear form $a(\cdot,\cdot)$ on Ω_i, $i = 1, 2$, and $(\cdot,\cdot)_i$ denotes the scalar product on $L^2(\Omega_i)$.

Initialization

Let $g^0 \in \Phi_h$ be given and let $\sigma_0 = 0$, $\tau_0 = 0$. Then solve

$$u_{1,h}^0 \in w_{1,h} + S_0^{1,h} : a_1(u_{1,h}^0, v_h) = (f, v_h)_1 \ \forall v_h \in S_{00}^{1,h}, \ \gamma_0 u_{1,h}^0 = g^0 \ , \tag{3.9}$$

$$\tilde{u}_{1,h}^0 \in S_0^{1,h} : a_1(\tilde{u}_{1,h}^0, v_h) = 0 \ \forall v_h \in S_{00}^{1,h}, \ \gamma_0 \tilde{u}_{1,h}^0 = \gamma_0 u_{1,h}^0 \ , \tag{3.10}$$

$$u_{2,h}^0 \in w_{2,h} + S_0^{2,h} : \begin{cases} a_2(u_{2,h}^0, v_h) &= (f, v_h)_2 \ \forall v_h \in S_{00}^{2,h} \\ a_2(u_{2,h}^0, \rho_{2,h}\phi_h) &= (f, \rho_{2,h}\phi_h)_2 \ \forall \phi_h \in \Phi_h \end{cases} , \tag{3.11}$$

$$\tilde{u}_{2,h}^0 \in S_0^{2,h} : a_2(\tilde{u}_{2,h}^0, v_h) = 0 \ \forall v_h \in S_{00}^{2,h}, \ \gamma_0 \tilde{u}_{2,h}^0 = \gamma_0 u_{2,h}^0 \ . \tag{3.12}$$

Step n, $n \geq 1$.

First, solve

$$z_{1,h}^n \in S_0^{1,h} : a_1(z_{1,h}^n, v_h) = 0 \ \forall v_h \in S_{00}^{1,h}, \ \gamma_0 z_{1,h}^n = \gamma_0 \tilde{u}_{2,h}^{n-1} \ . \tag{3.13}$$

Hence, determine α_n, σ_n, τ_n and θ_n according to

$$\alpha_n = \|z_{1,h}^n\|_1^2 / \|\tilde{u}_{2,h}^{n-1}\|_2^2 \ , \tag{3.14}$$
$$\sigma_n = \max(\sigma_{n-1}, \alpha_n) \ , \tag{3.15}$$
$$\tau_n = \max(\tau_{n-1}, 1/\alpha_n) \ , \tag{3.16}$$
$$\theta_n = (\tau_n + 1)/(\sigma_n^2 \tau_n + \tau_n + 2) \ , \tag{3.17}$$

and take

$$u_{1,h}^n = u_{1,h}^{n-1} + \theta_n(z_{1,h}^n - \tilde{u}_{1,h}^{n-1}) \ , \tag{3.18}$$
$$\tilde{u}_{1,h}^n = \tilde{u}_{1,h}^{n-1} + \theta_n(z_{1,h}^n - \tilde{u}_{1,h}^{n-1}) \ . \tag{3.19}$$

In order to compute $u_{2,h}^n$, first solve

$$\xi_{2,h}^n \in S_0^{2,h} : \begin{cases} a_2(\xi_{2,h}^n, v_h) &= 0 \ \forall v_h \in S_{00}^{2,h} \\ a_2(\xi_{2,h}^n, \rho_{2,h}\phi_h) &= -a_1(u_{1,h}^n, \rho_{1,h}\phi_h) \\ &= +(f, \rho_{1,h}\phi_h)_1 \ \forall \phi_h \in \Phi_h \end{cases} \tag{3.20}$$

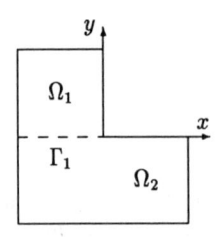

Figure 1: Decomposition of L-shape

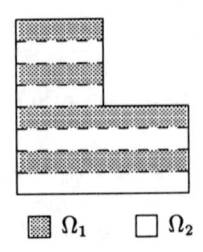

Figure 2: Decomposition suited for parallelism

Then take

$$u_{2,h}^n = u_{2,h}^0 + \xi_{2,h}^n ,\qquad (3.21)$$

$$\tilde{u}_{2,h}^n = \tilde{u}_{2,h}^{n-1} + u_{2,h}^n - u_{2,h}^{n-1} . \qquad (3.22)$$

$$(3.23)$$

The iteration is terminated if a certain stopping criterion is satisfied, as for instance

$$\|u_{1,h}^n - u_{1,h}^{n-1}\|_1 + \|u_{2,h}^n - u_{2,h}^{n-1}\|_2 \leq \varepsilon , \qquad (3.24)$$

where ε is sufficiently small. Otherwise, go back to (3.13).

IRM may be viewed as a variant of the Schur complement method. With a decomposition of an L-shaped region, as shown in Figure 1, the corresponding block Gaußian elimination step applied to A is described by

$$\begin{pmatrix} A_{11} & 0 & A_{13} \\ 0 & A_{22} & A_{23} \\ A_{13}^T & A_{23}^T & A_{33} \end{pmatrix} \rightarrow \begin{pmatrix} A_{11} & 0 & A_{13} \\ 0 & A_{22} & A_{23} \\ 0 & 0 & S \end{pmatrix}.$$

Here the subscripts $i = 1, 2$ denote contributions from finite element functions centered in Ω_i, the subscript 3 denotes contributions from FEM functions centered on Γ_{12}, and

$$S = A_{33} - A_{13}^T A_{11}^{-1} A_{13} - A_{23}^T A_{22}^{-1} A_{23} \qquad (3.25)$$

is the Schur complement matrix. Similarly, the Schur complement methods for subproblems defined on Ω_i, $i = 1, 2$, with Dirichlet boundary conditions on $\partial \Omega_i \cap \Gamma$ and Neumann boundary conditions on Γ_{12} read

$$\begin{pmatrix} A_{11} & A_{13} \\ A_{13}^T & A_{33}^{(1)} \end{pmatrix} \rightarrow \begin{pmatrix} A_{11} & A_{13} \\ 0 & S^{(1)} \end{pmatrix}, \quad \begin{pmatrix} A_{22} & A_{23} \\ A_{23}^T & A_{33}^{(2)} \end{pmatrix} \rightarrow \begin{pmatrix} A_{22} & A_{23} \\ 0 & S^{(2)} \end{pmatrix},$$

where

$$\begin{aligned} A_{33} &= A_{33}^{(1)} + A_{33}^{(2)} \\ S^{(1)} &= A_{33}^{(1)} - A_{13}^T A_{11}^{-1} A_{13} \\ S^{(2)} &= A_{33}^{(2)} - A_{23}^T A_{22}^{-1} A_{23} \end{aligned}$$

and $A_{33}^{(i)}$, $i = 1,2$ denote the contributions from integration over Ω_i.

Then, IRM acts on Γ_{12} like a Richardson method for the Schur complement system with matrix S as given by (3.25) and preconditioner $S^{(2)}$. Thus, the explicit assembly of S, which involves the computation of the inverses A_{11}^{-1} and A_{22}^{-1}, is avoided, and h-independent convergence is achieved, since S and $S^{(2)}$ are spectrally equivalent [2].

Moreover, if the subdomains are arranged as shown in Figure 2, the submatrices A_{11} and A_{22} will each have block-diagonal structure. Thus, the different blocks of each submatrix may be inverted in parallel, with the degree of parallelism being equal to half the number of strips (see Figure 2). [5]

4   An inner-outer iteration method

At step (2.3) of the BGS-algorithm M_{22} is inverted by an *inner-outer iteration* (INO) method designed by Axelsson and Vassilevski [1].
Outer iteration: Starting with $x^{n-1/2} = (u_f^0, u_b^0, \psi^0)$ at the k-th step of algorithm (INO) for given (u_b^{k-1}, ψ^{k-1}) we compute the residual

$$\begin{pmatrix} r_1^{k-1} \\ r_2^{k-1} \end{pmatrix} = \begin{pmatrix} r_{fc} - B^T u_f^0 \\ r_b \end{pmatrix} - \begin{pmatrix} C+W & I-K \\ I-K^T & -V \end{pmatrix} \begin{pmatrix} u_b^{k-1} \\ \psi^{k-1} \end{pmatrix},$$

its norm $R = \{\|r_1^{k-1}\|^2 + \|r_2^{k-1}\|^2\}^{1/2}$ and then a correction $(\delta u_b^{k-1}, \delta \psi^{k-1})$ to (u_b^{k-1}, ψ^{k-1}) by approximately solving the system

$$\begin{pmatrix} C+W & I-K \\ I-K^T & -V \end{pmatrix} \begin{pmatrix} \delta u_b^{k-1} \\ \delta \psi^{k-1} \end{pmatrix} = \begin{pmatrix} r_1^{k-1} \\ r_2^{k-1} \end{pmatrix}. \tag{4.1}$$

Inner iteration: We apply a gradient method to the Schur complement of (4.1)

$$(C + W + (I-K)V^{-1}(I-K^T))\delta u_b^{k-1} = r_1^{k-1} + (I-K)V^{-1}r_2^{k-1} \tag{4.2}$$

until the residual of the approximation of δu_b^{k-1} is less than $\epsilon_2 \cdot R$, ϵ_2 given.
The direct inversion of V in (4.2) is avoided. We use the cg–method to compute V^{-1} until the residual is less than $\epsilon_1 \cdot R$, ϵ_1 given.
Then the residuals r_1^{k-1}, r_2^{k-1} of the *outer iteration* satisfy

$$\{\|r_1^k\|^2 + \|r_2^k\|^2\}^{1/2} \le (\epsilon_1^2 + \epsilon_2^2)^{1/2} \{\|r_1^{k-1}\|^2 + \|r_2^{k-1}\|^2\}^{1/2}. \tag{4.3}$$

Algorithm INO:
Initialization: $(u_f^0, u_b^0, \psi^0) = x^{n-1/2}$;
For $k = 1, 2, \ldots$ until convergence compute:

$$r_1^{k-1} = r_{fc} - B^T u_f^0 - ((C+W)u_b^{k-1} + (I-K)\psi^{k-1});$$

$$r_2^{k-1} = r_b - ((I-K^T)u_b^{k-1} - V\psi^{k-1});$$

$$R = \{\|r_1^{k-1}\| + \|r_2^{k-1}\|\}^{1/2};$$

Solve $S\delta u_b = r_1^{k-1} + (I-K)V^{-1}r_2^{k-1}$ approximately,
by performing a number of steepest decent steps:

Initialization: $y_0 = 0$;
For $l = 1, 2, \ldots$ compute:

$$w = r_2^{k-1} - (I-K^T)y^{l-1};$$

$$v = V^{-1}w;$$

Here we perform a sufficient number of inner iteration steps until $\|w - Vv\| \le \epsilon_1 R$

$$s^{l-1} = r_1^{k-1} - (P+W)y^{l-1} - (I-K^T)v;$$

If $\|s^{l-1}\| \le \epsilon_2 R$ set

$$\delta u_b^{k-1} = v, \ \delta\psi^{k-1} = y^{l-1}, \text{ and goto C, otherwise}$$

$$\tilde{s}^{l-1} = (C+W)s^{l-1} + (I-K)V^{-1}(I-K^T)s^{l-1};$$

$$\alpha_{l-1} = (s^{l-1}, \tilde{s}^{l-1})/(\tilde{s}^{l-1}, \tilde{s}^{l-1});$$

$$y^l = y^{l-1} + \alpha_{l-1}s^{l-1};$$

C: Compute

$$u_b^k = u_b^{k-1} + \delta u_b^{k-1};$$

$$\psi^k = \psi^{k-1} + \delta\psi^{k-1};$$

and return to the computation of r_1^k, r_2^k and so on.

5 Numerical results and complexity

Our numerical experiments were carried out on the transmission problem (1.1), Ω being the L-shaped region of Figure 1, and the exact solutions known to be

$$u_1(r, \varphi) = r^{2/3} \sin[\frac{2}{3}(2\pi - \varphi)]$$
$$u_2(\bar{r}, \varphi) = \log \bar{r}, \quad \bar{r} = |(x_1, x_2) - (-s/2, -s/2)|,$$

in polar coordinates. The L-shape is scaled with factor $s < 1$ for the single layer potential operator V to be positive definite. Four variants of the BGS-algorithm for the approximate solution of a system of coupled FEM-BEM-equations (1.2) resulting from the h-version on a uniform mesh — piecewise bilinear functions in $\bar{\Omega}$, piecewise constant functions on Γ — are tested:

1. The Cholesky method is used for step (2.2), and the Gauß-Jordan algorithm is used for step (2.3).

2. Step (2.2) is carried out approximately using IRM (Section 3), and (2.3) is solved directly.

3. Step (2.2) is carried out directly, and (2.3) is done approximately by using INO (Section 4).

4. Step (2.2) and Step (2.3) are carried out approximately by using IRM and INO, respectively.

The rates of convergence for all the variants with various meshsizes are plotted in Figure 3. The main result is summarized in

Conjecture 1 *The rate of convergence ρ of the Block Gauß Seidel method applied to (2.1) behaves like*

$$\rho = 1 - C_1 h \sim 1 - C_2 N^{-1/2} \tag{5.1}$$

with positive constants C_1 and C_2 independent of the mesh parameter h, and N being the total number of unknowns.

It is therefore clear that the viability of BGS for FEM-BEM equations lies only with the parallelization of its modules. Let *cpi* be the number of arithmetic floating point operations (flops) needed per step of an iterative method, and let *eff* be the number of iterations which are needed to force the error due to the iterative method below the error resulting from the discretization. Then the computational complexity of the method is *cc*=*cpi*×*eff*. Let us look, for instance, at IRM. Suppose that the number of strips per subdomain in Figure 2 is k, and choose k such that $k^2 = N^{1/2}$. It follows that $cpi = (N/k)^2 = N^{3/2}$ for IRM. Since IRM has rate of convergence independent of N, also $eff = \log N$. This yields $cc = N^{3/2} \log N$ for the complexity of IRM as a module of BGS. As BGS has $eff = N^{1/2} \log N$, this gives rise to the statement that BGS with parallel IRM has complexity

$$cc(BGS) = N^2 \log N.$$

Here a (theoretical) speedup which is equal to the number of parallel processors used is assumed. Similar arguments seem to be valid for BGS with INO. Our numerical results indicate that the INO-algorithm when applied to (2.3) rather than the whole system of coupled FEM-BEM equations (1.2) has rate of convergence independent of the meshsize h. More precisely, if the error control parameters ϵ_1 and ϵ_2 are given certain fixed values,

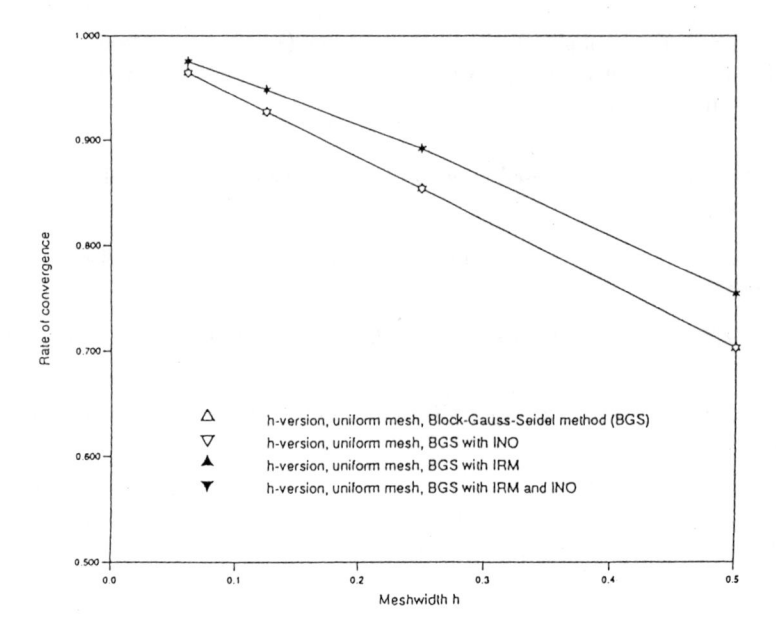

Figure 3: Linear transmission problem (Laplacian) with L-shaped interface; h-version, uniform mesh

which determine the contraction number $(\epsilon_1^2 + \epsilon_2^2)^{1/2}$, the number of inner and outer iterations needed to achieve convergence remains roughly the same. In addition, INO-algorithm can be parallelized at loop level.

Acknowledgement. The authors have been partly supported by the DFG research group Ste 238/25.

References

[1] O. Axelsson, P.S. Vassilevski: Construction of variable–step preconditioners for inner–outer iteration methods. *Proceedings of the IMACS* April 2–4 (1988) 1–14.

[2] P. Bjørstad, O.B. Widlund: Iterative methods for the solution of elliptic problems on regions partitioned into substructures. *SIAM J. Numer. Anal.* **23** (1986), 1097–1120.

[3] M. Costabel, E.P. Stephan: Coupling of finite element and boundary element methods for an elasto-plastic interface problem. *SIAM J. Numer. Anal.* **27** (1990) 1212-1226.

[4] L.D. Marini, A. Quarteroni: A relaxation procedure for domain decomposition methods using finite elements. *Numer. Math.* **55** (1989), 575-598.

[5] A. Quarteroni: Domain Decomposition Algorithms for the Stokes Equations. In: Chan, T., Glowinski, R., Périaux, J., Widlund, O. (eds.): *International symposium on domain decomposition methods for partial differential equations* **2** (1989), 431-442.

Numerical Simulations of Compressible Navier-Stokes Equations with Nonequilibrium Chemistry

M. Hilgenstock, J. Groenner and E. von Lavante
University of Essen, Lehrstuhl fuer Stroemungsmaschinen, Essen, Germany

The research and development of several hypersonic flight vehicles in the USA and Europe has brought many new computational methods for the prediction of hypersonic flowfields. The consideration of new physical phenomena in this regime, like chemical reactions and vibrational excitation, leads to the solutions of inhomogenous Navier-Stokes equations with a source term and additional equation of mass conservation for each chemical species. Several investigators have presented methods that work more or less reliably in two, and sometimes in three dimensions. The numerical efficiency and accuracy of these algorithms is still in question. The difficulties in numerically solving the resulting system of governing equations arise from the extreme stiffness of the equation system, due to the very short characteristic time scales associated with the chemistry, and the uncertainties about the material, physical and chemical properties of the participating species.

Usually main attention was paid to the analysis of hypersonic flows, while the problem of simulation of flows in typical combustion chambers was treated less extensively. The main goal of the present research was to investigate the possibilities of efficient simulation of complex supersonic flow with chemical reactions, that can be regarded as representation of combustion chambers.

Algorithm

The governing equations were in this case the compressible Navier-Stokes equation for n_s species with densities ρ_i. They are in general, body fitted coordinates, written in weak conservation law form:

$$\frac{\partial \hat{Q}}{\partial t} + \frac{\partial \hat{F}}{\partial \xi} + \frac{\partial \hat{G}}{\partial \eta} = \frac{S}{J} \qquad (1)$$

where

$$\hat{F} = \frac{1}{J}\begin{Bmatrix} \rho U \\ \rho u U + \xi_x p + \xi_x \tau_{xx} + \xi_y \tau_{xy} \\ \rho v U + \xi_y p + \xi_x \tau_{xy} + \xi_y \tau_{yy} \\ (e+p)U + \xi_x \Omega_x + \xi_y \Omega_y \\ \rho_1 U + \rho_1(\xi_x \tilde{u}_1 + \xi_y \tilde{v}_1) \\ \vdots \\ \rho_{n_s-1} U + \rho_{n_s-1}(\xi_x \tilde{u}_{n_s-1} + \xi_y \tilde{v}_{n_s-1}) \end{Bmatrix}$$

$$\tau_{xx} = -2\mu \frac{\partial u}{\partial x} + \frac{2}{3}\mu\left(\frac{\partial u}{\partial x} + \frac{\partial v}{\partial y}\right)$$
$$\tau_{yy} = -2\mu \frac{\partial v}{\partial y} + \frac{2}{3}\mu\left(\frac{\partial u}{\partial x} + \frac{\partial v}{\partial y}\right)$$
$$\tau_{xy} = \tau_{yx} = -\mu\left(\frac{\partial u}{\partial y} + \frac{\partial v}{\partial x}\right)$$
$$\Omega_x = \tau_{xx}u + \tau_{xy}v - k\frac{\delta T}{\delta x} + \sum_{i=1}^{n} h_i \rho_i \tilde{u}_i$$
$$\Omega_y = \tau_{xy}u + \tau_{yy}v - k\frac{\delta T}{\delta y} + \sum_{i=1}^{n} h_i \rho_i \tilde{v}_i$$

$$U = u\xi_x + v\xi_y \qquad J = \xi_x \eta_y - \xi_y \eta_x \ .$$

F and G are the flux vectors in the corresponding ξ and η direction.
In this work the flow was assumed to be compressible, viscous and mixture of thermally

perfect species. Due to the relatively low temperature and high pressure the gas-mixture was treated as in vibrational equilibrium. A simple model for the binary diffusion coefficient was used, along with the Sutherland equation for the viscous coefficient. The chemical reactions were realized with an 8-reaction model of Evans and Schexnayder [3], for the H_2-air combustion.

Three different flux-splitting methods were tested in the upwind-formulation. These methods are based on the work of Roe (flux-difference splitting) [9], van Leer (flux-vector-splitting) [6] and Liou (advection-upstream-splitting) [7]. The new method of Liou separates the convective and the pressure terms. The original Liou scheme was modified to:

$$(\rho M_\xi)_{1/2} = M_{\xi L}^+ \rho_L + M_{\xi R}^- \rho_R$$

$$\left(\hat{F}_{1/2}\right)_{inv} = (\rho M_\xi)_{1/2} \begin{Bmatrix} a \\ au \\ av \\ ah \\ a\frac{\rho_1}{\rho} \\ \vdots \\ a\frac{\rho_{n_s-1}}{\rho} \end{Bmatrix}_{L,R} + \begin{Bmatrix} 0 \\ \xi_x P_{1/2} \\ \xi_y P_{1/2} \\ 0 \\ 0 \\ \vdots \\ 0 \end{Bmatrix}$$

$$\hat{F}_{1/2} = \left(\hat{F}_{1/2}\right)_{inv} + F_v$$

where a is the local speed of sound; here,

$$M_\xi^+ = \begin{cases} \frac{1}{2}(M_\xi + |M_\xi|) , & |M_\xi| \geq 1.0 \\ \frac{1}{4}(M_\xi + 1)^2 , & |M_\xi| < 1.0 \end{cases}$$

In a similiar way,

$$P_{\frac{1}{2}} = P_L^+ + P_R^-$$

$$P_{L,R}^\pm = \begin{cases} p(1 \pm sgn(u))/2 , & |M_\xi| \geq 1.0 \\ p(M \pm 1)^2 (2 \mp M)/4 , & otherwise. \end{cases}$$

This scheme is automatically positivity preserving in the sence of Larrouturou [5] and Einfeldt [2].

The governing equations were integrated with a semi-implicit method, with different multistages Runge-Kutta types used for the explicit operator of the fluid-dynamics part. Following an idea of Bussing and Murmann [1], only the chemical source terms were treated implicitely.

$$\left\{I - \Delta t \, \Theta \, \frac{\partial S^n}{\partial Q^n}\right\} \frac{\partial Q^n}{\partial t \, J} = \hat{S}^n - \frac{\partial \hat{F}^n}{\partial \xi} - \frac{\partial \hat{G}^n}{\partial \eta} . \qquad (2)$$

with the relaxation parameter Θ. In most of the present computations, a two stage Runge-Kutta procedure with $\Theta = 1$. proposed to be the best choice.

The numerical effort to invert the Matrix $B = I - \Delta t \, \Theta \, \frac{\partial S^n}{\partial Q^n}$ depends on the formulation of the Jacobian of the chemical source terms. Several different forms of the Jacobian matrix, with increasing complexity and accuracy, were compared. The most obvious choice is to invert the full $n_s \times n_s$ matrix B. This, however, is a problem from the numerical point of view, since the inversion is CPU time consuming, and the matrix B usually illconditioned. This approach worked, but was rather inefficient. The next possibility to simplify the matrix B was proposed by Bussing and Murmann [1] and consists of dropping all the off diagonal terms, while keeping only the diagonal terms. In their case of only two reactions with two species, this simplification did not pay off, as the rate of convergence was also reduced. In our case of eight reactions with seven species, it turned out to be an effective means of accelerating the convergence in real time, as there was no apparent increase of the overall number of iterations. The simplest possibility is to replace the original Jacobian matrix by a diagonal matrix with the spectral radius on its diagonal. Although this simplification worked sometimes in the case of perfect gas, it was hopelessly unstable when combined with nonequilibrium chemistry.

Using a multi-block approach results in a flexible code with the possibilty of working with different chemical models (nonequilibrium, equilibrium, frozen) in different blocks (zones). Besides, some of the blocks were selectively refined, depending on the developing results. The present geometrical treatment of the computational domain was simple, yet flexible enough.

Results

1. supersonic diffusor

In this configuration, presented already by Bussing and Murmann [1], the supersonic flow of a premixed hydrogen-air mixture is investigated. The geometry is chosen such that the oblique shock is strong enough to ignite the mixture. The flow is steady, so that the reaction zone remains stationary, aligned with the shock region. The air enters the diffusor at a Machnumber $M = 2.5$, a temperature of $T = 900K$, the equivalence ratio was set to $\Phi = 0.1$. The pressure contours for the three different schemes are shown in Fig. 1.a-c. Clearly, the best results were obtained for the Roe flux-difference-splitting, although the AUSM scheme of Liou gave similiar results. The only difference are slight oscillations at the shock. The scheme of van Leer is the most dissipative one and gives thicker boundary layers, causing significant leading edge shocks. These shocks (Fig.2) lead to unphysical solutions for the chemical reaction calculations. The resulting H_2O contours given by the Roe- and Liou-schemes are practically identical, so that only the results from the Roe scheme are shown here. They indicate that the combustible mixture is ignited by the oblique shock, which agrees with the predictions by Bussing and Murmann [1].

2. transverse hydrogen jet in a supersonic airstream

Here, a hydrogen jet is injected into a two-dimensional channel with parallel walls. The inflow Mach number was $M = 4.0$, the pressure was $p = 0.1 MPa$ and the temperature was $T = 1000K$. The hydrogen jet enters at sonic conditions, the pressure of the jet was $p = 0.4 MPa$ and the temperature was $T = 600K$. This case is of particular interest, since it has been frequently used in numerical simulations by several authors and was (and is still being) extensively experimentally investigated. This allows a comparison between the numerical results and experimental data.

For this test case the AUSM scheme of Liou failed, because of its insufficient damping characteristics. Therefore we have compared only the results of van Leer's flux-vector-splitting scheme with Roe's flux-difference-splitting scheme. In Fig. 3, the pressure contours indicate the position of the oblique shock due to the leading edge of the upper plate, a strong oblique shock due to the blockage effect of the jet, it's reflection from the lower wall as well as the location of the H_2 jet. The H_2O contours in Fig. 4 show the position of the reaction zone. In accordance with the experimental measurements, the H_2 is carried upstream of the injection opening by recirculating fluid in the boundary layer. Downstream of the region, where most of the H_2O production occures, the flow is basically chemically frozen, with H_2O being convected. A small part of the produced water is convected into the boundary layer ahead of the jet by the recirculation, present at this location. Similar conclusions can be made considering the OH concentration contours. The formation of OH is the neccessary condition for the production of H_2O. The location of its highest concentration coincides with the highest rate of production and concentration of H_2O.

Fig.5 shows a detailed plot of the velocity vectors in the injection zone. Two recirculation zones are generated due to the transverse jet. A recirculation zone is located upstream of the injection, a smaller one can be found downstream of the injection. Again the Roe-scheme gave the best results. In contrast to the results of the Roe scheme the results of the van Leer scheme are influenced by the more dissipative character of the scheme. In particular, this behaviour can be recognized in Fig. 5., which shows smoother gradients of the velocity vectors. The present results agree well with the results obtained by Shuen [10]. Preliminary comparisons with available experimental measurements support qualitatively the present numerical predictions.

Conclusions

In the present work, aimed at finding a simple numerical method for simulations of compressible, viscous flows in chemical nonequilibrium, three different spatial discretization schemes were tested on a few supersonic and hypersonic configurations with increasing complexity. The van Leer FVS scheme, known to be very dissipative, was in some cases too inaccurate to be of practical use. It provided however, an reasonable initial distribution of the flow variables. The prolific Roe scheme gave consistantly good results on adequate computational grids, but displayed a distinct lack of robustnes. It failed to provide any results on grids with insufficient resolution. The relatively untested AUSM scheme, with the proper modifications, gave some promising results, while being simple and comparatively reliable. The CPU time savings due to its simplicity are, however, negligible compared with the effort spent on the modelling of the other real gas effects.

Surprisingly, the simplest possibility of removing the exponential stiffnes of the governing equations by implicit formulation, seemed to be the most effective one. The primitive diagonal form of the jacobian matrix $\frac{\partial S}{\partial Q}$ worked well, making CFL numbers, based on the acoustic wave speeds, of O(1) possible.

The multi-block, structured computational grid layout was flexible enough for the present computations. The selective, blockwise, grid refinement provided an additional means of improving the results. Still, the physics of the chemically reacting flows, especially in turbulent flows, is not well understood. The resolution of the grid would have to be greatly increased in order to provide a detailed inside into the complex flow phenomena typically present in these configurations. Currently, the FAS multigrid procedure is being tested with respect to possible convergence acceleration. The R-K semi-implicit procedure has, however, an insufficient numerical damping of the high frequency errors. Therefore, an fully implicit approach will be necessary in order to take an full advantage of the multigrid acceleration. Clearly, more work and more computer power is needed in this area of numerical research.

References

[1] Bussing, T.R.A. and Murmann, E.M., " Finite Volume Method for the Calcultion of Compressible Chemically Reacting Flows ", AIAA Paper 85-0331, Jan. 1985.

[2] Einfeldt, B., Munz, C.D., Roe, P.L. and Sjoegren, B., " On Godunov-Type Methods near Low Densities ", J. Comp. Phys., vol. 92, pp. 273-295, 1991.

[3] Evans, J.S., Schexnayder, C.J., " Influence of Chemical Kinetics and Unmixedness on Burning in Supersonic Hydrogen Flames", AIAA-Journal, Febr. 1980, pp. 188-193.

[4] Grossmann, B., and Cinella, P., " Flux-Split Algorithms for Flows with Non-equilibrium Chemistry and Vibrational Relaxation ", J. Comp. Phys., vol. 88, pp. 131-168, 1990.

[5] Larrouturou, B., and Fezoui, L. " On the Equations of Multi-Component Perfect or Real Gas Inviscid Flow ", Nonlinear Hyperbolic Problems, Lecture Notes in Mathematics, 1402, Springer Verlag, Heidelberg 1989.

[6] van Leer, B., " Flux-Vector-Splitting for the Euler Equations ", Institue for Computer Applications in Science and Engineering, Hampton, VA, Rept. 82-30, Sept. 1982.

[7] Liou, M.S., " On a New Class of Flux Splitting ", Proceedings , 13th International Conference on Numerical Methods in Fluid Dynamics, Rome 1992, pp. 115-119.

[8] von Lavante, E. and Yao, J. " Simulation of Flow in Exhaust Manifold of an Reciprocating Engine ", AIAA-93-2954.

[9] Roe, P.L., Pike, J., " Efficient Construction and Utilisation of Approximate Riemann Solutions ", Computing Methods in Applied Sciences and Engineering, VI, pp. 499-516, INRIA, 1984.

[10] Shuen, Jiang-Shun, " Upwind Differencing and LU Factorization for Chemical Nonequilibrium Navier-Stokes Equations ", Journal of Comp. Phys., vol. 99, pp. 213-250, 1992.

[11] Wada, Y. Ogawa, S. and Ishiguro, T., " A Generalized Roe's Approximate Riemann Solver for Chemically Reacting Flows ", AIAA 89-0202.

Liou Advection-Upstream-Splitting

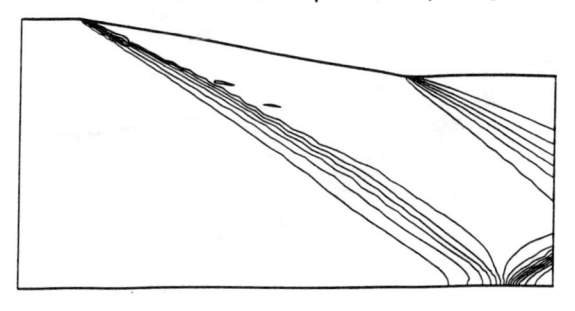

Roe Flux-Difference-Splitting

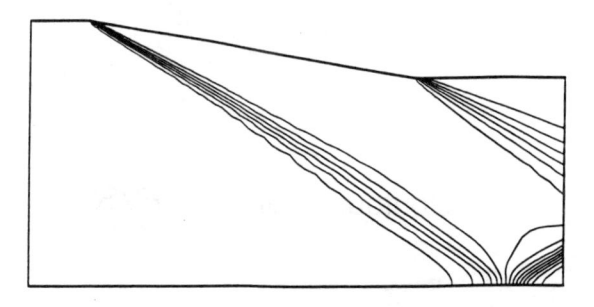

van Leer Flux-Vector-Splitting

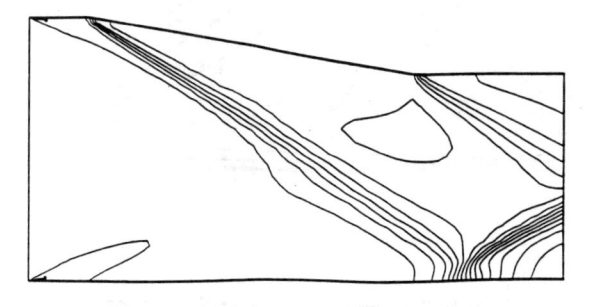

Fig. 1: Pressure Contours in Supersonic Diffusor

H_2O Contours, Roe FDS

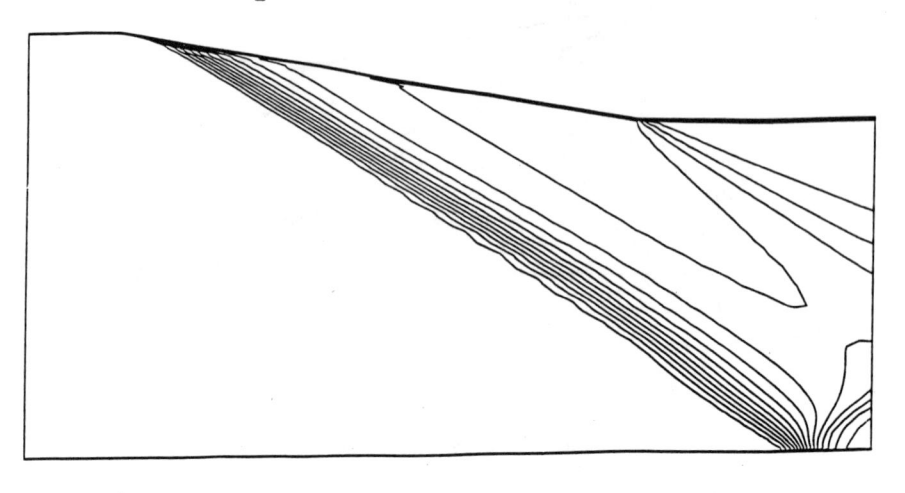

H_2O Contours, van Leer FVS

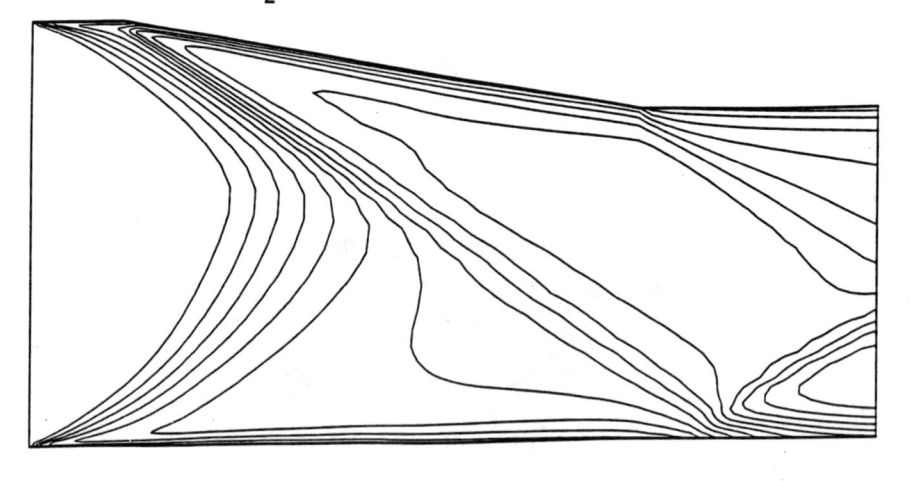

Fig. 2: H_2O Contours in Supersonic Diffusor

Pressure Contours, Roe FDS

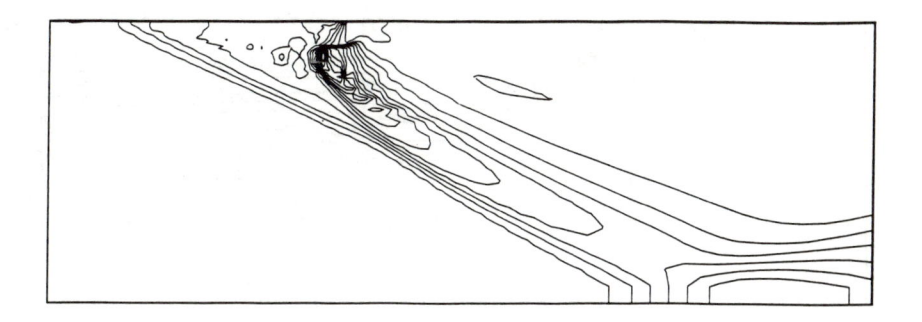

Pressure Contours, van Leer FVS

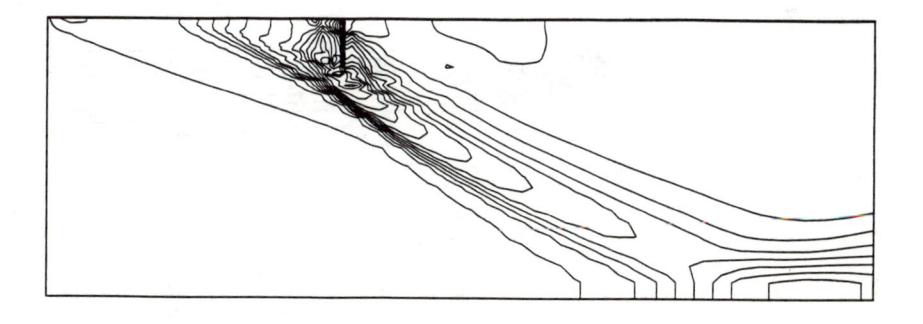

Fig. 3: Pressure Contours, Transverse Jet in Supersonic Airstream

H_2O Contours, Roe FDS

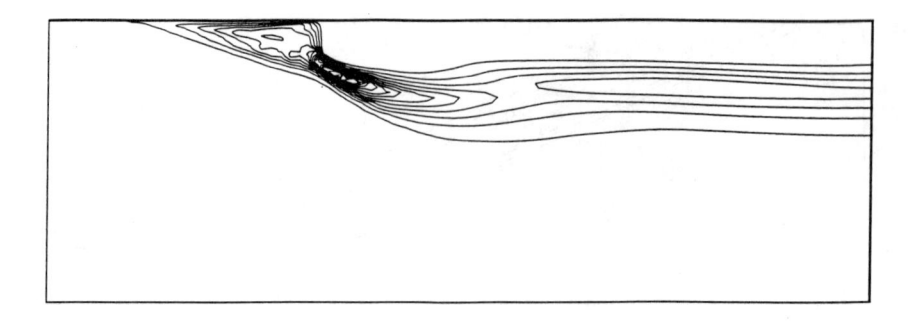

H_2O Contours, van Leer FVS

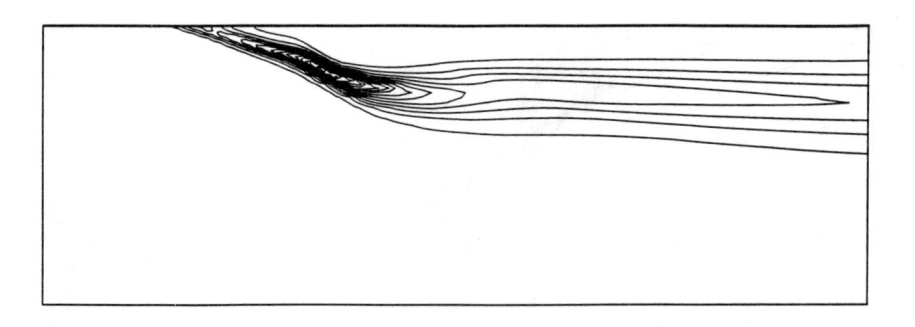

Fig. 4: H_2O Contours, Transverse Jet in Supersonic Airstream

Velocity Vectors, Roe FDS

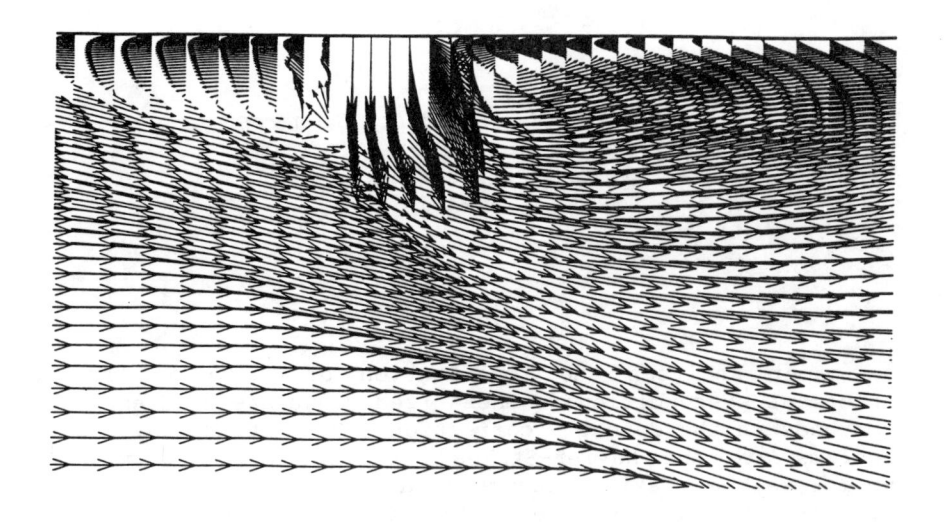

Velocity Vectors, van Leer FVS

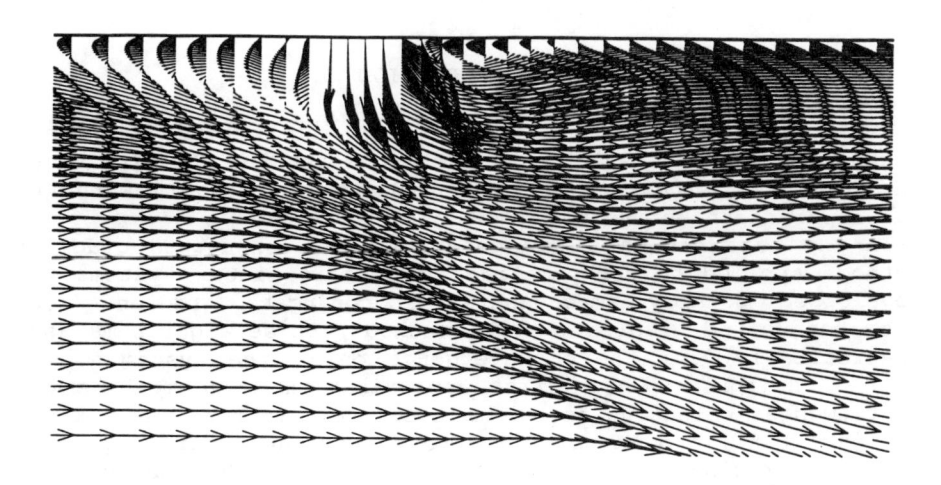

Fig. 5: Velocity Vectors, Transverse Jet in Supersonic Airstream

PRECONDITIONED CG - LIKE METHODS AND DEFECT CORRECTION FOR SOLVING STEADY INCOMPRESSIBLE NAVIER - STOKES EQUATIONS

Gh. Juncu
POLITEHNICA University Bucharest
Catedra de Inginerie Chimica
Polizu 1, 78126 Bucharest
ROMANIA

Summary

The paper analyses the implantation of three of the most performant preconditioned generalized conjugate gradient (PGCG) methods in the defect - correction (DC) iteration. The test problems are the Navier-Stokes equations which describe the incompressible, axisymmetric steady flow past a circular cylinder and past a sphere. The PGCG methods, being linear solvers a Newton process must be used in the DC iteration. The necessary number of Newton steps in a DC step is analysed. The preconditioner used is based on the ILU of the Jacobi matrix. The quantities which monitored the computational behaviour are the average reduction factor (ρ) and the efficiency (τ). The computational results are compared with those provided by the multigrid (MG) - DC method.

1. INTRODUCTION

The DC iteration [13] is a well known indirect way to obtain high order accurate solutions of the discretized PDE. Its usage in connection with the iterative methods, especially with MG, was studied exhaustively.

In the last decade a new generation of conjugated gradient methods applied to nonsymmetric algebraic linear systems was developped. Usually named generalized conjugated gradient (GCG) or conjugated gradient - like methods, these algorithms ask, theoretically, for their convergence only the system matrix to be nonsingular. In a first approximation these methods can be viewed as belonging to two main classes: Petrov - Galerkin - Krylov (PGK) and Lanczos. For very large and sparse systems, as those arising from the PDE discretization, the employing of the PGK algorithms full version becomes inefficient and, from this reason the truncated or restarted variants are used. For convection dominated convection - diffusion equations, discretized with high order accuracy schemes, the convergence of the preconditioned GCG (PGCG) algorithms is not allways satisfactory. If ILU preconditioner is used, Elman [4] shows (for the ORTHOMIN class of PGCG) that the preconditioner is the cause for the lack of convergence. Other preconditioners, proposed by Elman and Schultz [3] and tested in [8], [9] do not restore the convergence for any value of the parameter which quantifies the convection-diffusion ratio.

One of the problem's keys seems to be the usage of the adequate preconditioner. The most robust and used preconditioner still remains the ILU preconditioner. To restore the ILU stability the upwind schemes must be used. In this case the accuracy of the solutions becomes unsatisfactory. The remedy is the DC iteration. The combination PGCG-DC is analysed and compared with the MG-DC iteration in this paper, using as test problems steady incompressible Navier-Stokes equations.

Note that a different manner to couple the conjugated gradient with the deferred-correction iteration was proposed by Rubin and Khosla [10]. This method is not taken into consideration here.

2. TEST PROBLEMS

The selected test problems are the Navier-Stokes equations which describe the incompressible steady axisymmetric flow past a sphere and past a circular cylinder. The references considered, for problem formulation and results too, are Clift et al. [2] for sphere and Fornberg [5] for cylinder.

The variables used to describe the flow are the vorticity and the stream function. Using the polar spherical coordinates $z = \ln(r)$ and θ the dimensionless form of the Navier-Stokes equations for the flow past a sphere are

- stream function (Ψ)

$$\frac{\partial^2 \Psi}{\partial z^2} + \frac{\partial^2 \Psi}{\partial \theta^2} - \frac{\partial \Psi}{\partial z} - \cot\theta \frac{\partial \Psi}{\partial \theta} = \zeta \exp(3z)\sin\theta, \qquad (1a)$$

- vorticity (ζ)

$$\frac{Re}{2}\left(\frac{\partial}{\partial \theta}(e^z V_\theta \zeta) + \frac{\partial}{\partial z}(e^z V_R \zeta)\right) = \frac{\partial^2 \zeta}{\partial z^2} + \frac{\partial^2 \zeta}{\partial \theta^2} + \frac{\partial \zeta}{\partial z} + \cot\theta \frac{\partial \zeta}{\partial \theta} - \frac{\zeta}{\sin^2\theta} \qquad (1b)$$

where

$$V_R = -\frac{1}{r^2 \sin\theta}\frac{\partial \Psi}{\partial \theta}, \qquad V_\theta = \frac{1}{r^2 \sin\theta}\frac{\partial \Psi}{\partial z}$$

and $Re = U_\infty a / \nu$ is the particle Reynolds number. The equations (1) and (2) have been made dimensionless by dividing the radial coordinate r by the sphere radius a and the velocity by the free steam value U_∞. The boundary conditions considered are

$$\theta = 0, \pi: \quad \Psi = 0, \qquad \zeta = 0, \qquad (2a)$$

$$z = 0: \quad \Psi = 0, \quad \frac{\partial \Psi}{\partial z} = 0, \quad \zeta = \frac{1}{\sin \theta} \frac{\partial^2 \Psi}{\partial z^2}, \tag{2b}$$

$$z = \infty: \quad \Psi = 0.50 \exp(2z) \sin^2 \theta, \quad \zeta = 0. \tag{2c}$$

The limit $z = \infty$ was stated at $z = \pi$. This location assures accurate results for the Re values considered. The domain in which the equations are solved is the square $[0,\pi] \times [0,\pi]$.

According to Fornberg's recommandations [5], it is more convenient to solve the cylinder problem working in the deviation from the uniform flow ψ, $\psi = \Psi - y$, Ψ being the stream function, and (x,y) the cartezian coordinate system. Using the polar coordinate system defined by the conformal transformation

$$z + i\eta = \frac{1}{\pi} \ln(x + iy)$$

the dimensionless Navier-Stokes equations in terms of ψ and the vorticity ζ take the form

$$\Delta \psi = -\pi^2 r^2 \zeta \tag{3a}$$

$$\Delta \zeta = \frac{Re}{2} (v_1 \frac{\partial \zeta}{\partial z} - v_2 \frac{\partial \zeta}{\partial \eta}) \tag{3b}$$

where

$$\Delta = \frac{\partial^2}{\partial z^2} + \frac{\partial^2}{\partial \eta^2}$$

$$v_1 = \frac{\partial \psi}{\partial \eta} + \pi \exp(\pi z) \cos(\pi \eta)$$

$$v_2 = \frac{\partial \psi}{\partial z} + \pi \exp(\pi z) \sin(\pi \eta)$$

and Re is the Reynolds number based on the cylinder diameter. The variables z and η are connected to the usual polar coordinate r and θ by $r = \exp(\pi z)$ and $\theta = \pi\eta$. The boundary conditions used are

$$\eta = 0,1: \quad \psi = \zeta = 0, \tag{4a}$$

$$z = \infty: \quad \frac{\partial \psi}{\partial z} = \frac{\partial \zeta}{\partial z} = 0, \tag{4b}$$

$$z = 0: \quad \psi = -\sin(\pi \eta). \tag{4c}$$

The choice $z = \infty$ is made according to the Re number values so that Fornberg's results [5] to be reproduced. For the Re values used, a good choice is $z = 1.0$.

An important quantity which is considered as a measure of the solution accuracy is the drag coefficient C_D. The formulas for computing C_D are
- for the sphere - the relations (8) to (12) of [7];
- for the cylinder - the relation (29) of [5].

The flow equations were discretized according to the finite difference methods. In both tangential and radial directions equal step sizes were used. The two schemes employed are: the stable first order accurate upwind scheme and the second order centered scheme. The boundary conditions were implemented in the usual way [7], [8].

3. METHOD OF SOLUTION

Three PGCG algorithms, which seem to be the most important, are tested as inner iteration in the DC process: CGS [12], GMRES [11] and GCG-LS [1]. CGS is a modification of the biconjugate gradient iteration (BCG) that appears to outperform BCG consistently. GMRES and GCG-LS are the most robust of the Krylov space orthogonalization and residual minimization methods. From the experience accumulated until now it seems that none of the other iterations proposed significantly outperform CGS, GMRES and GCG-LS.

The concrete form of these algorithms used is briefly presented above, considering the general linear nonsymmetric system
$$Ax = b$$
preconditioned at the left with the matrix C. Note that the notations used for the algorithms presentation are independent from the other notations employed in the paper.

CGS
  initialization: choose x^0
$$r^0 = Ax^0 - b$$
$$pr^0 = C^{-1} r^0$$
    choose ar^0
$$p^{-1} = 0$$
$$q^0 = 0$$
$$\rho^{-1} = 1$$
$$k = 0$$
  control: if satisfied stop criterion out;
  iteration loop:
$$\rho^k = <ar^0, pr^k>$$
$$\beta^k = \rho^k / \rho^{k-1}$$
$$u^k = pr^k + \beta^k q^k$$
$$p^k = u^k + \beta^k (q^k + \beta^k p^{k-1})$$
$$\sigma^k = <ar^0, C^{-1} Ap^k>$$
$$\alpha^k = \rho^k / \sigma^k$$
$$q^{k+1} = u^k - \alpha^k C^{-1} Ap^k$$
$$x^{k+1} = x^k - \alpha^k (u^k + q^{k+1})$$
$$r^{k+1} = r^k - \alpha^k A (u^k + q^{k+1})$$
$$pr^{k+1} = C^{-1} r^{k+1}$$
$$k = k+1$$
    go to control

The choice used for ar^0 is r^0.

GMRES (m)
  1. initialization: choose x^0 and the dimension m of the Krylov subspaces;

2. Arnoldi process: $r^0 = C^{-1} (b - Ax^0)$, $\beta = \| r^0 \|$ and
$v_1 = r^0 / \beta$;
for (j=1; j <= m; j++){
$h_{i,j} = < C^{-1} Av_j, Av_i >$, i=1,...,j
$va_{j+1} = C^{-1} Av_j - \sum_{1 \le i \le j} h_{i,j} v_i$
$h_{j+1,j} = \| va_{j+1} \|$
$v_{j+1} = va_{j+1} / h_{j+1,j}$
}
3. find the vector y_m which minimizes the function
$J(y) = \| \beta e_1 - H_m y \|$,
where $e_1 = [1,0,..,0]$, among all vectors of R^m. Compute
$x^m = x^0 + v_m y_m$
4. restart: if satisfied stop criterion out, else set
$x^0 \leftarrow x^m$ and go to 2.

GCG-LS(m)
initialization: choose x^0
$r^0 = Ax^0 - b$
$pr^0 = C^{-1} r^0$
$d^0 = -pr^0$
k = 0
control: if satisfied stop criterion out
iteration loop:
$\alpha^k = - < r^k, Ad^k > / < Ad^k, Ad^k >$
$x^{k+1} = x^k + \alpha^k d^k$
$r^{k+1} = r^k + \alpha^k Ad^k$
$pr^{k+1} = C^{-1} r^{k+1}$
$h^{k+1} = A pr^{k+1}$
$\beta_{j,k} = <h^{k+1}, Ad^j>/<Ad^j, Ad^j>$,
j=max(0,k-m+1),...,k
$d^{k+1} = - pr^{k+1} + \sum_j \beta_{j,k} d^j$

k = k + 1
go to control.

The convergence criterion of the PGCG iteration is
$\| r^k \| < 1.E-05 \| r^0 \|$.

Denote by $N_{h,1}$ and $N_{h,2}$ the first order accurate respectively, the second order accurate discretization of the nonlinear equations (1) and (3). The DC iteration can be written as

$$N_{h,1} u_h^{(i+1)} = N_{h,1} u_h^{(i)} - R_{h,2}^{(i)} \qquad (5)$$

where $R_{h,2}$ is the residual of the second order discretization. The starting approximation u_h^1 is the first order accurate solution. A DC step needs solving the nonlinear system (5). The practice of the DC iteration shows that one or more steps of a given nonlinear iterative method can replace the exact solving of (5). The PGCG algorithms are designated as linear solvers. Their implantation in the DC method needs a linearization technique. The linearization algorithm used is of the Newton type. The Newton algorithm is the inner iteration in the DC method. The PGCG algorithms are the linear solvers of the Newton method.

The DC-MG algorithms used reproduce those presented in detail in [7] and [8]. Here, only the main elements of these algorithms are briefly mentioned: one V-FAS-MG [6] cycle in a DC step; prolongation by bilinear interpolation [6]; restriction by full-weighting [6]; smoothing with point Gauss-Seidel (for sphere) and line Gauss Seidel (eq. 3a) - point Gauss Seidel (eq. 3b) (for cylinder) [6].

4. RESULTS

The results are reported for a mesh having 33x33 points. The quantities which monitored the algorithms' numerical performances are the average reduction factor (ρ) and the efficiency (τ) [6]. Both quantities are computed for the DC iteration. The DC convergence is measured by the discrete euclidean norm of the second order approximation residuals $R_{h,2}$. The problems analysed are:
- the number of Newton steps in a DC step;
- the usage of a Newton or simplified Newton algorithm;
- the role of the preconditioner.

The practice of the DC-MG algorithm shows that the most efficient variant is that with one MG cycle in a DC step. In these conditions it seems naturally to try, for the DC-PGCG methods, the variant with one Newton step in a DC step. The other elements of the algorithm are considered, for the begining, in their simpler form, i.e.:
- the Jacobi matrix corresponds to the quasi-linearization of the convective terms in the vorticity equation;
- the preconditioner is the ILU(5) [12] of the Jacobi mat matrix without the term which refers to the coupling be between stream function and vorticity; the variant of van der Vorst [14] is employed.

For the flow past the sphere, at Re = 20, 50 and 100 the algorithm works very well. The values obtained for ρ and τ are depicted in table 1. Graphically, the evolution of the $\| R_{h,2} \|$ and C_D during the DC iteration is presented in figures 1 and 2. To avoid the aglomeration in the figure 1, only the results for Re=50 are depicted. The abscissa in figure 1 is the work (i.e. the number of elementary operations) per grid point and in figure 2 the DC iteration number. In the figure 2 only three curves are depicted because, for a given Re number, the results provided by the algorithms coincide. In table 1 and figures 1 and 2 the results of the DC-MG algorithm are also depicted. To compute the efficiency of the DC-PGCG algorithm the average number of PGCG iterations per Newton step was considered. Note that, the number of PGCG iterations per Newton step does not vary significantly with the Re number. Concerning the accuracy of the results it must be mentioned that the second order accurate C_D values were confirmed experimentally.

Table 1. Computational results for the flow past a sphere.

$\rho(\tau)$	DC - MG	DC - CGS	DC - GMRES	DC - GCG-LS
Re = 20.0	0.762(1093)	0.760(6387)	0.76(10931)	-
Re = 50.0	0.824(1511)	0.82(8818)	0.82(15117)	-
Re = 100.0	0.846(1720)	0.84(10037)	0.84(17206)	-

The first observation which can be made is the lack of convergence for the GCG-LS algorithm. The maximum order tried was 20. The GCG-LS method was not able to solve the linear system of the Newton step with the convergence criterion addopted. The order of the GMRES (9) is close to the minimum order (8) for which the convergence occurs. Without preconditioning both CGS and GMRES (for an order up to 20) do not converge. Table 1 shows that the convergence rate of the DC-MG, DC-CGS and DC-GMRES(9) algorithms differs insignificantly. In terms of work, DC-CGS outperforms DC-GMRES(9) and both DC-PGCG algorithms are less efficient than the DC-MG method.

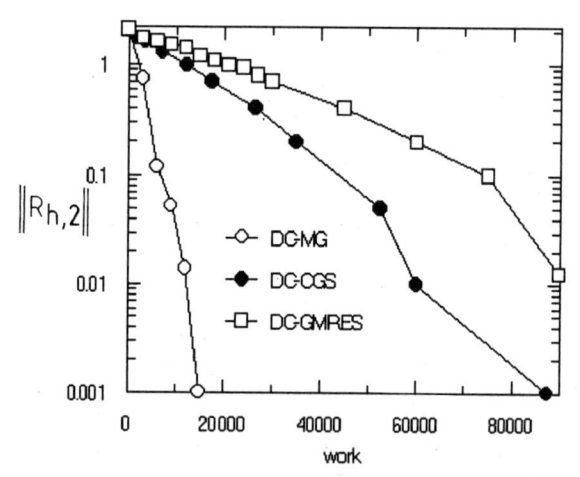

Figure 1. The evolution of the $\| R_{h,2} \|$ during the DC process: the flow past a sphere at Re = 50.

Replacing the approximate Jacobi matrix with the exact Jacobi matrix in the Newton method, the numerical performances of the algorithm do not improve. Also, using a more evolved ILU the effect is the same. The variants with more Newton steps in a DC step improve slightly the convergence rate but deteriorates the efficiency.

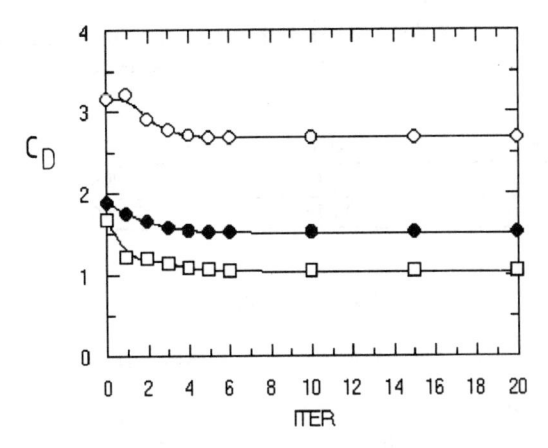

Figure 2. The evolution of the drag coefficient during the DC iteration: O - Re = 20; ● - Re = 50; □ - Re = 100.

For the cylinder problem, a different situation was encountered. The Reynolds number values employed are 2, 40, 100. The previous algorithm, on the 33x33 mesh does not work. The first modification made was the usage of the exact Jacobi matrix. The convergence was restored only for Re = 2. The next step was the replacement of ILU(5) with ILU(9) [12]. The convergence rate increases but results for Re = 40 and Re = 100 were not obtained. In [8] results for Re = 40 and Re = 100 were obtained with the DC - MG algorithm.

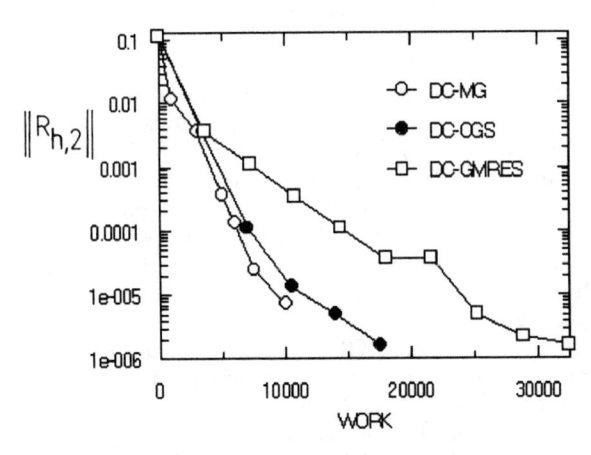

Figure 3. The evolution of the $\| R_{h,2} \|$ during the DC process: the flow past a cylinder at Re = 2.

The computational results for Re = 2 are depicted in the table 2 and figure 3. As in the previous case, the GCG-LS algorithm does not converge. The minimum order for GMRES convergence is 13. The results presented were obtained using the full Jacobi matrix and ILU(5). Table 2 and figure 3 show that, at low Re number, DC-CGS algorithm can be a concurrent for the DC-MG algorithm.

Table 2. Computational results for the flow past a cylinder

$\rho(\tau)$	DC - MG	DC - CGS	DC - GMRES(15)	DC - GCG-LS
Re = 2	0.686(1351)	0.10(760)	0.10(1563)	-

5. REFERENCES

1. Axelsson, O., A generalized conjugate gradient least square method, Numer. Math. 51, 209 (1987).
2. Clift, R., Grace, J.R., and Weber, M.E., *Bubbles, Drops and Particles*, Academic Press, New York, 1978.
3. Elman, H.C., and Schultz, M.H., Preconditioning by fast direct methods for nonself-adjoint nonseparable elliptic equations, SIAM J. Numer. Anal. 23, 44 (1986).
4. Elman, H.C., Relaxed and stabilized incomplete factorizations for non-self-adjoint linear systems, BIT 29, 890 (1989).
5. Fornberg, B., A numerical study of steady viscous flow past a circular cylinder, J. Fluid. Mech. 98, 819 (1980).
6. Hackbusch, W., *Multi-Grid Methods and Applications*, Springer, Berlin, 1985.
7. Juncu, Gh., and Mihail, R., Numerical solution of the steady incompressible Navier-Stokes equations for the flow past a sphere by a multigrid defect correction technique, Int. J. Num. Meth. Fluids 11, 379 (1990).
8. Juncu, Gh., Conjugated gradient and multigrid methods for solving the steady incompressible Navier-Stokes equations, paper submited to the 8th Int. Conf. Numerical Methods in Laminar and Turbulent Flow, Swansea, 1993.
9. Juncu, Gh., and Iliuta, I., Preconditioned CG-like methods for solving nonlinear convection-diffusion equations, Int. J. Numer. Meth. Heat & Fluid Flow, in press (1994).
10. Khosla, P.K., and Rubin, S.G., A conjugate gradient iterative method, Lecture Notes in Physics 141, Springer, Berlin, 1981, pp 248-253.
11. Saad, Y., and Schultz, M.H., GMRES: A generalized minimal residual algorithm for solving nonsymmetric linear systems, SIAM J. Sci. Statist. Comput. 7, 856 (1987).
12. Sonneveld, P., CGS A fast Lanczos-type solver for nonsymmetric linear systems, SIAM. J. Sci. Statist. Comput. 10, 36 (1989).
13. Stetter, H.J., The defect correction principle and discretization methods, Numer. Math. 29, 425 (1979).
14. van der Vorst, H.A., Iterative solution methods for certain sparse linear systems witha non-symmetric matrix arising from PDE-problems, J. Comput. Phys. 44, 1 (1981).

A Domain Decomposition Method for Singularly Perturbed Elliptic Problems

Andrei Kapurkin
Otto von Guericke University of Magdeburg
GKMBI
Postfach 4120
D-39016 Magdeburg
Germany
and
G. Lube
Georg-August-University Göttingen
Institute of Numerical
and Applied Mathematics
Lotzestr. 16-18
D-37083 Göttingen

Summary

We consider a modified Schwarz method for solving scalar singularly perturbed elliptic boundary value problems in a two-dimensional domain with an underlying stable discretization (of upwind or Galerkin/Least-squares type). The method is fully parallelizable and allows for minimal overlap.

A convergence proof of the decomposition method is available for the continuous problem in the turning-point free case. The method has been implemented on a transputer system. Numerical experiments for problems with different kinds of boundary and interior layer confirm the robustness of the method and the ability to solve some large scale problems.

1 Introduction

The resolution of fine structures of incompressible flow problems requires the solution of very large systems of linear algebraic equations. Domain decomposition methods are a very promising tool to handle such problems but there exist only a few papers considering the convection dominated case (e.g. [BS], [C], [CGK], [CR], [FQ], [G], [TA]).

In this note, we consider a domain decomposition method for solving singularly perturbed elliptic boundary value problems with an underlying stable discretization method. In particular, we analyse as a model problem the scalar convection–diffusion–reaction problem with $0 < \varepsilon \ll 1$

$$L_\varepsilon u_\varepsilon \equiv -\varepsilon \Delta u_\varepsilon + \sum_{j=1}^{2} b_j(x) \frac{\partial u_\varepsilon}{\partial x_j} + c(x)u_\varepsilon = f(x) \quad \text{in } \Omega \subseteq \mathbb{R}^2 \tag{1.1}$$

$$u_\varepsilon = g(x) \quad \text{on } \partial\Omega. \tag{1.2}$$

(1.1) is the Fourier–Kirchhoff equation modelling the energy balance in nonisothermal, incompressible flow problems.

In the limit case ($\varepsilon = 0$), information is propagated along the characteristics of the first order hyperbolic operator L_0, i.e. the solution of

$$\dot{x}(\tau) = \mathbf{b}(x(\tau)), \quad x(0) = x_0 \in \overline{\Omega}. \tag{1.3}$$

Such an "upwind character" is also reflected in domain decomposition methods. Let us consider the following

Example: Let $u_\varepsilon = \exp(x_1, x_2) \sin(\pi x_1) \sin(\pi x_2)$ be the solution of (1.1), (1.2) with $\varepsilon = 10^{-4}$, $\mathbf{b} = (1, 0)^T$ in the unit square Ω. Results are given for a modified Schwarz method (cf. Chap. 4) on a) an equidistant "isotropic" macro–mesh and "anisotropic" macro–meshes b) and c) (cf. Fig. 1):

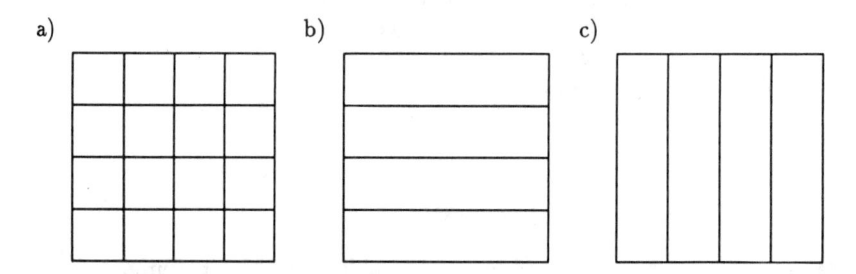

a)   b)   c)

Fig. 1: a) "Isotropic" macro–mesh, b) "Anisotropic" macro–mesh I,
c) "Anisotropic" macro–mesh II

The number of iteration steps of the decomposition method required for the stopping criterium is given in Table 1.

Table 1: Number of decomposition iteration steps depending on the number N_p of sub-domains (= of processors) for Example 1 ($\varepsilon = 10^{-4}$, $h = 1/128$)

	N_p mesh-type	2	4	8	16	32	64
a)	"isotropic"	–	3	–	8	–	15
b)	"anisotropic" I	6	6	6	6	6	6
c)	"anisotropic" II	3	6	10	17	32	64

The "upwind character" of information propagation from subdomain to subdomain is clearly to be seen for mesh c). □

The key problem of domain decomposition methods is to find the unknown function u_ε at interfaces between subdomains. For overlapping or non–overlapping subdomains, such methods consist in passing from one subdomain to the neighbouring ones Dirichlet or Neumann data (or convex combinations of both) on interior interfaces. In this note, we prefer a modified Schwarz method with non–overlapping subdomains where the information transfer is accomplished via "small" subproblems around the interfaces (following an idea in [BS] for $1D$ problems).

This paper is organized as follows: In Chap. 2 we derive a basic stability estimate for the continuous problem. In Chap. 3 we describe a domain decomposition method for the continuous problem and give a convergence result for the turning–point free case $\mathbf{b} \neq 0$. The domain decomposition method for the discrete problem is considered in Chap. 4. Several numerical test problems are shown in Chap. 5.

2 A Stability Estimate for the Continuous Problem

We consider the singularly perturbed problem (1.1), (1.2) with the small "viscosity" parameter ε, $0 < \varepsilon \ll 1$ and denote by Γ_-, Γ_+ and Γ_0, respectively, the parts of boundary $\partial\Omega \in C^{0,1}$ where the scalar product $(\mathbf{b} \cdot \nu)(x)$ is negative, positive or zero. ν is the outward pointing unit normal vector on $\partial\Omega$. Furthermore we introduce (for simplicity) the following assumptions:

(H.1) There exist no closed characteristics of the first order system (1.3) and no "turning points" (i.e. $\mathbf{b}(x) \neq 0$) in $\overline{\Omega}$.

(H.2) All points in $\overline{\Omega}$ "upstream a point" in Γ_0 belong to Γ_0, i.e. all characteristic solutions $\xi(\tau)$ of (1.3) with $\xi(0) \in \Gamma_0$ do not enter domain Ω.

(H.3) $|(\mathbf{b} \cdot \nu)(x)| \geq \beta > 0$, $x \in \overline{\Gamma_- \cup \Gamma_+}$

(H.4) $b_i, c, f \in C(\overline{\Omega})$; $g \in C(\partial\Omega)$; $\partial\Omega \in C^{0,1}$, piecewise C^2.

As a result of (H.1) – (H.4), the solution of (1.1), (1.2) with $0 \leq \varepsilon \ll 1$ admits boundary layers of width $0(\varepsilon|ln\varepsilon|)$ at Γ_+ and $0(\sqrt{\varepsilon}|ln\varepsilon|)$ at Γ_0 and possibly interior layers emanating from singular points of Γ_- or discontinuities of boundary data g on Γ_-.

In order to simplify the presentation, we assume that $b_i \in C^2$ in (H.4). As a consequence of (H.1) – (H.4), the field $\mathbf{b}(x)$ is diffeomorphic to the field $(b(x), 0)$ with $b(x) > 0$ and equation (1.1) is transformed into

$$L_\varepsilon u := -\varepsilon L_1 u + b(x) \frac{\partial u}{\partial x_1} + c(x) u = f(x) \quad \text{in } \Omega \tag{2.1}$$

with a uniformly elliptic operator

$$L_1 u := \sum_{i,j=1}^{2} a_{ij}(x) \frac{\partial^2 u}{\partial x_i \partial x_j} + \sum_{j=1}^{2} a_j(x) \frac{\partial u}{\partial x_j} + a_0(x) u \tag{2.2}$$

and

$$\Omega := \{(x_1, x_2) \in \mathbb{R}^2 : \eta_1(x_2) < x_1 < \eta_2(x_2), \ 0 < x_2 < L\} \tag{2.3}$$

with Lipschitz–continuous and piecewise C^2 functions η_i.

Lemma 1: Let $U_1, U_2 \in C(\overline{\Omega}) \cap C^2(\Omega)$ be two solutions of (2.1) – (2.3). Then there holds for $U := U_1 - U_2$ in the strip $\overline{\Omega} \cap K_\delta$ with $K_\delta := K_\delta(x_2^0) : 0 \leq x_2^0 - \delta \leq x_2^0 \leq$

$x_2^0 + \delta \leq L$

$$|U(x)| \leq \exp(\gamma x_1) \cdot \left\{ \sup_{\Gamma_- \cap K_\delta} |U(x)| + \sup_{\Gamma_+ \cap K_\delta} |U(x)| \cdot \exp\left[\frac{\alpha}{\varepsilon}(x_i - \eta_2(x_2))\right] \right.$$
$$+ \sup_{x_2 = x_2^0 - \delta} |U(x)| \cdot \exp\left[\frac{\beta}{\sqrt{\varepsilon}}((x_2^0 - \delta) - x_2)\right]$$
$$\left. + \sup_{x_2 = x_2^0 + \delta} |U(x)| \cdot \exp\left[\frac{\beta}{\sqrt{\varepsilon}}(x_2 - (x_2^0 + \delta))\right] \right\}$$
(2.4)

with ε-independent constants $\alpha > 0$, $\beta > 0$ and γ.

Proof: The proof of (2.4) is based on the maximum principle: Let $L_\varepsilon S \geq |L_\varepsilon U|$ in $G \subseteq \Omega$, $S \geq |U|$ on $\partial G \in C^{0;1}$, then $S \geq |U|$ in $\overline{G}$. A straightforward calculation with barrier function S and suitable constants α, β, γ as the r.h.s. of (2.4) gives

$$L_\varepsilon S \geq 0 = |L_\varepsilon U| \text{ in } \Omega \cap K_\delta, \qquad S \geq |U| \text{ on } \partial(\Omega \cap K_\delta)$$

which completes the proof of (2.4). □

Remark 2.1: In [KL] we will give a proof of Lemma 1 in the general case (H.1) – (H.4). Furthermore we will give extensions concerning assumptions (H.1) – (H.3). One can also argue with the weak maximum principle [G] and therefore use a weaker assumption on the data as (H.4). □

Remark 2.2: Lemma 1 reflects the boundary and interior layer behaviour of the solution of (2.1) – (2.3) and will be the basis of the convergence proof of the domain decomposition method in Chap. 3. □

3 Domain Decomposition of the Continuous Problem

We start with some notation. Let $\{M_j\}$ with $\overline{\Omega} = \bigcup_{j=1}^{I} \overline{M}_j$ be a partition of $\overline{\Omega}$ into non-overlapping subdomains M_i ("macroelements") with Lipschitz-continuous and piecewise C^2 boundary ∂M_i and outward pointing unit normal vector ν^i. Furthermore let be ∂M_i^+, ∂M_i^- and ∂M_i^0, respectively, the parts of ∂M_i where $(\mathbf{b} \cdot \nu^i)(x)$ is positive, negative or zero.

$\mathcal{O}_{ij}$ denotes an "overlap domain" between adjacent macroelements M_i and M_j such that $\partial M_i \cap \partial M_j \subseteq \overline{\mathcal{O}_{ij}}$. Set

$$B_i := \{x \in M_i \cap \bigcup_{i,j} \mathcal{O}_{ij} : \partial M_i \cap \partial M_j \neq 0\} \quad (3.1)$$

and

$$ovlp_i^- := \min_{M_i \setminus B_i} \text{dist}(x, \partial M_i^-) \quad ovlp_i^+ := \min_{M_i \setminus B_i} \text{dist}(x, \partial M_i^+)$$
$$ovlp_i^0 := \min_{M_i \setminus B_i} \text{dist}(x, \partial M_i^0). \quad (3.2)$$

Furthermore we introduce crosspoints C as the vertices of macroelements M_i and "crosspoint regions" $U(C)$ with

$$U(C) := \{x \in \bigcap_{i,j=1}^{M} \mathcal{O}_{ij} : C \in \partial M_i \cap \partial M_j\}. \tag{3.3}$$

We propose the following domain decomposition method for the continuous problem (1.1), (1.2):

1. Solve (parallel) subproblems on M_i, $i = 1, \ldots, I$:

$$L_\varepsilon u^n = f \quad \text{in} \quad M_i; \qquad u^n = u^{n-1} \quad \text{on} \quad \partial M_i. \tag{3.4}$$

2. Solve (parallel) subproblems on $U(C)$:

$$L_\varepsilon z^n = f \quad \text{in} \quad U(C); \qquad z^n = u^n \quad \text{on} \quad \partial U(C). \tag{3.5}$$

Set

$$u^n = z^n \quad \text{on} \quad \partial M_i \cap U(C). \tag{3.6}$$

3. Solve (parallel) subproblems on $\mathcal{O}_{ij}$:

$$L_\varepsilon w^n = f \quad \text{in} \quad \mathcal{O}_{ij}; \qquad w^n = u^n \quad \text{on} \quad \partial \mathcal{O}_{ij}. \tag{3.7}$$

Set

$$u^n = w^n \quad \text{on} \quad \partial M_i \cap \partial M_j. \tag{3.8}$$

4. Check the stopping criterion for given tolerance TOL:

$$q^n := |u^n - u^{n-1}| < \text{TOL}. \tag{3.9}$$

Set $n := n + 1$ in case of $q^n \geq \text{TOL}$ and repeat steps (1) – (4), otherwise stop.

The following result states convergence of the iteration method (3.4) – (3.9) in the singularly perturbed case of (1.1), (1.2) without turning–points considered in Chap. 2. Therefore we consider the transformed problem (2.1) – (2.3).

Theorem 3.1 *Assume (H.1) – (H.4) and for each macroelement M_i:*

(H.5) All points in $\overline{M_i}$ "upstream a point in ∂M_i^0 " belong to ∂M_i^0.

(H.6) $|(\mathbf{b} \cdot \nu^i)(x)| \geq \beta_i > 0$, $x \in \overline{\partial M_i^-} \cup \overline{\partial M_i^+}$.

(H.7) $ovlp_i^+ \geq \beta_i^{-1} \varepsilon |\ln \text{TOL}|$, $ovlp_i^- \geq \beta_i^{-1} \varepsilon |\ln \text{TOL}|$

$$ovlp_i^0 \geq \gamma^{-1} \sqrt{\varepsilon} |\ln \text{TOL}|. \tag{3.10}$$

For given tolerance TOL there exists an index $N < \infty$ and a positive constant K (both independent of ε and TOL) such that after N iteration steps (3.4) – (3.9) holds

$$\sup_\Omega |(u_\varepsilon - u^N)(x)| \leq K \cdot \text{TOL}. \tag{3.11}$$

Remark 3.1: We assume (H.5) – (H.7) only for clearness of the presentation. Note also that, for given vector field b, the choice of the macroelement mesh is at our hand. □

Remark 3.2: It follows from the proof that N depends linearly on the maximal number of macroelements which a characteristic curve of (1.3) passes in $\overline{\Omega}$ starting from a point on $\overline{\Gamma}_-$. □

Remark 3.3: In case of $c(x) \geq 0$ in Ω, there holds $K \leq 1$ in estimate (3.11). In the general case, the constant K is characterized by the term $\exp(\gamma x_1)$ in estimate (2.4). Under our assumptions (H.1) – (H.4) it is always possible to transform (1.1) to the case $c(x) \geq 0$, hence we consider this case in the following proof. □

Proof of Theorem 3.1: We can assume by a scaling argument that $|g(x)| \leq 1$ on $\partial\Omega$. The maximum principle yields $|u_\varepsilon(x)| \leq 1$ in Ω. So it is natural to assume for the initial guess that $|u^0(x)| \leq 1$ in $\overline{\Omega}$. Furthermore we assume for simplicity that at most four macroelements form a common crosspoint (cf. Remark 3.4). According to Chap. 2 we consider the transformed problem (2.1) – (2.3).

We prove the result by induction and consider the following situation (cf. Fig. 2 for some macroelement M_i after $n-1$ iteration steps of (3.4) – (3.9).

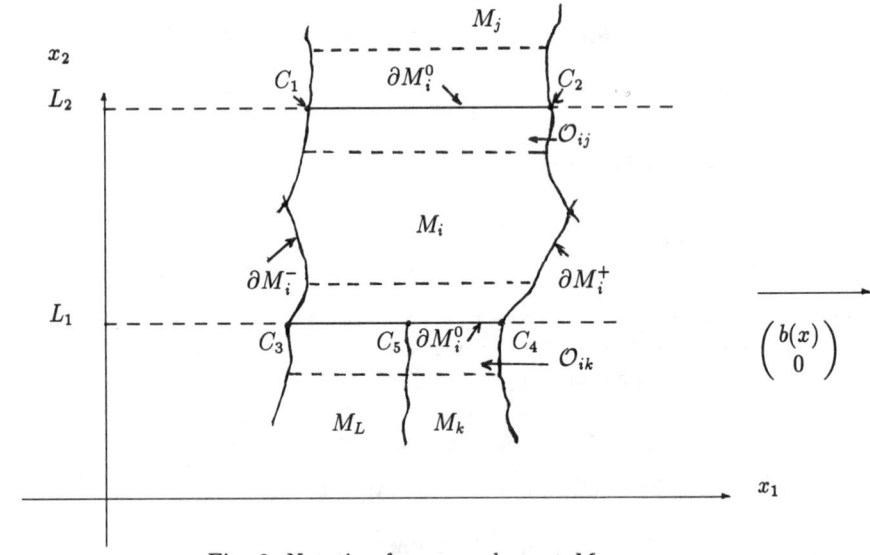

Fig. 2: Notation for macroelement M_i

We start with the following induction assumption:

(H) Suppose that there holds $|(u_\varepsilon - u^{n-1})(x)| < $ TOL on the inflow part ∂M_i^- of M_i and in crosspoint regions $U(C)$ around crosspoints on the closure of ∂M_i^-.

Let us assume first that M_i is an "interior" macroelement s.t. $\partial M_i \cap \overline{\Gamma_0 \cup \Gamma_+} = \emptyset$ (cf. (vi)).

(i) We conclude from (H), Lemma 1 and the overlap condition (3.10) that there holds after step (1)
$$|u_\varepsilon - u^n| < \forall x \in \overline{M_i} : \text{dist}(x, \partial M_i^0) \geq ovlp_i^0 , \text{dist}(x, \partial M_i^+) \geq ovlp_i^+ .$$

(ii) If ∂M_i^- and ∂M_i^+ form a common crosspoint C, we have furthermore $|u_\varepsilon - u^n| <$ TOL in $\overline{M_i} \cap U(C)$ because of $|u_\varepsilon - u^{n-1}| = |u_\varepsilon - u^n| <$ TOL on $\partial M_i^+ \cap U(C)$ by (3.4) and (3.6) and (H).

(iii) Let us assume now that a connecting component of ∂M_i^0 contains at most two crosspoints $C_1 = \partial M_i^0 \cap \partial M_i^-$ (cf. arc $C_1 C_2$ at $x_2 = L_2$ in Fig. 2. By (i) we know that $|u_\varepsilon - u^{n-1}| = |u_\varepsilon - u^n| <$ TOL on $\partial M_j^- \cap U(C_1)$ if M_j is an adjacent macroelement of M_i. Then step (3) yields in the overlap domain $\mathcal{O}_{ij}$ that $|u_\varepsilon - w^n| <$ TOL $\forall x \in \overline{M_i} \cap \overline{\mathcal{O}_{ij}}$, $\text{dist}(x, \partial M_i^+) \leq ovlp_i^0$, hence by (3.8) $|u_\varepsilon - u^n| <$ TOL $\forall x \in M_i \cap \partial M_j$, $\text{dist}(x, \partial M_i^+) \geq ovlp_i^0$.

(iv) Suppose now that $|u_\varepsilon - u^n| <$ TOL $\forall x \in \overline{M_i}$ such that $\text{dist}(x, \partial M_i^+) \geq ovlp_i^+$ and that a similar estimate is valid on adjacent macroelements. We correct the data $u^n = z^n$ in (somewhat smaller) crosspoint regions $U(C)$ around crosspoints on the closure of ∂M_i^+ in at least two steps of type (2). Then we can update u^n by step (3) on ∂M_i^+ with smooth transition to the values of u^n on ∂M_i^+ around crosspoints on the closure of ∂M_i^+. Note that we have also an correct update of u^n in crosspoint regions on the closure of ∂M_i^- of macroelements M_j following M_i "downstream".

(v) Consider now the case that additional crosspoints are located on ∂M_i^0 (cf. C_5 on the arc $C_3 C_4$ at $x_2 = L_1$ in Fig. 2). It can be easily seen that a correct update of u^n in a neighbourhood of ∂M_i^0 is given after a finite number of steps (2) and (3).

(vi) Finally we consider the case that M_i is not an "interior" macroelement such that $\partial M_i \cap (\Gamma_+ \cup \Gamma_0) \neq \emptyset$. Note that by construction holds $u_\varepsilon - u^{n-1} = 0$ on $\partial M_i \cap \partial \Omega$. Hence we find the correct update u^n in M_i whenever $|u_\varepsilon - u^{n-1}| <$ TOL on $\partial M_i / \partial \Omega$.

(vii) As a consequence of (i) – (vi), we can proceed with the update of u^n in macroelements which are located "downstream" w.r.t. M_i starting again with (i). The number N of iteration steps (1) – (4) is finite because the number of macroelements and therefore of crosspoints is finite. □

Remark 3.4: The assumption in the proof that at most four macroelements form a common crosspoint is not essential. As a rule, the number of iteration steps (1) – (4) may increase with the number of adjacent macroelements at a crosspoint. □

4 Domain Decomposition of a Stable Discrete Problem

The assumption of the (possibly) singularly perturbed problem (1.1), (1.2) requires a stable discretization method. It is well known that standard central difference schemes or the standard Galerkin finite element method produce nonphysical solutions unless the mesh is sufficiently fine.

Upwind or stabilized Galerkin schemes are candidates for a stable method. Recent results on local L^∞-error estimates for a hybrid finite element [Ri] or the streamline diffusion method [ZR] allow to formulate discrete analogues of Lemma 1 and Theorem 3.1 with possibly enlarged overlap and crosspoint regions (cf. [KL]).

Let us describe the domain decomposition method for a discrete problem

$$L_h u_h = f_h \qquad (4.1)$$

based on a stable difference scheme on a fine mesh $\overline{\Omega}_h$ on $\overline{\Omega}$. We assume for simplicity that the macromesh Ω_H formed by the crosspoints of $\{M_j\}$ with characteristic mesh width $0(H)$ is contained in the fine mesh with width $0(h)$, $h \ll H$. We introduce notation

$$M_i^h := M_i \cap \Omega_h, \quad \mathcal{O}_{ij}^h := \mathcal{O}_{ij} \cap \Omega_h, \quad U^h(C) := U(C) \cap \Omega_h \qquad (4.2)$$

and similarly ∂M_i^h, $\partial \mathcal{O}_{ij}^h$ and $\partial U^h(C)$.

Initialization: Solve a coarse grid problem

$$L_H u_H^0 = f_H \quad \text{on } \Omega_H; \qquad u_H^0 = g_H \quad \text{on } \partial\Omega_H. \qquad (4.3)$$

and interpolate u_H^0 on ∂M_i^h.

Iteration: Let u_h^{n-1}, $n = 1, 2, \ldots$ be given.

(1) Solve parallel subproblems on M_i^h, $i = 1, \ldots, I$:

$$L_h u_h^n = f_h \quad \text{in } M_i^h; \qquad u_h^n = u_h^{n-1} \quad \text{on } \partial M_i^h. \qquad (4.4)$$

(2) Solve parallel subproblems on $U^h(C)$:

$$L_h z_h^n = f_h \quad \text{in } U^h(C); \qquad z_h^n = u_h^n \quad \text{on } \partial U^h(C). \qquad (4.5)$$

Set

$$u_h^n = z_h^n \quad \text{on } \partial M_i^h \cap U^h(C). \qquad (4.6)$$

(3) Solve (parallel) subproblems on $\mathcal{O}_{ij}^h$:

$$L_h w_h^n = f_h \quad \text{in } \mathcal{O}_{ij}^h; \qquad w_h^n = u_h^n \quad \text{on } \partial \mathcal{O}_{ij}^h. \qquad (4.7)$$

Set

$$u_h^n = w_h^n \quad \text{on } \partial M_i^h \cap \partial M_j^h. \qquad (4.8)$$

(4) Check the stopping criterion for given tolerance TOL:

$$d_h^n := \|f_h - L_h u_h^n\| < \text{TOL}; \qquad q_h^n := \|u_h^n - u_h^{n-1}\| < \text{TOL}. \qquad (4.9)$$

Set $n := n + 1$ in case of $d_h^n \geq \text{TOL}$ or $q_h^n \geq \text{TOL}$ and repeat steps (1) – (4), otherwise stop.

Note that the iteration steps (1) – (4) are fully parallelizable. Only local communication across interfaces $\partial M_i \cap \partial M_j$ and in crosspoint regions is required. In the singularly perturbed case, steps (2) and (3) are very cheap due to the very small values $ovlp_i^+$, $ovlp_i^-$ and $ovlp_i^0$.

5 Numerical Results

The domain decomposition method of Chap. 4 has been implemented on a T 800–transputer system GCeL 128. Each macroelement M_i is assigned to a processor P_i. Subproblems in overlap domains $\mathcal{O}_{ij}$ are solved redundantly on processors P_i and P_j, similarly the subproblems in crosspoint regions.

The discretization method (4.1) is an upwind method proposed in [Ro].

The following examples are based on problem (1.1), (1.2) in the unit square $\Omega = (0,1) \times (0,1)$. We use equidistant coarse and fine meshes with characteristic parameters H and h and denote by $\Delta = \max\{ovlp_i^+; ovlp_i^-; ovlp_i^\circ\}$ the diameter of overlap domains perpendicular to the interfaces $\partial M_i \cap \partial M_j$.

The iteration (4.3) – (4.9) was terminated with TOL $= 10^{-5} d_h^0$.

Example 1: (Smooth solution)
Let $u = \exp(x_1 x_2) \sin(\pi x_1) \sin(\pi x_2)$ be the solution of (1.1), (1.2) with $b = (1,1)^T$, $c = 0$, $f = L_\varepsilon u$ and $g = 0$ in the unit square $\Omega = (0,1) \times (0,1)$ and on $\partial \Omega$, respectively.

The number of iterations is given in Table 2 in dependence of the small parameter ε, the coarse mesh parameter H and the overlap width Δ for $h = 1/128$.

Table 2: Number of iterations in Example 1 ($h = 1/128$)

	$H = 1/4$			$H = 1/8$		
$-\log(\varepsilon)$	$\Delta = h$	$\Delta = 2h$	$\Delta = 3h$	$\Delta = h$	$\Delta = 2h$	$\Delta = 3h$
3	13	12	11	17	16	15
4, 5, 6, 7	8	8	8	15	15	15

It is clearly to be seen that the method is stable for moderate small and very small values of the singular perturbation parameter ε. Furthermore it is possible to minimize the overlap width to $\Delta = h$. This result is stable w.r.t. to the fine mesh width h (cf. Table 3).

Table 3: Number of iterations in Example 1 ($\Delta = h$, $\varepsilon = 10^{-i}$, $i = 4, 5, 6, 7$)

H	$h = 1/128$	$h = 1/256$	$h = 1/1024$
1/2	3	3	–
1/4	8	7	8
1/8	15	15	15

Furthermore we compare in Table 4 the number of required iteration steps for the proposed modified Schwarz method (MSM) and an additive Schwarz method with small overlap (cf. [CGK]).

Table 4: Comparison of iteration steps for
ASM – additive Schwarz method (cf. [CGK]) and
MSM – modified Schwarz method (present)

method	$\varepsilon = 10^{-3}$		$\varepsilon = 10^{-4}$		H
	$\Delta = h$	$\Delta = 2h$	$\Delta = h$	$\Delta = 2h$	
ASM	17	17	17	17	1/4
MSM	13	12	8	8	
ASM	22	20	23	19	1/8
MSM	17	16	15	15	

□

Example 2: (Outflow boundary layer)
Let be $b = -(1, 1/2)^T$, $c = 0$, $g = 1$ if $x_1 + x_2 \leq 1$, $g = 0$ on $\partial\Omega$ else. We assume that $u_\varepsilon = \Phi + U_\varepsilon$ with the smooth part

$$\Phi = \sin(k\pi x_1)\sin(k\pi x_2), \quad k \in \mathbb{N}$$

and the boundary layer term U_ε as the solution of $L_\varepsilon U_\varepsilon = 0$ in Ω and $U_\varepsilon = g$ on $\partial\Omega$. In Table 5 we present the number of iterations depending on ε, H and k for fixed $h = 1/100$ and $\Delta = h$.

Table 5: Number of iterations in Example 2 ($h = 1/100$, $\Delta = h$)

	H	$k = 1$			$k = 5$		
$-\log(\varepsilon)$		1/2	1/5	1/10	1/2	1/5	1/10
3		5	10	21	5	11	22
4, 5, 6, 7		4	9	19	4	10	20

The results are stable w.r.t. the fine mesh width h (cf. Table 6).

Table 6: Number of iterations in Example 2 ($h = 1/1000$, $H = 1/10$, $\Delta = h$)

$-\log(\varepsilon)$	$k = 1$	$k = 5$
4	21	23
5,6	19	20

□

Next we will show that the proposed modified Schwarz method works also when the assumptions (H.1) – (H.3) of Lemma 1 are not valid.

Example 3: (Interior shock line)
Let be $b = (1/2 - x_1, 0)^T$, i.e. $x_1 = 1/2$ consists of turning points. Furthermore let $c = f = 0$, $g = 0$ at $\partial\Omega$ with $x_2 = 0$ or $x_2 = 1$, $g = \sin(\pi x_2)$ at $\partial\Omega$ with $x_1 = 0$ and $g = \sin(5\pi x_2)$ at $\partial\Omega$ with $x_1 = 1$, hence there is an interior shock layer at $x_1 = 1/2$.

Table 7: Number of iterations in Example 3 ($h = 1/100$, $\Delta = h$)

$-\log(\varepsilon)$	$H = 1/2$	$H = 1/5$	$H = 1/10$
3	32	31	42
4	23	15	24
5	10	8	14
6	9	7	13
7	9	6	13

Obviously the number of iterations depends in a more severe way of the size of ε, decreasing with decreasing ε. Furthermore it is interesting that the results are much better if the shock line is not contained in the macro mesh (cf. $H = 1/5$). □

Example 4: (Rotating flow field)
We consider the unit square with a slit:

$$\Omega_1 = \Omega \setminus \Sigma, \quad \Omega = (0,1)^2, \quad \Sigma = \{1/2\} \times (0, 1/2].$$

and the field $\mathbf{b} = (1/2 - x_2, x_1 - 1/2)^T$ which generates (formally) closed characteristics of (1.3). Let be $c = f = 0$ and $g = 0$ on $\partial\Omega$ and $g = \Phi(x_2)$ at Σ with

a) $\quad\quad\quad \Phi(x_2) = \sin(4\pi x_2) \quad\quad\quad$ – smooth inflow data

b)
$$\Phi(x_2) = \begin{cases} 0 & , \text{if } |x_2 - 1/4| > 1/8 \\ 1 & , \text{if } |x_2 - 1/4| \leq 1/8 \end{cases}$$
– discontinuous inflow data.

Note that interior layers in Ω emanate in case b) from the discontinuities of the inflow data.

The results in Table 8 show that the number of iterations is not influenced by the smoothness of the inflow data at Σ.

Table 8: Number of iterations in Example 4 ($h = 1/100$, $\Delta = h$)

$-\log(\varepsilon)$	$H = 1/2$		$H = 1/4$		$H = 1/10$	
	a	b	a	b	a	b
3	42	43	51	52	63	64
4	15	14	21	21	42	40
5, 6, 7	13	13	22	21	42	40

Remark 5.1: Results for moderate values of $\varepsilon \geq 10^{-2}$ are not given above. We found in our experiments that the proposed method still works for $\varepsilon = 10^{-1}$ and 10^{-2} but we had to increase the overlap width Δ depending on the coarse grid Peclet number $\varepsilon^{-1}\|\mathbf{b}\| \cdot H$. The method is obviously not designed for diffusion–dominated problems as the standard Poisson problem. □

6 Concluding Remarks

We presented a modified Schwarz method for solving scalar elliptic boundary value problems of singularly perturbed type. A convergence proof for the method is available in the turning-point free case. Numerical results confirm the robustness of the method for high Peclet numbers $\|\mathbf{b}\| \cdot h \cdot \varepsilon^{-1}$, for coarse grids up to 100 macroelements, fine grids with up to 10^6 points and different kinds of boundary and interior layers. Further research will be concentrated on overlap–free variants, mesh adaptation in boundary and interior layers and on extensions to nonlinear problems as the incompressible Navier–Stokes problem.

References

[BS] Boglaev, I.P.; Sirotkin, V.V.: Domain decomposition technique for singularly perturbed problems and its parallel implementation, in: Proc. 13. IMACS (J.J.H. Miller, R. Vichnevetsky, eds.), 52–523, Dublin 1991

[C] Cai, X.C.: Some domain decomposition algorithms for nonselfadjoint and parabolic partial differential equation, Ph.D. thesis, Courant Inst. 1989

[CGK] Cai, X.C.; Gropp, W.D.; Keyes, D.E.: A comparison of some domain decomposition and ILU preconditioned iterative methods for nonsymmetric elliptic problems (submitted to J. Numer. Lin. Al. Applics.)

[CR] Canuto, C.; Russo, A.: On the elliptic–hyperbolic coupling. I: The advection–diffusion equation via the χ–formulation, Math. Mod. Meths. Appl. Sc. 3 (1993) 2, 145–170

[FQ] Frati, A.; Quarteroni, A.: Spectral approximation to advection–diffusion problems by the fictious interface method, J. Comp. Phys. 107 (1993), 201–212

[G] Gastaldi, L.: A domain decomposition method associated with the streamline diffusion FEM for linear hyperbolic systems, Appl. Numer. Math. 10 (1992), 357–380

[KL] Kapurkin, A.; Lube, G.: A modified Schwarz method for singularly perturbed elliptic problems (in preparation)

[L] Lions, P.L.: On the Schwarz alternating method III: A variant for nonoverlapping subdomains, in: Proc. II, Int. Conf. Domain Decomposition Methods for Partial Differential Equations, SIAM, Philadelphia 1989

[Ri] Risch, U.: An upwind finite element method for singularly perturbed elliptic problems and local estimates in the L^∞-norm, Math. Model. Anal. Numer. 24 (1990) 2, 235–264

[Ro] Roos, H.G.: Necessary convergence conditions for upwind schemes in the two–dimensional case, IJNME 21 (1985), 1459–1469

[TA] Tezduyar, T.E., Aliabadi, S.; Behr, M, Johnson, A.; Mittal, S.: Massively parallel finite element computations of $3D$ flow problems, AHPCRC Preprint 92-119 (1992)

[ZR] Zhou, G.; Rannacher, R.: Mesh orientation and anisotropic mesh refinement in the streamline diffusion method, Preprint 93-57 (SFB 359), Universität Heidelberg 1993

A parallel subspace decomposition method for elliptic and hyperbolic systems

Edgar Katzer

Graduiertenkolleg "Modellierung, Berechnung und Identifikation mechanischer Systeme", Otto-von-Guericke-Universität, 39016 Magdeburg, Germany

Summary

Robust and efficient algorithms for solving linear hyperbolic and elliptic systems of first order are presented. A vertex-centered discretisation of the conservation equations on a triangular grid is applied. The discrete system is obtained by minimizing the flux residuals on the triangles. The resulting normal equations are solved by a subspace decomposition approach. A new sequence of four prolongations defines the subspaces.

Additive and multiplicative Schwarz iterations modified by smoothing steps solve the equations. Robust relaxation schemes with uniformly bounded error reduction rates for anisotropic elliptic and hyperbolic problems are obtained. The combination of a conjugate gradient method with modified additive Schwarz iteration as preconditioner improves the convergence.

The additive variants are perfectly suited for parallel processing. A parallel efficiency up to 0.8 is obtained with nine processors.

1 Introduction

The efficient solution of flow problems is still a challenging numerical task. The Euler equations which govern steady inviscid compressible flows are a nonlinear first order system of composite elliptic/hyperbolic character. This complicates their numerical solution considerably. Time marching methods are therefore very popular but time consuming. The situation is even worse for the Navier-Stokes equations which, for high Reynolds numbers, comprise a singular pertubation of the Euler equations.

A direct approach for solving the steady equations is more economical. Excellent results were obtained with multigrid methods by Jameson [1], Hemker and Spekreijse [2], Shaw and Wesseling [3], Dick [4], Mulder [5] and Ta'asan [6]. The efficiency of the approach depends on carefully designed relaxation schemes and/or on multiple coarse grid corrections (as e. g. in Mulder [5]). Usually the accuracy is restricted to first order and higher order accuracy is obtained with defect correction. Robustness of the algorithm is not guaranteed, that means that the convergence rate depends on flow characteristics, e. g. the flow direction, or deteriorates for hypersonic flows and a special treatment of these cases is necessary (see e. g. Koren and Hemker [7]).

A new approach to robust multigrid methods, the frequency decomposition multigrid method, was introduced by Hackbusch [8, 9]. The usage of several coarse grid corrections is similar to the semi-coarsening approach of Mulder [5].

This paper presents a new robust method for the parallel solution of linear elliptic and hyperbolic systems. It is a modification of the frequency decomposition approach. Efficient algorithms for such model problems are a prerequisite for the fast solution of the Euler and Navier-Stokes equations.

Hyperbolic model problem:

A simple hyperbolic problem is given by a scalar advection equation with a constant coefficient:

$$q\,\partial_x u + \partial_y u = 0 \;, \tag{1}$$

and x-periodic inflow and boundary conditions on the unit square:

$$u(x,0) = u_0(x) \;,$$

$$u(x+1,y) = u(x,y) \;, \quad \forall \; 0 \leq y \leq 1 \;, \; x \in \mathbb{R} \;.$$

Model systems:

A simple model system which allows a varying type is given by:

$$A_1\,\partial_x u + A_2\,\partial_y u = 0 \;, \tag{2}$$

where:

$$A_1 = q \begin{bmatrix} 0 & 1 \\ \eta & 0 \end{bmatrix} \;, \quad A_2 = I \;.$$

In the elliptic case ($\eta = -1$, $q > 0$), we specify periodic conditions at all boundaries and impose zero mean values. The Cauchy-Riemann equations are obtained for ($\eta = -1$, $q = 1$). For the hyperbolic system ($\eta = 1$, $q \geq 0$), we specify x-periodic Dirichlet conditions at $y = 0$.

A conservative integral formulation is given by Stoke's theorem:

$$\int_{\partial \tau} A_1\,u\,dy - A_2\,u\,dx = 0 \;, \tag{3}$$

for sufficiently smooth control elements τ with boundary $\partial \tau$.

A finite volume discretisation on a triangular grid and a minimization problem are defined in section 2. In section 3 we introduce subspace decomposition iteration schemes. Convergence rates for hyperbolic and elliptic model problems are shown in section 4 and results for parallel procedures are discussed in section 5.

(x_{j-1}, y_k) (x_j, y_k)

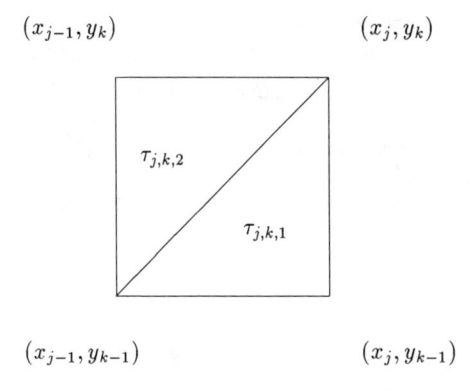

$\tau_{j,k,2}$

$\tau_{j,k,1}$

(x_{j-1}, y_{k-1}) (x_j, y_{k-1})

Figure 1: Subdivision of a square into two triangles

2 Discretisation

The integral formulation (3) is approximated on a structured triangular grid which is defined by an equidistant grid with step size $h = 1/n'$ (n' even):

$$\Omega_h = \{(x_j, y_k) \mid x_j = jh, \ y_k = kh, \ 1 \leq j, k \leq n'\} \ .$$

The triangulation is given by dividing the cells into two triangles as shown in Figure 1.

The fluxes (3) across triangle edges are approximated by the trapezoidal rule. As there are twice the number of triangles than vertices, the discrete system $Lu = f$ is overspecified and no solution of the flux equations exists in the general case. Therefore a minimization problem based on the squared Euclidean norm $E(u) = \|Lu - f\|_2^2$ is defined by:

$$E(u^*) \leq E(u) \ , \quad \forall \, u \in U \ . \tag{4}$$

The well known solution of this problem is given by the normal equations:

$$Au^* = L^*Lu^* = L^*f = b \ . \tag{5}$$

Matrix A is positive definite but not an M-matrix. The minimization approach is an asymptotic limit of the streamline diffusion method (SUPG) of Hughes et al. [10, 11].

3 Subspace decomposition approach

3.1 Schwarz Iteration

Modifications of additive and multiplicative Schwarz iteration are defined here as relaxation schemes. In contrast to domain decomposition techniques based on a geometrical decomposition of the domain, the solution space U is decomposed into subspaces:

$$U = \sum_{\kappa=1}^{K} U_\kappa \ .$$

The subspaces, which are not necessarily disjoint, are given in section 3.3. The present approach is a modification of the frequency decomposition method of Hackbusch [8, 9].

The minimization problems on the subspaces defines corrections v_κ of an approximate solution u^n by:

$$E(u^n + v_\kappa) \leq E(u^n + v) \ , \quad \forall \, v \in U_\kappa \ , \tag{6}$$

and correction operators N_κ on the residuals $r^n = b - Au^n$ by:

$$N_\kappa(r^n) =: v_\kappa \ .$$

A linear combination of these corrections yields the *additive Schwarz iteration:*

$$\Phi^{AS}(u^n) = u^n + \sum_{\kappa=1}^{K} \alpha_\kappa N_\kappa(r^n) \ ,$$

with suitable coefficients α_κ, and the *multiplicative Schwarz iteration:*

$$\Phi^{MS} = \prod_{\kappa=1}^{K} G_\kappa \ ,$$

where subspace corrections G_κ are given by $G_\kappa(u) = u + N_\kappa(b - Au)$.

3.2 Smoothing steps

As the subspace corrections v_κ are not sufficiently smooth, it was found advantageous to apply additional smoothing steps (usually one per subspace correction). For the multiplicative variant, few iterations of the gradient method are applied:

$$S(u) = u + \lambda r \ ,$$

where $\lambda = <r, r> \, / \, <Ar, r>$ and $r = b - Au$.

A simple smoothing step for the additive variant is given by adding correction operators based on Matrix A:

$$R_\kappa = AN_\kappa \ , \quad 1 \leq \kappa \leq K \ , \tag{7}$$

and a symmetric variant is obtained with:

$$S_\kappa = AN_\kappa + N_\kappa A \ , \quad 1 \leq \kappa \leq K \ . \tag{8}$$

The Schwarz iteration with smoothing steps is now given as a **modified multiplicative Schwarz** variant:

$$\Phi^{MMS} = \prod_{\kappa=1}^{K} G_\kappa S^\nu \tag{9}$$

with $\nu \times K$ smoothing steps. We consider only the cases $\nu \in \{0, 1\}$. We also define a **nonsymmetric additive Schwarz** variant:

$$\Phi^{NAS}(u^n) = u^n + \sum_{\kappa=1}^{K} (\alpha_\kappa N_\kappa(r^n) + \bar{\alpha}_\kappa R_\kappa(r^n)) \tag{10}$$

and a **symmetric additive Schwarz** variant:

$$\Phi^{SAS}(u^n) = u^n + \sum_{\kappa=1}^{K}(\alpha_\kappa N_\kappa(r^n) + \bar{\alpha}_\kappa S_\kappa(r^n)) \ . \tag{11}$$

The coefficients α_κ and $\bar{\alpha}_\kappa$ are optimized with respect to the Euclidean norm, e. g.:

$$E\left(\Phi^{SAS}(u^n)\right) \to \min \ , \quad \forall \ \alpha_\kappa \, , \ \bar{\alpha}_\kappa \ .$$

The optimization of these coefficients requires the solution of a small minimization problem of dimension K (or $2K$ when smoothing steps are applied).

The numerical efficiency is considerably increased when the symmetric additive Schwarz variant Φ^{SAS} is used as a preconditioner for a conjugate gradient method Φ^{PCG}. Convergence rates for the variety of iteration schemes are presented in section 4.

3.3 The subspaces

The modified Schwarz iterations of section 3.2 are totally determined by the subspaces. These subspaces are defined by carefully constructed linear prolongations:

$$p_\kappa : V_\kappa \to U \ , \quad U_\kappa = \text{Range}\,(p_\kappa) \ ,$$

on suitable coarse grid spaces V_κ. The correction operators are then given by:

$$N_\kappa = p_\kappa (p_\kappa^* A p_\kappa)^{-1} p_\kappa^* \ .$$

A standard coarse grid is given by a grid spacing of $2h$:

$$\Omega_0 = \{(x_j, y_k) \,|\, x_j = 2hj \, , \ y_k = 2hk \, , \ 1 \leq j, k \leq n'/2\} \ .$$

Prolongation p_0 is the seven point prolongation (see [12]) with stencil notation given below. Three additional prolongations, $p_1, \ldots, p_3$, are defined in analogy to the frequency decomposition approach ([8, 9]). They are based on p_0 and use alternating signs of the coefficients. The prolongations are given in stencil notation by:

$$p_0 = \tfrac{1}{2}\begin{bmatrix} 0 & 1 & 1 \\ 1 & 2 & 1 \\ 1 & 1 & 0 \end{bmatrix} \ , \quad p_1 = \tfrac{1}{2}\begin{bmatrix} 0 & 1 & -1 \\ -1 & 2 & -1 \\ -1 & 1 & 0 \end{bmatrix}$$

$$p_2 = \tfrac{1}{2}\begin{bmatrix} 0 & -1 & -1 \\ 1 & 2 & 1 \\ -1 & -1 & 0 \end{bmatrix} \ , \quad p_3 = \tfrac{1}{2}\begin{bmatrix} 0 & -1 & 1 \\ -1 & 2 & -1 \\ 1 & -1 & 0 \end{bmatrix} \ .$$

These prolongations operate on corresponding shifted coarse grids:

$$\Omega_1 = \{(x_j, y_k) \,|\, x_j = 2hj - h \, , \ y_k = 2hk \, , \quad 1 \leq j, k \leq n'/2\}$$
$$\Omega_2 = \{(x_j, y_k) \,|\, x_j = 2hj - h \, , \ y_k = 2hk - h \, , 1 \leq j \leq n'/2, \ 1 \leq k \leq 1 + n'/2\}$$
$$\Omega_3 = \{(x_j, y_k) \,|\, x_j = 2hj \, , \ y_k = 2hk - h \, , \quad 1 \leq j \leq n'/2, \ 1 \leq k \leq 1 + n'/2\} \ .$$

Table 1: Robust convergence for scalar hyperbolic problems

q	-4.0	-2.0	-1.0	-0.5	0.0	0.25	0.5	0.75	1.0	2.0	4.0
Φ^{TG}	0.64	0.54	0.55	0.53	0.44	0.50	0.50	0.52	0.44	0.56	0.59
Φ^{MMS}	0.75	0.72	0.76	0.75	0.51	0.69	0.71	0.72	0.49	0.75	0.78
Φ^{NAS}	0.85	0.85	0.79	0.79	0.66	0.79	0.80	0.79	0.50	0.81	0.85
Φ^{SAS}	0.86	0.86	0.83	0.82	0.77	0.82	0.84	0.82	0.59	0.86	0.86
Φ^{PCG}	0.59	0.59	0.57	0.54	0.52	0.55	0.58	0.53	0.35	0.59	0.58

Notice, that the grids Ω_2 and Ω_3 have a grid line outside of Ω_h on $y = 1 + h$ and the dimensions of the coarse grid spaces V_κ are different. It was found, that these additional degrees of freedom were essential in the hyperbolic case for obtaining robust iterations using only four coarse grid corrections. A preceding approach of the author [13] based on coarse grids without these additional grid line, needed four additional subspace corrections for obtaining robustness. Apparently, the additional degrees of freedom improve the treatment of the errors near the "outflow boundary" of the coarse grid. This is essential, because hyperbolic problems allow undamped transport of errors from the boundary into the field. Poor error reduction near the boundaries may then pollute the solution in the whole field.

4 Numerical results

Robustness of the subspace decomposition relaxations proposed in section 3.2 is analysed for hyperbolic and elliptic model problems. The grid step size for all presented results is $h = 1/32$. Further numerical investigations revealed, that the grid spacing has only minor influence on the convergence for smaller step sizes.

Asymptotic error reduction rates for the scalar hyperbolic model problem (1) are given in Table 1 for the variety of relaxation schemes defined in section 3.2. As the convergence rates depend only slightly on the "flow direction" parameter q, the algorithms are robust. It is only the case $q = 1$ which converges considerably faster. Convergence rates are uniformly bounded by 0.9 for the additive variants and by 0.8 for the multiplicative variant of the Schwarz iteration with smoothing steps. The multiplicative variant Φ^{MMS} is nearly twice as efficient as the additive variants Φ^{NAS} and Φ^{SAS}.

The symmetric additive variant Φ^{SAS} is used as a preconditioner for a CG-iteration Φ^{PCG} which converges even faster than the multiplicative Schwarz variant. Mean error reduction rates are uniformly bounded by 0.6.

For comparison, results of a twogrid iteration Φ^{TG} obtained by a preceding version [13] are also shown. This preceding version is the modified multiplicative Schwarz variant, with *eight* subspace corrections. Convergence is faster than the present approach with four subspace corrections, but the numerical work is also doubled. Eight subspace corrections were necessary for robustness of the preceding approach. Main advantage of the present algorithm is the robustness obtained with only four subspace corrections. This enables a future extension to a V-cycle type multigrid iteration with $n \log n$ operations per cycle.

Table 2: Robust convergence for hyperbolic systems

q	0.0	0.1	0.25	0.5	1.0	2.0	4.0	10.0
Φ^{MMS}	0.51	0.70	0.72	0.74	0.76	0.78	0.77	0.74
Φ^{NAS}	0.66	0.76	0.83	0.85	0.82	0.87	0.85	0.83
Φ^{SAS}	0.77	0.81	0.86	0.89	0.86	0.88	0.86	0.84
Φ^{PCG}	0.53	0.53	0.58	0.65	0.62	0.63	0.61	0.55

Table 3: Robust convergence for elliptic systems

q	0.1	0.25	0.5	1.0	2.0	4.0	10.0
Φ^{MMS}	0.34	0.21	0.14	0.12	0.19	0.23	0.37
Φ^{NAS}	0.50	0.48	0.45	0.39	0.44	0.48	0.50
Φ^{SAS}	0.50	0.49	0.46	0.40	0.46	0.49	0.50
Φ^{PCG}	0.30	0.28	0.25	0.23	0.25	0.27	0.32

Similar results are obtained for hyperbolic systems in Table 2. The asymptotic error reduction rate is uniformly bounded for all analysed schemes. Mean error reduction rates of the PCG-iteration are bounded by 0.65 independent of the characteristic directions. For the simple model systems analysed here, the relaxation schemes seem to be robust with respect to the number of equations involved.

Convergence rates for elliptic systems in Table 3 are much better than for hyperbolic systems. In the isotropic case of the Cauchy-Riemann equations, convergence is comparable to that of multigrid methods for the Poisson problem. Error reduction rates for the anisotropic case are bounded by 0.5.

The influence of the smoothing step is shown in Table 4 for the additive and in Table 5 for the multiplicative Schwarz variants. Smoothing improves convergence because the subspace corrections are optimised in a larger space. The relative speedup based on the number of iterations:

$$\sigma_{sm} = \frac{\text{number of iterations without smoothing}}{\text{number of iterations with smoothing}} \cong \frac{\log(\rho(\Phi^{SAS};\text{with}))}{\log(\rho(\Phi^{SAS};\text{without}))}.$$

is also given in Tables 4 and 5. Smoothing reduces the number of iterations by a factor of 1.3 to 1.9 for the elliptic case and 1.1 to 2 in the hyperbolic case.

The numerical results show, that smoothing is not really necessary. But as smoothing is cheap for the multiplicative Φ^{MMS} and the nonsymmetric additive variant Φ^{NAS}, it can always be recommended. The situation is different when we consider the symmetric additive variant Φ^{SAS}. Then the smoothing steps require the expensive solution of coarse grid problems and the numerical work is twice the work without smoothing. This is not efficient on sequential computers. But symmetric smoothing could be advantegous

Table 4: Influence of smoothing step for additive variant Φ^{SAS}

q		0.0	0.1	0.5	1.0	2.0	4.0
elliptic	no smooth.	–	0.67	0.57	0.55	0.57	0.62
	with sm.	–	0.50	0.46	0.40	0.46	0.49
	σ_{sm}	–	1.7	1.4	1.5	1.4	1.5
hyperbolic	no smooth.	0.87	0.90	0.92	0.92	0.91	0.91
	with sm.	0.77	0.81	0.89	0.86	0.88	0.86
	σ_{sm}	1.9	2.0	1.5	1.7	1.4	1.6

Table 5: Influence of smoothing step for multiplicative variant Φ^{MMS}

q		0.0	0.1	0.5	1.0	2.0	4.0
elliptic	no smooth.	–	0.45	0.34	0.33	0.27	0.41
	with sm.	–	0.34	0.14	0.12	0.19	0.23
	σ_{sm}	–	1.4	1.8	1.9	1.3	1.6
hyperbolic	no smooth.	0.66	0.76	0.79	0.80	0.80	0.79
	with sm.	0.51	0.70	0.74	0.76	0.78	0.77
	σ_{sm}	1.6	1.3	1.3	1.2	1.1	1.1

on parallel computers, when it is done simultaneously and needs no additional time. Symmetric relaxations with or without smoothing should be combined with a conjugate gradient method which is highly efficient and easily implemented on parallel computers.

5 The parallel algorithm

The variants of the additive Schwarz iterations (Φ^{NAS}, Φ^{SAG}, Φ^{PCG}) are perfectly suited for parallel processing. The subspace corrections N_κ and the smoothing operators S_κ and R_κ may be calculated in parallel. The number of processors involved is $1 + 2K$ (or, without smoothing $1 + K$), where K is the number of subspaces. A parallel procedure for the additive variant with symmetric smoothing steps is given as follows:

1. The approximate solution u^n is stored on processor zero. The residual r^n is calculated and distributed to all processors.

2. Processor κ calculates $N_\kappa(r^n)$ and processor 2κ calculates $N_\kappa A(r^n)$ and communicates the result to processor κ. S_κ is calculated on processor κ and the results are communicated to processor zero. The calculation steps and parts of the communication steps are done parallel.

3. Processor zero calculates the coeffecients α_κ and $\bar{\alpha}_\kappa$ and updates the approximate solution u^{n+1}. Work continues with step 1. At the moment this step is sequential. Of course parts of this step could be done in parallel on several processors, but this would require additional communication.

The most expensive parts are the calculations of the subspace corrections N_κ and S_κ in step 2 because this involves the solution of linear systems on the coarse grids. At the moment this is done by conjugate gradient iterations. As the coarse grid systems have different condition numbers, the number of CG-iterations is different on each processor which leads to some load imbalance. This load imbalance could be overcome by using alternative solution procedures on the coarse grids, as e. g. noniterative procedures or multigrid algorithms which are not so dependent on the condition numbers. A simple remedy is the application of a fixed number of CG-iterations on each processor. The coarse grid corrections are then solved with different accuracy on different subspaces. Clearly, this reduces the numerical efficiency of the Schwarz iteration but increases the load balance and thus the parallel efficiency. Preliminary tests with a fixed number of ten CG-iterations on the coarse grid revealed a minor decrease of the convergence rate. The overall effect of this approach should be further analysed.

Further load imbalance is associated with the sequential calculation of the coefficients α_κ and $\bar{\alpha}_\kappa$ in step 3. Although this step comprises only a small part of the total numerical work, it is of nearly the same cost as the solution of the subspace corrections. This load imbalance reduces the parallel efficiency to 0.5 with eight processors. Improved load balance could either be achieved by using a fixed set of coefficients, or the coefficients are optimized with data of the previous iterations, as in the present approach. The optimization and subspace corrections are performed simultaneously. The extrapolation of the coefficients requires some damping of the coefficients to prevent instabilities. Coefficients at the current iteration n are average value of coefficients at iteration $n-2$ and the optimal coefficients at level $n-1$. The convergence rate deteriorates with the nonoptimal choice of extrapolated coefficients α_κ, $\bar{\alpha}_\kappa$ from 0.46 to 0.65 for an elliptic problem ($q = 0.5$) on a 32×32 grid. The estimation of the coefficients α_κ, $\bar{\alpha}_\kappa$ should be improved for faster convergence.

We define the parallel efficiency of a parallel algorithm Φ_P by the quotient of computation time needed for solving the problem on one processor divided by p times the time $T(p, \Phi_P)$ needed by the *same* algorithm on p processors:

$$\eta_{par}(\Phi_P) = \frac{T(1, \Phi_P)}{p\, T(p, \Phi_P)} \; .$$

Notice, that this is not identical with the speedup for solving the linear system on a parallel computer. For assessing the enhancement obtainable with parallel computation, we should compare the parallel procedure with "the best" sequential method. But as the "best sequential method" is not clearly defined and depends on the class of methods considered, we should only compare with a given "good sequential algorithm" Φ_S. (See Bertsekas [14], p. 15, for a discussion of this problem.) This influence could be taken into account by the *relative efficiency* of the two algorithms on *one* processor:

$$\eta_{rel}(\Phi_P, \Phi_S) = \frac{T(1, \Phi_S)}{T(1, \Phi_P)} \simeq \frac{C_S \, \log(\rho(\Phi_P))}{C_P \, \log(\rho(\Phi_S))} \; .$$

Table 6: Parallel efficiency of Φ^{SAS} for elliptic problem, (32×32) grid

	no load balancing			with load balancing		
	1 proc.	5 proc.	9 proc.	1 proc.	5 proc.	9 proc.
CPU [sec]	78.1	36.7	18.5	51.4	12.9	7.4
Speed	–	2.1	4.2	–	4.0	7.0
η_{par}	–	0.43	0.47	–	0.80	0.77

Here $\log(\rho(\Phi_S))^{-1}$ measures the number of iterations required for a given error bound and C_S and C_P accounts for the different numbers of numerical operations of Φ_S and Φ_P. The various modified Schwarz iterations discussed here have different operation counts, because S_κ is as expensive as N_κ whereas the smoothing steps S and R_κ are cheap and their costs could be neglected in Φ^{NAS} and Φ^{MMS}.

The *relative efficiency* of a parallel algorithm Φ_P on p processors with respect to a sequential algorithm Φ_S may be defined as:

$$\eta_{rel,\,par}(\Phi_P, \Phi_S) = \frac{T(1, \Phi_S)}{p\,T(p, \Phi_P)} \cong \eta_{rel}(\Phi_P, \Phi_S)\,\eta_{par}(\Phi_P)\ .$$

It should be understood, that all CPU-times are measured on equivalent hardware. Otherwise, a hardware influence factor η_{hard} should be included.

Parallel efficiencies on a transputer cluster PARYTEC GCel-128 are given in Table 6. We compare computation times on one to nine processors needed for one iteration of the modified symmetric additive Schwarz iteration Φ^{SAS} for an elliptic problem ($q = 0.5$) on a 32×32 grid. No efforts have been made in efficient implementation of the algorithms. This explains the rather high computation time for one iteration. Parallel efficiency is 0.4 – 0.5 without load balancing. Load balance is improved with a fixed number of 10 iterations on the coarse grids combined with extrapolation of the coefficients α_κ, $\bar\alpha_\kappa$. Then parallel efficiency reaches 0.8 for five and nine processors.

6 Conclusion

A new subspace decomposition method for solving hyperbolic and elliptic systems of first order partial differential equations is presented. The conservation equations are discretized on a triangular grid by a new vertex-centered scheme which avoids the use of a second (dual) grid. The conservation equations are overspecified and the discrete system is solved by minimizing the flux balance for all triangles.

The system of normal equations are solved by a new subspace decomposition method based on a modification of the frequency decomposition approach of Hackbusch [8, 9]. The subspaces are defined by carefully constructed prolongations. A special treatment of the prolongations near the farfield boundary in the hyperbolic case is essential for robustness. Now we obtain a robust method based on only four subspace corrections in contrast to eight corrections required by a preceding version of the algorithm [13].

The subspace decompositon method consists of additive or multiplicative Schwarz iterations modified by smoothing steps. Asymptotic error reduction rates for multiplicative and additive variants with symmetric and nonsymmetric smoothing steps are presented. Uniformly bounded error reduction rates independent of the step size and characteristic direction are obtained.

Error reduction rates for elliptic systems are even better than for the hyperbolic cases. The multiplicative Schwarz iteration with smoothing has convergence rates of 0.12 for the isotropic and 0.4 for the anisotropic problem. Convergence is comparable with that of twogrid iterations for the Poisson problem. Robustness is obtained with convergence rates depending only slightly on the anisotropy. Convergence of the hyperbolic problem is slower but we have to keep in mind, that relaxations for hyperbolic problems usually show poor convergence and a lack of robustness. So robustness is most important in that case.

The symmetric modified additive Schwarz variants are useful as preconditioners. Preconditioned CG-iterations reveal mean error reduction rates bounded by 0.65.

The additive variants are perfectly suited for parallel processing. At the moment we use four subspace corrections and four smoothing operations, a maximum of nine processors could work simultaneously. Together with communication overhead and sequential parts of the algorithm, we obtain a parallel efficiency up to 0.8 on nine processors.

The present results provides a basis for efficient parallel solution of the Euler equations. Further work concerns the extension to nonlinear problems and to more processors.

Acknowledgement

Many fruitful and inspiring discussions with Dr. Grambow and Prof. Tobiska, Magdeburg, and Prof. Hackbusch, Kiel, are gratefully acknowledged. The present work was supported by the "Deutsche Forschungsgemeinschaft" (German Research Society) and the "Land Sachsen-Anhalt" (Saxony Anhalt).

References

[1] Jameson, A.: *Solution of the Euler equations for two dimensional transonic flow by a multigrid method.* Applied Math. Computation 13 (1983) 327–356.

[2] Hemker, P. W.; Spekreijse, S. P.: *Multiple grid and Osher's scheme for the efficient solution of the Euler equations.* Applied Numer. Math. 2 (1986) 475–493.

[3] Shaw, G.; Wesseling, P.: *A multigrid method for the Navier–Stokes equations.* Delft University of Technology, Dept. Mathematics and Informatics. Report No.: 86–13, 1986.

[4] Dick, E.: *Multigrid formulation of polynomial flux-difference splitting for steady Euler equations.* J. Computational Physics 91 (1990) 161–173.

[5] Mulder, W. A.: *A high resolution Euler solver based on multigrid, semi-coarsening, and defect correction.* J. Computational Physics 100 (1992) 91–104.

[6] Ta'asan, Sholmo: *Optimal multigrid solvers for steady state inviscid flow problems.* Fourth European Multigrid Conference, EMG '93, Amsterdam, 1993.

[7] Koren, P. W.; Hemker, P. W.: *Damped, direction-dependent multigrid for hypersonic flow computation.* Applied Numerical Mathematics 7 (1991) 309–328.

[8] Hackbusch, W.: *A new approach to robust multi-grid solvers.* First. Int. Conf. on Industrial and Applied Mathematics. ICIAM '87. Proceedings, Philadelphia, PA: SIAM 1988, pp. 114–126.

[9] Hackbusch, W.: *The frequency decomposition multi-grid method. Part I: Application to anisotropic equations.* Numerische Mathematik 56 (1990) 229-245.

[10] Hughes, Th. J. R.: *Recent progress in the development and understanding of SUPG methods with special reference to the compressible Euler and Navier-Stokes equations.* Int. J. for Numer. Methods in Fluid Dynamics 7 (1987) 1261–1275.

[11] Hughes, Th. J. R.; Franca, L. P.; Hulbert, G. M.: *A new finite element formulation for computational fluid dynamics: VIII. The Galerkin/Least-Squares method for advective-diffusive equations.* Computer Methods in Applied Mechanics and Engineering 73 (1989) 173–189.

[12] Hackbusch, W.: *Multi-grid methods and applications.* Berlin: Springer, 1985.

[13] Katzer, E.: *A subspace decomposition twogrid method for hyperbolic equations.* Fourth European Conference on Multigrid Methods, EMG '93. In: Hemker, P. W.; Wesseling, P. (esd.): Contributions to Multigrid, pp. 87–98. CWI Tract 103. Amsterdam: Centrum voor Wiskunde en Informatica, 1994.

[14] Bertsekas, D. P.; Tsitsiklis, J. N.: *Parallel and distributed computation.* Englewood Cliffs, NJ: Prentice Hall, 1989.

IMPLICIT TIME-DISCRETIZATION OF THE NONSTATIONARY INCOMPRESSIBLE NAVIER-STOKES EQUATIONS

S. Müller, A. Prohl, R. Rannacher, S. Turek
Institut für Angewandte Mathematik
Universität Heidelberg
INF 293, D-69120 Heidelberg

This paper deals with implicit time-discretization of the nonstationary incompressible Navier-Stokes equations. The emphasis is on the so-called "fractional-step-θ-scheme" in which the incompressibility constraint is treated either fully implicitly or semi-implicitly by employing operator splitting (projection method). Recently a complete mathematical analysis has become available for methods of this type which supports current computational experience. The fractional-step-θ-scheme has proven to be a robust and accurate time-stepping method which is in several respects superior to the traditional Crank-Nicolson scheme. Combined with operator splitting it is particularly suitable for the long time computation of complex flows.

1. THE NAVIER-STOKES PROBLEM

We consider the incompressible Navier-Stokes problem

$$v_t - \nu\Delta v + v\cdot\nabla v + \nabla p = f, \quad \nabla\cdot v = 0 \quad \text{in } \Omega\times I, \tag{1.1}$$

$$v_{|t=0} = v^0, \quad v_{|\Gamma_0} = 0, \quad v_{|\Gamma_{in}} = b, \quad \nu\partial_n v - pn_{|\Gamma_{out}} = 0, \tag{1.2}$$

where $\Omega \subset \mathbf{R}^d$ (d=2 or 3) is a bounded domain with sufficiently regular boundary, I = [0,T] is a bounded time-interval, Γ_0, Γ_{in} and Γ_{out} are the rigid, the inflow and the outflow part of the boundary, respectively, and ν is the viscosity. Assuming the characteristic length and speed to be normalized to one, the problem may be parametrized by the Reynolds number $Re \equiv 1/\nu$. This note concerns laminar flows related to values $Re \approx 10^0\text{-}10^5$ depending on the problem. The numerical schemes considered are supposed to resolve all relevant spatial and temporal structures of the flow. Accordingly, all numerical results presented are 2D, though our numerical techniques have also been applied in 3D (see [23]) The aspects of small scale modeling or turbulence as necessary for larger Reynolds numbers will not be touched.

Below, we will use the notation $(\cdot,\cdot)$ and $\|\cdot\|$ for the inner product and norm of $L^2(\Omega)$. Vector-valued functions and corresponding function spaces are written with bold face type, e.g., $\mathbf{L}^2(\Omega) = [L^2(\Omega)]^d$, while no distinction is made in the notation of the corresponding inner products and norms.

2. SPATIAL DISCRETIZATION

For setting up a finite element model of the Navier-Stokes problem, one starts from its variational formulation: Find $v \in H+b$ and $p \in L^2(\Omega)$, such that

$$(v_t,\phi) + \nu(\nabla v, \nabla \phi) + (v \cdot \nabla v, \phi) - (p, \nabla \cdot \phi) = (f,\phi) \quad \forall\, \phi \in H, \tag{2.1}$$

$$(\chi, \nabla \cdot v) = 0 \quad \forall\, \chi \in L^2(\Omega). \tag{2.2}$$

The choice of the function space $H \subset H^1(\Omega)$ depends on the specific boundary conditions chosen for the problem to be solved. We note that in working with the above "free-stream" outflow boundary condition, it is not advisable to write the diffusive term with the deformation tensor $D[v] = (\nabla v + \nabla v^T)/2$, or to "symmetrize" the convective term according to $v \cdot \nabla v \to \{v \cdot \nabla v + \nabla \cdot (v^T v)\}/2$, as is sometimes proposed in the literature. Both modifications may lead to nonphysical flow behavior accross the outlets (see [10] or [19] for more details on this matter).

On a finite mesh T_h (triangulation, etc.) covering the domain Ω, with (local) element width h, one defines spaces $H_h \times L_h$ "$\subset$" $H \times L(\Omega)$ of piecewise polynomial ansatz functions. The spaces H_h may be "nonconforming", i.e., the discrete velocities v_h are required to be continuous across the interelement boundaries and to vanish along the rigid boundary only in an approximate sense. In such a case, the nabla operator ∇ used below is to be understood in the "piecewise" sense. The discrete analogues of (1) then read: Find $v_h \in H_h + b_h$ and $p_h \in L_h$, such that

$$(v_{h,t},\phi) + \nu(\nabla v_h, \nabla \phi) + (v_h \cdot \nabla v_h, \phi) - (p_h, \nabla \cdot \phi) = (f,\phi) \quad \forall\, \phi \in H_h, \tag{2.3}$$

$$(\chi, \nabla \cdot v_h) = 0 \quad \forall\, \chi \in L_h, \tag{2.4}$$

where b_h is an appropriate approximation to b. In order to ensure that (2.3), (2.4) is a numerically stable approximation to (2.1), (2.2), as $h \to 0$, the spaces $H_h \times L_h$ should satisfy the so-called Babuska-Brezzi condition (see, e.g., [6]),

$$\inf_{q_h \in L_h/R} \sup_{w_h \in H_h} \left(\frac{(q_h, \nabla \cdot w_h)}{\|q_h\| \, \|\nabla w_h\|} \right) \geq \gamma > 0. \tag{2.5}$$

Many stable pairs of finite element spaces $\{H_h, L_h\}$ have been proposed in the literature (see, e.g., [6]). Below, two particularly simple examples of quadrilateral elements will be described (see Figure 1 for the corresponding nodal points in 2D) which have satisfactory approximation properties and are applicable in two as well as three space dimensions. In both cases the incompressibility constraint (2.2) is treated in a completely implicit way.

1) The first example uses piecewise "rotated" bi- or tri-linear shape functions (on the reference element) for the velocities, spanned by $\{1,x,y,x^2-y^2\}$ (in 2D) or by $\{1,x,y,z,x^2-y^2,x^2-z^2\}$ (in 3D), and piecewise constant pressures. The nodal values are the mean values of the velocity vector over the element edges and the mean values of the pressure over the elements rendering this ansatz "nonconforming". This $\widetilde{Q}_1/P_0$-element is the natural quadrilateral analogue of the well-known triangular Stokes element of Crouzeix/Raviart (see [6]). A convergence analysis is given in [20] and very promising computational results are reported in [23] and [27].

2) The second example is the conforming Q_1/Q_1-Stokes element which uses continuous bi-linear (in 2D) or tri-linear (in 3D) shape functions for both the velocity and the pressure approximations where the nodal values are the function values of the velocity and the pressure at the vertices of the mesh. This element would be unstable, i.e., would violate condition (2.5), if used together with formulation (2.3), (2.4). In order to get a stable discretization, one adds certain least-squares terms in the continuity equation (2.4) (pressure stabilization method of Hughes, et al., [11]),

$$(\chi_h, \nabla \cdot v_h) + \frac{\sigma}{\nu} \sum_{K \in T_h} h_K^2 (\nabla \chi_h, \nabla p_h)_K = \frac{\sigma'}{\nu} \sum_{K \in T_h} h_K^2 (\nabla \chi_h, f + \nu \Delta v_h - v_h \cdot \nabla v_h)_K .$$

The correction terms on the right hand side make this modification fully consistent, as (for $\sigma = \sigma'$) the additional terms cancel out if the solution $\{v,p\}$ is inserted. Normally, one takes $\sigma \approx 0.1$ and $\sigma' = \sigma$, or simply $\sigma' = 0$.

These quadrilateral elements have several important features. They admit simple upwind strategies which lead to matrices with certain M-matrix properties. Further, efficient multigrid solvers are available which work satisfactorily over the whole range of relevant Reynolds numbers, $1 \leq Re < 10^5$ (see [8], [22], and [27]), and also on non-uniform and highly stretched meshes (for the latter aspect see [1]).

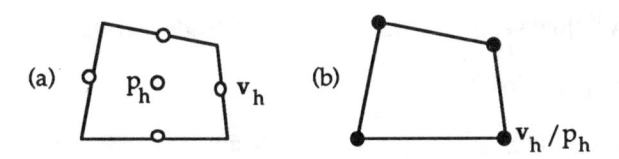

Figure 1. Nodal points of a) the nonconforming "rotated" $\widetilde{Q}_1/P_0$- and b) the conforming (stabilized) Q_1/Q_1-Stokes element

We introduce the auxiliary function $u_h := v_h - b_h \in H_h$. Then, using the same symbols u_h and p_h also for the coefficient vectors in the nodal bases representation of u_h and p_h, problem (2.3), (2.4)) may be written as a differential-algebraic system of the form

$$M_h u_{h,t} + A_h u_h + N_h(u_h)u_h + B_h p_h = g_h, \quad (2.6)$$

$$B_h^T u_h + \sigma C_h p_h = \sigma' c_h(u_h), \quad (2.7)$$

with the "mass" matrix M_h, the "diffusion" matrix A_h, the "convection" matrix $N_h(u_h)$, the "gradient" matrix B_h, the "Laplace-Neumann" matrix C_h, the correction vector $c_h(u_h)$, and finally the "load" vector g_h, the latter containing also the contribution from the nonhomogeneous inflow boundary condition. For dominant transport, the convection matrix $N_h(u_h)$ may include some upwind mechanism.

3. TIME DISCRETIZATION

3.1. GENERAL ASPECTS

The DAE system (2.6), (2.7) is highly stiff, with a stiffness ratio of order $O(vh^{-2})$, or even $O(vh^{-4})$ in the case of a divergence-free FE formulation (see [27]), essentially depending on the smallest mesh-size h_{min}. Further stiffening is caused by the pressure-velocity coupling (2.7). Therefore, in the choice of time-stepping methods for solving this system, one is limited to implicit schemes. In the past, explicit time-stepping schemes have been commonly used in nonstationary flow calculations, mainly for managing the transition to steady-state limits. Because of the severe stability problems inherent in this approach (for moderately sized Reynolds numbers and non-uniform meshes), the required very small time-steps prohibited the accurate solution of really time-dependent flows. Since implicit methods became feasible thanks to more efficient linear system solvers, the schemes most frequently used were either the simple first-order backward Euler (BE) scheme or the second-order Crank-Nicolson (CN) scheme. These two methods belong to the group of so-called "one-step θ-schemes". Dropping from now on the subscript h, these schemes read as follows (k = time step):

$$[M+\theta k \overline{A}^{n+1}]u^{n+1} + Bp^{n+1} = [M-(1-\theta)kA^n]u^n + \theta k g^{n+1} + (1-\theta)kg^n, \quad (3.1)$$

$$Bu^{n+1} + \sigma C p^{n+1} = \sigma' \overline{c}^{n+1}, \quad (3.2)$$

where $A^n = A+N(u^n)$, $g^n = g(t_n)$, $\overline{A}^n = A+N(\overline{u}^n)$ and $\overline{c}^n = c(\overline{u}^n)$, with some prediction $\overline{u}^n \approx u^n$ obtained, e.g., by linear extrapolation from the preceding time levels. The CN scheme (i.e., θ=0.5) occasionally suffers from unexpected instabilities because of the only weak damping properties of this scheme (not *strongly* A-stable). For a discussion of this issue, see [17] and [9; Part 4]. This defect can in principle be cured by an adaptive step-size selection, but this may necessitate the use of an unreasonably small time step, thereby increasing the computational cost.

Alternative schemes of higher order are based on the (diagonally) implicit Runge-Kutta formulae or the backward differencing multi-step formulae, being well-known from the ODE literature. These schemes, however, have not yet found wide applications in flow computations, mainly because of their higher complexity and storage requirements compared with the traditional CN scheme.

3.2. THE FRACTIONAL-STEP-θ-SCHEME

There is another method which seems to have the potential to excel in this competition, the so-called "fractional step θ-scheme" (in short FS$_\theta$ scheme). Applied to the model equation $y' = \lambda y$, this 3-level one-step method derives from a third-order rational approximation of the exponential function, i.e.,

$$R_\theta(\lambda) \equiv \frac{(1+\alpha\theta'\lambda)(1+\beta\theta\lambda)^2}{(1-\alpha\theta\lambda)^2(1-\beta\theta'\lambda)} = e^\lambda + O(|\lambda|^3), \quad \lambda < 0, \tag{3.3}$$

where $\theta = 1-\sqrt{2}/2 = 0.29289321...$, $\theta'=1-2\theta$, and $\alpha \in (0.5,1]$, $\beta=1-\alpha$. This choice of parameters insures the second-order accuracy and, additionally, the strong A-stability of the scheme,

$$\overline{\lim}_{\lambda \to -\infty} |R_\theta(\lambda)| = \beta/\alpha < 1. \tag{3.4}$$

Further, for Re $\lambda \to 0$, the amplification factor $R_\theta(\lambda)$ becomes nearly one (e.g., $R_\theta(0.8i) \approx 0.9998$), which means that the resulting time-stepping scheme contains only very little numerical dissipation. Applied to the Navier-Stokes system (2.6), (2.7) the FS$_\theta$ scheme reads:

$$[M+\alpha\theta k\overline{A}^{n+\theta}]u^{n+\theta}+Bp^{n+\theta} = [M-\beta\theta kA^n]u^n+\theta kg^n, \tag{3.5}$$

$$B^T u^{n+\theta}+\sigma Cp^{n+\theta} = \sigma'c^n, \tag{3.6}$$

$$[M+\beta\theta'k\overline{A}^{n+1-\theta}]u^{n+1-\theta}+Bp^{n+1-\theta} = [M-\alpha\theta'kA^{n+\theta}]u^{n+\theta}+\theta'kg^{n+1-\theta}, \tag{3.7}$$

$$B^T u^{n+1-\theta}+\sigma Cp^{n+1-\theta} = \sigma'\overline{c}^{n+1-\theta}, \tag{3.8}$$

$$[M+\alpha\theta k\overline{A}^{n+1}]u^{n+1}+Bp^{n+1} = [M-\beta\theta kA^{n+1-\theta}]u^{n+1-\theta}+\theta kg^{n+1-\theta}, \tag{3.9}$$

$$B^T u^{n+1}+\sigma Cp^{n+1} = \sigma'c^{n+1-\theta}. \tag{3.10}$$

The use of the FS$_\theta$ scheme for solving the Navier-Stokes equations was first proposed by Glowinski, et al. ([3]), for the case $\sigma = \sigma' = 0$, in the form of an operator-splitting scheme separating the two problems "nonlinearity" and "incompressibility" within each cycle $t_n \to t_{n+\theta} \to t_{n+1-\theta} \to t_{n+1}$. However, this scheme deserves also attention as a mere time-stepping method as will become clear by the following discussion.

3.3. THEORETICAL ASPECTS

Being a strongly A-stable scheme the FS_θ scheme possesses the full smoothing property, which is important in the case of rough initial or boundary data. Further, it contains only very little numerical dissipation, which is crucial in the computation of non-enforced temporal oscillations in a flow (for a more detailed discussion of these aspects see [17]). A rigorous theoretical analysis of the FS_θ scheme applied to the Navier-Stokes problem, in the case $\sigma = \sigma' = 0$, has recently been given in [14]. The main result is a second-order error estimate of the form

$$\|v_h^n - v(\cdot, t_n)\| \leq C(t_n)(h^2 + \tau_n^{-1} k^2), \tag{3.11}$$

where $v_h^n = u_h^n + b_h^n$, $v = u + b$, and $\tau_n = \min\{1, t_n\}$. The proof is based on a tricky use of "energy arguments" where the main problem consists in a proper combination of the three substeps (3.7)-(3.9) in order to obtain second-order truncation error terms. The result (3.11) has several aspects:

- It shows the stability of the scheme for sufficiently small time step-size (independent of the spatial mesh size!) and additionally its actual second-order convergence on finite time-intervals. Previously, only qualitative stability results had been known (see [4] and [13]).

- It guarantees second-order accuracy at times $t \geq t_0 > 0$, even for generic initial data satisfying only $v^0 \in H \cap H^2(\Omega)$, $\nabla \cdot v = 0$. A related "smoothing property" for the CN scheme has been proven in [9, Part 2].

- The error constant $C(t)$ remains bounded for $t \to \infty$, if the boundary data and the body force are bounded and if the exact solution v is exponentially stable. A general discussion of this aspect for finite element semi-discretizations and particularly for the CN scheme has been given in [9, Part 2], while the most important question of the dependence of $C(t)$ on the viscosity ν has recently been considered in [12].

3.4. SOLUTION METHODS

For the solution of the linear systems arising in each substep of the FS_θ-scheme, efficient multigrid methods are available (see [22], [23], and [27]). This is particularly important in long-time simulations when the time step has to be taken rather large. But, though the convergence of these algorithms is sufficiently fast (convergence rates $\rho \approx 0.1$-0.4) for very regular situations, they may drastically slow down on perturbed meshes, e.g., those with large cell aspect ratios ($h_{max}/h_{min} > 5$). This lack of robustness can be cured by special tricks. These are oriented at the experience with the related problem for scalar diffusion problems where the use of ILU-based or line-oriented Gauß-Seidel smoothers leads to remarkably robust multigrid algorithms. However, the effective implementation of these techniques for the indefinite Navier-Stokes problem is difficult in general situations (for more details on this particular aspect see [21] and [1]). The way out of this dilemma seems to be the operator splitting approach. The FS_θ scheme may be combined with any of the projection methods of Chorin-type for simplifying the treatment of the incompressibility constraint (see [18] and the literature cited therein). This approach will be described in more detail below.

3.5. ACCURACY AND ROBUSTNESS

Next, we consider in some more detail the accuracy and robustness of the FS_θ scheme in comparison with the backward Euler and the CN scheme. A time step in the FS_θ scheme consists of a cycle of three substeps with total step-length $k = \theta k+(1-2\theta)k+\theta k$, and each substep is approximately of the same computational complexity as one step of the standard CN scheme. Here, we assume that in both schemes the nonlinear terms are fully updated in each time (sub-)step. We note that in the case of an autonomous system the special choice $\alpha = (1-2\theta)/(1-\theta) = 0.58578...$, results in identical coefficient matrices in all three substeps of the FS_θ-scheme. Hence, in order to make a fair comparison, we have to compare one step (cycle) of the FS_θ scheme with total length k with three steps of the CN (or the BE) scheme with length $k/3$. All comparisons described below have been made on this basis of an "effective" time-step length k. Further, the values $\bar{v}^m$ in the convective terms are predicted by linear extrapolation from the two preceding time levels and then, if necessary, successively improved by a defect correction iteration (with $L \geq 1$ correction steps).

The test cases are the van Kármán vortex shedding behind a plate, and the flow in a Venturi pipe. Figures 2 and 3 show the respective coarse grids from which the computational meshes are generated by uniform refinement (4-6 refinement steps) and corresponding sample results.

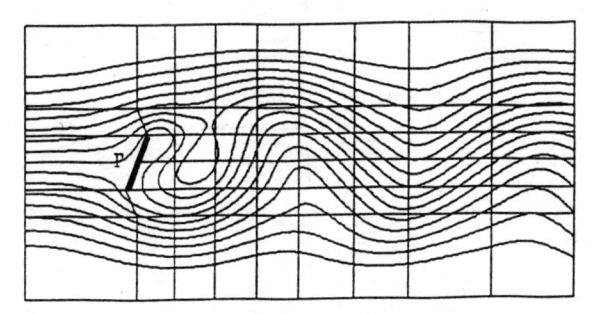

Figure 2. Coarsest mesh and streamline plot at Re = 500
for the van Kármán vortex shedding

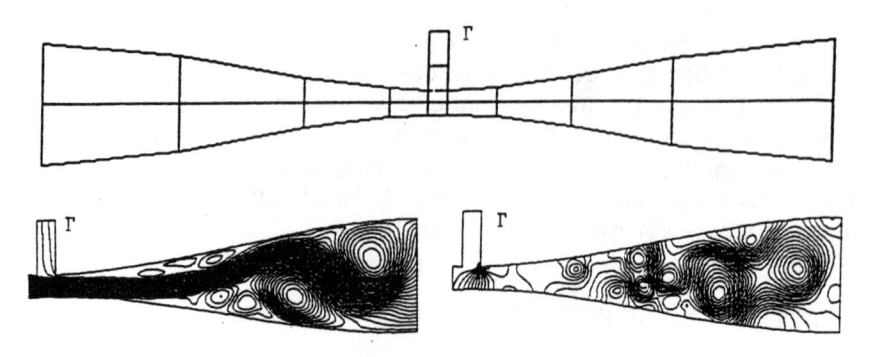

Figure 3. Coarsest mesh and streamline and pressure contour plots for Re = 5,000 for the Venturi pipe flow

a) Accuracy Test

First we examine the ability of the methods to capture the large scale dynamics of the flow. Figures 4 shows (relative) streamline plots of the flow around an inclined plate for Re = 500 at T = 10, computed with a) the CN scheme, b) the FS_θ scheme, and c) the BE scheme, all schemes combined with nonlinear defect correction (L = 2-4). It seems clear that at this test problem, the FS_θ scheme is as accurate as the CN scheme and that both are far superior to the (first-order) BE scheme which achieves a comparable accuracy only for a step size k = 0.11. Because of its poor performance the BE scheme will not further considered.

a) CN scheme (k = 0.33)   b) FS_θ scheme (k = 0.33)

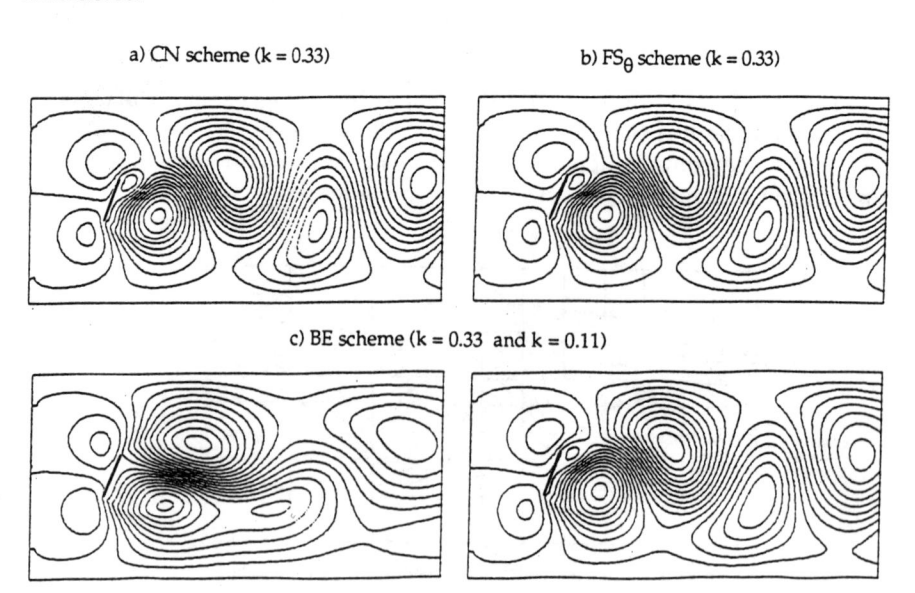

c) BE scheme (k = 0.33 and k = 0.11)

Figure 4. Relative streamlines of van Kármán vortex shedding for Re=500, at time T = 10

Next, we look at the fine scale of the flows. As a first example, Figure 5 shows the time behavior of the mean pressure drop accross the plate,

$$\Delta P = \int_{\Gamma_{up}} p \, ds - \int_{\Gamma_{down}} p \, ds ,$$

calculated by the CN and the FS_θ scheme.

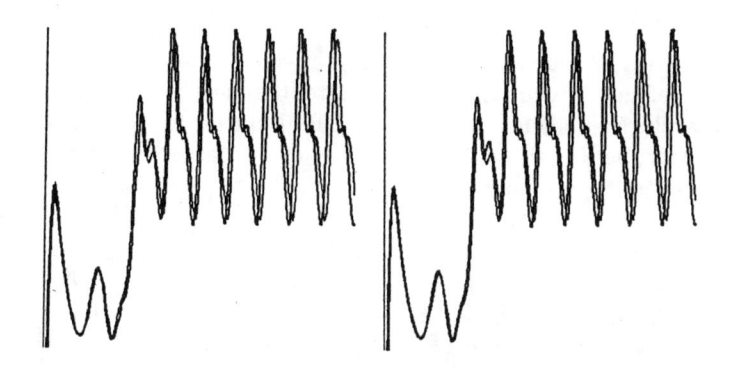

Figure 5. Mean pressure drop accross the plate at $Re = 500$ over $I = [0,60]$, calculated with the CN scheme (left) and the FS_θ scheme (right), with $k=0.33$; reference solution thick line

Further, Figure 6 shows the time behavior of the total flux through the upper inlet Γ of the Venturi pipe, again calculated by the CN and the FS_θ scheme. The corresponding "reference solutions" have been calculated on the same spatial mesh but with a much smaller time step. These results confirm that both schemes have about the same accuracy for realistic step-sizes.

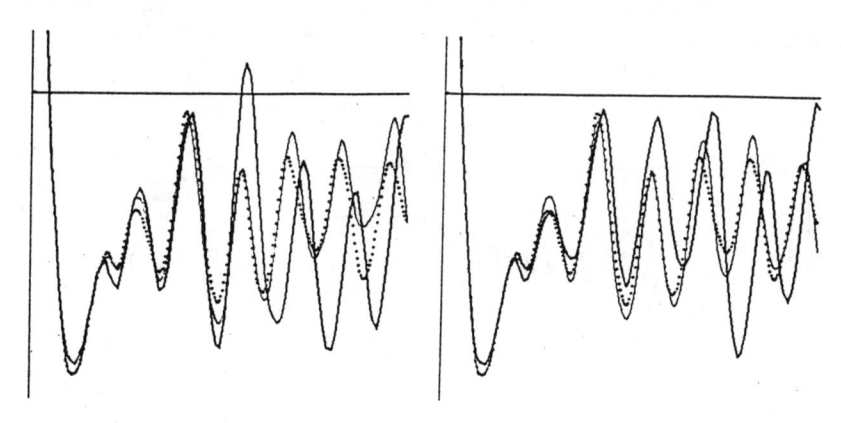

Figure 6. Total flux through the upper inlet Γ of the Venturi pipe at $Re = 5,000$, over $I = [0,30]$, calculated with the CN scheme (left) and the FS_θ scheme (right) with $k = 0.33$ (thick line) and $k = 0.11$ (thin line); reference solution dotted line

b) Robustness Test

We now consider the stability properties of the two second-order schemes. Since the Crank-Nicolson scheme is only A-stable we expect that it may have problems in coping with high frequency perturbations caused by rough initial or boundary data or by errors in the prediction (time extrapolation) of the transport coefficient which are amplified through nonlinear mechanisms.

First, Figure 7 shows an instability effect of the CN scheme in the computation of the flow around a plate. Here, the mean pressure drop shows non-physical fluctuations for $k = 0.11$ which disappear for $k = 0.05$. For the same step size, the FS_θ scheme is stable. For even coarser step-size, $k \geq 0.33$, both schemes exhibit unstable behavior.

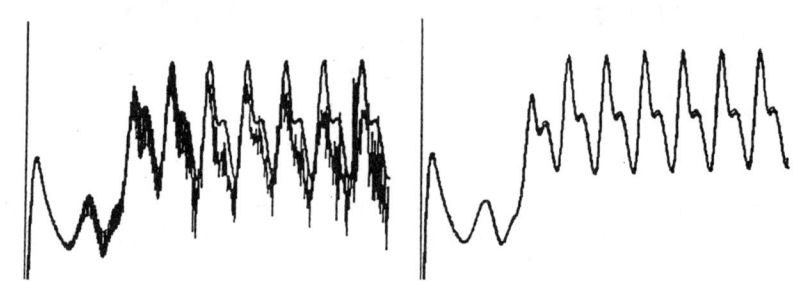

Figure 7. Mean pressure drop accross the plate at Re=500, over $I = [0,60]$, calculated with $k = 0.33$ by the CN scheme (left) and the FS_θ scheme (right), both with linear time-extrapolation (L = 0); reference solution thick line

Next, Figure 8 shows an defect of the CN scheme (combined with an operator splitting approach as described in Section 4, below) in the computation of the flow in a Venturi pipe. Here, for $k = 0.33$, it produces a nonphysical solution with positive outflow through the upper inlet Γ.

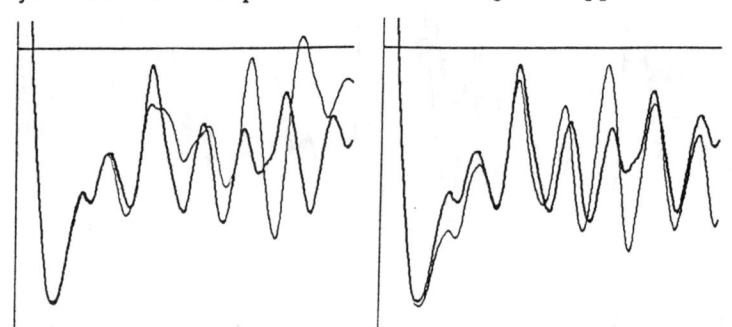

Figure 8. Total flux through the upper inlet of the Venturi pipe Re=5.000, over $I = [0,30]$, calculated with $k = 0.033$ by the CN scheme (left) and the FS_θ scheme (right), both with linear time-extrapolation (L = 0); reference solution thick line

4. PROJECTION METHOD

As mentioned above, the FS_θ-scheme may be combined with any of the projection (or pressure correction) methods of Chorin-type for simplifying the treatment of the incompressibility constraint (see [18] and the literature cited therein). We first consider Chorin's classical projection method. Suppose, for simplicity, that problem (1) is posed with zero Dirichlet boundary conditions, $v_{|\partial\Omega} = 0$. For a given initial value $v^0 \in J_1(\Omega)$, choose a time step k, and solve for $n \geq 1$:

i) Find $\tilde{v}^n \in H_0^1(\Omega)$, such that ("Burgers step")

$$\frac{1}{k}(\tilde{v}^n - v^{n-1}) - \nu\Delta\tilde{v}^n + \tilde{v}^n \cdot \nabla\tilde{v}^n = f^n, \quad \text{in } \Omega, \tag{4.1}$$

ii) Find $v^n = P\tilde{v}^n \in J_0(\Omega)$ by solving ("Projection step")

$$v^n = \tilde{v}^n - k\nabla p^n, \quad \nabla \cdot v^n = 0, \text{ in } \Omega, \quad n \cdot v^n_{|\partial\Omega} = 0. \tag{4.2}$$

Here, the spaces $J_0(\Omega)$ and $J_1(\Omega)$ are obtained through the completion of the subspace $\{\phi \in D(\Omega), \nabla\cdot\phi \equiv 0\}$ in $L^2(\Omega)$ and $H_0^1(\Omega)$, respectively. This time stepping scheme may now be combined with any spatial discretization method, e.g., the finite element methods mentioned above.

The intermediate approximations $\tilde{v}^n$ satisfy the correct boundary condition but violate the incompressibility constraint, while the v^n are divergence free but may have a non-zero tangential component. In view of the orthogonal splitting $L^2 = J_0 \oplus \{\nabla q, q \in H^1(\Omega)\}$ (see, e.g., [6]), the projection step (ii) can equivalently be written as $v^n = \tilde{v}^n - k\nabla p^n$, with some $p^n \in L_0^2(\Omega) \cap H^1(\Omega)$, which is determined through the Neumann problem

$$\Delta p^n = \frac{1}{k}\nabla \cdot \tilde{v}^n, \text{ in } \Omega \quad \partial_n p^n_{|\partial\Omega} = 0. \tag{4.2'}$$

This amounts to a Poisson equation for p^n with zero Neumann boundary conditions. It is this nonphysical boundary condition, $\partial_n p^n_{|\partial\Omega} = 0$, which has caused a lot of controversal discussion about the principle value of the projection method. Nevertheless, the method has proven to work well for representing the velocity field in many flow problems of physical interest (see, e.g., [7]). It is very economical as it requires in each time step only the solution of a (nonlinear) advection-diffusion system for v^n (d-dimensional Burgers problem) and a scalar Laplace-Neumann problem for p^n. Still, it was argued that the pressure p^n were a mere fictitious quantity without any physical relevancy (see, e.g., [26]). It remained the question: How can such a method work at all?

The first convergence results for the projection method was already given by Chorin himself, but concerned only cases with absent rigid boundaries (all-space or spatially periodic problems). Later on, qualitative convergence was shown even for the pressure, but in a measure theoretical sense, too weak for practical purposes, [25]). The first error estimate, yet still sub-optimal, has been given in [24],

$$\|v^n - v(t_n)\| + \left(k \sum_{m=0}^{n} \|p^m - p(t_m)\|_{L^2(\Omega)}^2 / R\right)^{1/2} = O(\sqrt{k}) . \quad (4.3)$$

This indicates that the quantities p^n are really reasonable approximations to the pressure $p(t_n)$. This result has then been sharpend in [18] to optimal order,

$$\|v^n - v(t_n)\| + \tau_n \|p^n - p(t_n)\|_{H^{-1}(\Omega)} = O(k) , \quad (4.4)$$

where the occurance of the time weight $\tau_n = \min\{1, t_n\}$ reflects a peculiar reduction in the regularity of the pressure at $t=0$. This finally confirmed that Chorin's original method is indeed a first order time stepping scheme for the incompressible Navier-Stokes problem. The key to this result is the re-interpretation of the projection method in the context of the so-called "pressure stabilization methods". To this end one inserts the quantity $v^{n-1} = \tilde{v}^{n-1} - k\nabla p^{n-1}$ into the momentum equation, obtaining

$$\frac{1}{k}(\tilde{v}^n - \tilde{v}^{n-1}) - \nu\Delta\tilde{v}^n + (\tilde{v}^n \cdot \nabla)\tilde{v}^n + \nabla p^{n-1} = f^n, \quad \tilde{v}^n|_{\partial\Omega} = 0 , \quad (4.5)$$

$$\nabla \cdot \tilde{v}^n - k\Delta p^n = 0 , \quad \partial_n p^n|_{\partial\Omega} = 0 . \quad (4.6)$$

This looks like an approximation of the Navier-Stokes equations with a first-order "pressure stabilization" term. Pressure stabilization methods for solving the Navier-Stokes equations have a long history. In these methods the continuity equation is supplemented by terms involving the pressure thereby giving it a similar appearance as in compressible flow models; this approach is therefore sometimes refered to as "pseudo-compressibility method". The observation is that the projection method can (at least on the continuous level) be viewed as a pressure stabilization method with a global stabilization parameter $\varepsilon = k$, and an explicit treatment of the pressure term. In [18], [15], and [16] it is shown how this may be used to derive optimal order error estimates for the projection method. One result of this analysis is that the pressure error is of better order in the interior of the domain Ω than at the boundary, i.e., for any strict subdomain $\Omega' \subset\subset \Omega$, there holds

$$\tau_n \|p^n - p(t_n)\|_{L^2(\Omega')} / R = O(k) . \quad (4.7)$$

Moreover, it turns out that the pressure error is actually confined to a small boundary strip of width $\delta \approx \max\{\sqrt{\nu k}, h\}$ and decays exponentially into the interior of Ω which is confirmed by numerical experiments (for a discussion of this matter see also [7]). Subsequently, it was shown that in fact

$$|p^n(x)-p(x,t_n)| \leq \gamma \exp(-\alpha \frac{d(x)}{\sqrt{vk}})\sqrt{k} + O(k), \tag{4.8}$$

where $d(x) = \text{dist}(x,\partial\Omega)$, see [16] for the details. This result is supported by numerical experiments for the pressure stabilization method applied to the stationary Stokes problem. In fact, it is even possible to recover the optimal order accuracy of the pressure at the boundary by simply extrapolating the pressure values from the interior of the domain (see [2], [18]).

The analysis described above can be extended to higher-order projection methods. One popular example is the method of Van Kan [28]:
Starting from some $v^0 \in J_1(\Omega)$, and $p^0 = 0$, and compute, for $n \geq 1$,

i) Find $\tilde{v}^n \in H_0^1(\Omega)$, such that (Burgers step): \hfill (4.9)

$$\frac{1}{k}(\tilde{v}^n-v^{n-1}) - \frac{v}{2}\Delta(\tilde{v}^n+v^{n-1}) + \frac{1}{2}(\tilde{v}^n\nabla\cdot\tilde{v}^n+v^{n-1}\nabla\cdot v^{n-1}) + \nabla p^{n-1} = f^{n-1/2}, \text{ in } \Omega,$$

ii) Find $q^n \in H^1(\Omega)$, such that (projection step):

$$-\Delta q^n = \frac{2}{k}\nabla\cdot\tilde{v}^n \text{ in } \Omega, \quad \partial_n q^n|_{\partial\Omega} = 0, \tag{4.10}$$

and set $v^n = \tilde{v}^n - 2k\nabla q^n$, $p^n = p^{n-1} + q^n$.

An examination of this scheme shows that it may again be interpreted as a certain pressure stabilization method using a stabilization of the form

$$\nabla\cdot v - \alpha k^2 \partial_t \Delta p = 0, \text{ in } \Omega, \quad \partial_n p|_{\partial\Omega} = 0, \quad \partial_n p|_{t=0} = 0. \tag{4.11}$$

This observation, in turn, then led to improved error estimates of optimal order analogously to (4.4) (see [16])

$$\|v^n-v(t_n)\| + \|\bar{p}^n-\bar{p}(t_n)\|_{-1} = O(k^2), \tag{4.12}$$

where ther bar in the pressure term refers to a certain mean value over time. Further, the following analogue of (4.8) holds true:

$$|\bar{p}^{n-1/2}(x)-\bar{p}(t_{n-1/2},x)| \leq c \exp(-\alpha \frac{d(x)}{vk})k + O(k^2). \tag{4.13}$$

This shows again that the effect of the nonphysical boundary condition in the projection step is restricted to a small boundary strip of width $O(\max\{vk,h\})$.

Beside its higher order accuracy the Van Kan scheme has an important advantage over Chorin's classical (only first-order) method. It can also be used for calculating stationary solutions as the limit $v^n \to v^\infty$ ($n\to\infty$), since it has the structure of a fixed point iteration.

These projection methods have discrete analogues in which the discretization in space is applied first followed by the operator splitting. Suppose that the spatial discretization is stable, i.e., satisfies the the condition (2.5). We exemplarily formulate the discrete analogue of Chorin's scheme:

i) Find $\tilde{v}_h^n \in H_h$, such that ("discrete" Burgers step):

$$\left(M + kA + kN(\overline{v}_h^n)\right)\tilde{v}_h^n = M v_h^{n-1} + k f_h^{n-1/2}. \tag{4.14}$$

ii) Find $q_h^n \in L_h$, such that ("discrete" projection step):

$$B^T M^{-1} B q_h^n = \frac{1}{k} B^T \tilde{v}^n, \quad v_h^n = \tilde{v}_h^n - k M^{-1} B q_h^n, \quad p_h^n = p_h^{n-1} + q_h^n. \tag{4.15}$$

Written in this form, the discrete projection method does not seem to inherit the questionable Neumann boundary condition, as all discrete velocities are computed in the finite element space H_h containing the no-slip boundary condition. The reason for this surprising fact may be that the matrix $C = B^T M^{-1} B$ does not correspond to a consistent discretization of the Lapalace-Neumann problem $-\Delta p = g$ in Ω, $\partial_n p_{|\partial\Omega} = 0$, as the matrix B is constructed in the space $H_h \times L_h$ containing zero Dirichlet boundary conditions for the velocities. A consistent approximation of the Laplace-Neumann problem based on a "mixed" variational formulation would have to use a slightly larger velocity space $\tilde{H}_h$ including the "no-penetration" boundary condition $n \cdot v_{h|\partial\Omega} = 0$. A more detailed discussion of this problem together with results of related numerical tests will be given in a forthcoming paper.

Clearly, the approach described in (4.14) and (4.15) may also be used in connection with the FS_θ scheme resulting in a splitting of each of the sub-steps into a "discrete" Burgers step and a "discrete" projection step. Then, for both types of sub-steps efficient and robust multigrid solvers are available (even on strongly nonuniform and anisotropic meshes). The achievable convergence rates for the momentum equation are nearly by one order of magnitude smaller than those obtained for the fully coupled problem (3.5)-(3.10). This leaves open the questions how the operator splitting scheme performs with respect to accuracy. Preliminary numerical tests indicate that, for larger Reynolds numbers and small time-step sizes, the splitting approach may indeed be superior with respect to total computational efficiency.

In Figures 9 and 10, we present exemplary results for the (time-peridic) flow around a plate and for the (chaotic) flow in a Venturi pipe which show that the FS_θ scheme combined with operator splitting described above results in a time stepping scheme which for only slightly smaller time step-sizes is (nearly) as accurate as its fully coupled counterpart.

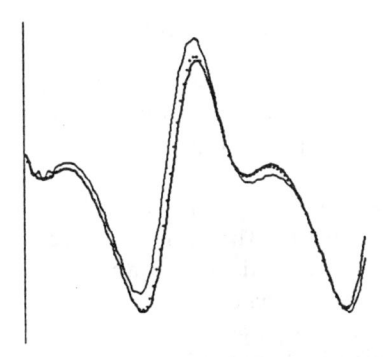

Figure 9. Mean pressure drop accross the plate at Re = 500, over I = [0,10], calculated by the FS_θ scheme in fully coupled form for k = 0.075 (thick line), and with operator splitting for k = 0.075 (thin line) and for k = 0.033 (dotted line)

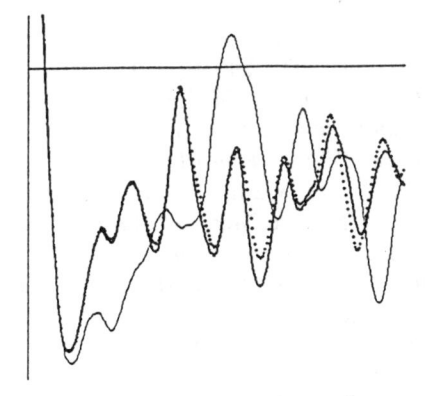

Figure 10. Total flux through the upper inlet of the Venturi pipe at Re=5.000, over I = [0,30], calculated by the FS_θ scheme in fully coupled form for k = 0.05 (thick line), and with operator splitting for k = 0.05 (thin line) and for k = 0.022 (dotted line)

Finally, we note that, due to the better convergence rate of the inner multigrid solver in the splitting scheme, the CPU-time (on a SUN Sparc 10/51) needed for calculating the solution at time T = 10 was only about 40% of that needed by the fully coupled method. A more detailed comparison of the solution efficiency of the fully coupled FS_θ scheme and its variants based on the concept of operator splitting will be presented in a forthcoming paper.

ACKNOWLEDGEMENT

This work has partly been supported by the Deutsche Forschungsgemeinschaft through the SFB 359 "Reaktive Strömungen, Diffusion und Transport" and the GK "Modellierung und Wissenschaftliches Rechnen in Mathematik und Naturwissenschaften" at the University of Heidelberg.

REFERENCES

[1] R. Becker, R. Rannacher, *Finite element discretization of the Stokes and Navier-Stokes equations on anisotropic grids*, Proc. 10th GAMM-Seminar, Kiel, January 14-16, 1994 (G. Wittum, W. Hackbusch, eds.), Vieweg, to appear.

[2] H. Blum, *Asymptotic Error Expansion and Defect Correction in the Finite Element Method*, Habilitation Thesis, Heidelberg, 1991.

[3] M.O. Bristeau, R. Glowinski, J. Periaux, *Numerical methods for the Navier-Stokes equations: Applications to the simulation of compressible and incompressible viscous flows*, Report UH/MD-4, Univ. of Houston, 1987, in Computer Physics Report 1987.

[4] E. F. Cara, M. M. Beltran, *The convergence of two numerical schemes for the Navier-Stokes equations*, Preprint, University of Sevilla, 1988.

[5] A.J. Chorin, *Numerical solution of the Navier-Stokes equations*, Math. Comp. 22, 745-762 (1968).

[6] V. Girault, P.A. Raviart, *Finite Element Methods for Navier-Stokes Equations*, Springer, Berlin-Heidelberg 1986.

[7] P.M. Gresho, *On the theory of semi-implicit projection methods for viscous incompressible flow and its implementation via a finite element method that also introduces a nearly consistent mass matrix. Part 1: Theory*, Int. J. Numer. Meth. Fluids 11, 621-659 (1990). Part 2: Implementation, Int. J. Numer. Meth. Fluids 11, 587-620 (1990).

[8] J. Harig, *A 3-d finite element upwind approximation of the stationary Navier Stokes equations*, IWR-Report, Univ. of Heidelberg 1991.

[9] J.G. Heywood, R. Rannacher, *Finite element approximation of the nonstationary Navier-Stokes problem*, Part 1, SIAM J. Numer. Anal. 19, 275-311 (1982), Part 2, ibidem, 23, 750-777 (1986), Part 3, ibidem, 25, 489-512 (1988), Part 4, ibidem, 27, 353-384 (1990).

[10] J. G. Heywood, R. Rannacher, S. Turek, *Artificial boundaries and flux and pressure conditions for the incompressible Navier-Stokes eqtuions*, Int. J. Numer. Math. Fluids, to appear.

[11] T.J.R. Hughes, L.P. Franca, M. Balestra, *A new finite element formulation for computational fluid mechanics: V. Circumventing the Babuska-Brezzi condition: A stable Petrov-Galerkin formulation of the Stokes problem accommodating equal order interpolation*, Comp. Meth. Appl. Mech. Eng., 59, 85-99 (1986).

[12] C. Johnson, R. Rannacher, M. Boman, *Numerics and hydrodynamic stability: Toward error control in CFD*, SIAM J. Numer. Anal., to appear.

[13] P. Kloucek, F. S. Rys, *On the stability of fractional step-θ-scheme for the Navier-Stokes equations*, SIAM J. Numer. Anal., to appear.

[14] S. Müller-Urbaniak, *Eine Analyse des Zwischenschritt-θ-Verfahrens zur Lösung der instationären Navier-Stokes-Gleichungen*, Doctor Thesis, Universität Heidelberg, 1993.

[15] A. Prohl, *Über die Chorinsche Methode zur Lösung der inkompressiblen Navier-Stokes-Gleichungen*, Diploma Thesis, Heidelberg, 1992.

[16] A. Prohl, R. Rannacher, *Analysis of some projection methods for the incompressible Navier-Stokes equations*, to appear.
[17] R. Rannacher, *Numerical analysis of nonstationary fluid flow. A survey*, in Applications of Mathematics in Industry and Technology (V. C. Boffi, H. Neunzert, eds.), pp. 34-53, B.G. Teubner, Stuttgart 1989.
[18] R. Rannacher, *On Chorin's projection method for the incompressible Navier-Stokes equations*, in "Navier-Stokes Equations: Theory and Numerical Methods" (R. Rautmann, et.al., eds.), Proc. Oberwolfach Conf., 19.-23.8.1991, Springer, 1992.
[19] R. Rannacher, *Numerical analysis of the Navier-Stokes equations*, Appl. Math. 38, 361-380 (1993).
[20] R. Rannacher, S. Turek, *Simple nonconforming quadrilateral Stokes element*, Numer. Meth. Part. Diff. Equ. 8, 97-111 (1992).
[21] H. Reichert, G. Wittum, *On the construction of robust smoothers for incompressible flow problems*, Proc. Workshop "Numerical Methods for the Navier-Stokes Equations", Heidelberg, Oct. 25-28, 1993, Vieweg, to appear.
[22] F. Schieweck, *A parallel multigrid algorithm for solving the Navier-Stokes equations on a Transputer system*, Impact Comput. in Sci. Engrg., 5, 345-378 (1993).
[23] P. Schreiber, S. Turek, *An efficient finite element solver for the nonstationary incompressible Navier-Stokes equations in two and three dimensions*, Proc. Workshop "Numerical Methods for the Navier-Stokes Equations", Heidelberg, Oct. 25-28, 1993, Vieweg, to appear.
[24] J. Shen, *On error estimates of projection methods for the Navier-Stokes equations: First order schemes*, SIAM J. Numer. Anal. (1991).
[25] R. Temam, *Sur l'approximation de la solution des equations de Navier-Stokes par la méthode des pas fractionaires II*, Arch. Rat. Mech. Anal., 33, 377-385 (1969).
[26] R. Temam, *Remark on the pressure boundary condition for the projection method*, Technical Note 1991, to appear in Theoretical and Computational Fluid Dynamics.
[27] S. Turek, *Tools for simulating nonstationary incompressible flow via discretely divergence-free finite element models*, Int. J. Numer. Meth. Fluids, 18, 71-105 (1994).
[28] J. Van Kan, *A second-order accurate pressure-correction scheme for viscous incompressible flow*, J. Sci. Stat. Comp., 7, 870-891 (1986).

Experiences with a parallel multiblock multigrid solution technique for the Euler equations.

C.W. Oosterlee, H. Ritzdorf, A. Schüller, B. Steckel

Gesellschaft für Mathematik und Datenverarbeitung,
P.O. Box 1316, 53731 St. Augustin.

Summary

The parallel solution of 2D steady compressible Euler equations with a multigrid method is investigated. The parallelization technique used is the grid partitioning strategy. The influence of splitting into many blocks on multigrid convergence rates is reduced with an extra interior boundary relaxation and an extra update of the overlap region. The finite volume discretization of the equations is based on the Godunov upwind approach, with Osher's flux difference splitting for the convective terms. Second order accuracy is obtained with defect correction. Solution times of the multigrid algorithms are presented for several parallel MIMD computers.

1 Introduction

The solution of partial differential equations on parallel MIMD distributed memory machines is starting to become very interesting for the industry. Several industrial codes made for vectorcomputers are now being parallelized ([5]), or are about to be parallelized. In the POPINDA project, a cooperation of DLR, Dornier, DASA, Deutsche Airbus, IBM, ORCOM and the GMD, 3 dimensional parallel compressible Navier-Stokes solvers will, amongst many other features, be developed, based on a 3 dimensional communications library. In the first stage the suitability of using parallelization strategies of already existing two dimensional codes for 3D solvers is investigated.
At GMD a two dimensional Euler code for block-structured grids has been developed as part of the LiSS package [15]. LiSS, developed in the eighties, is a program package to solve partial differential equations on general two-dimensional domains. From the beginning of the package development the parallel solution of equations has been focussed upon by constructing efficient parallel solution methods and parallelization tools, like a two-dimensional communications library, based on PARMACS ([1]). Next to efficiency the solution algorithm must be robust in order to solve complex flow problems. Therefore the resulting discretized equations are solved with a parallel multigrid algorithm. Multigrid ([2], [8]) has proved to be a very efficient solution technique for several (non)linear partial differential equations. However, the parallelization of multigrid, while keeping its favourable properties is not at all trivial. The parallelization strategy adopted is the grid partitioning method (explained in [13], [16]). In grid partitioning the grid is split into blocks, which are mapped to different processes. Each process is mapped to a processor of a parallel computer. We investigate the quality of this parallelized (standard) multigrid algorithm for steady compressible Euler equations. References to research in the field of parallel multigrid methods, where standard multigrid is changed in order to get parallelism are (amongst others) [7], [9], [16], [17], [19] and [21].
A flow problem around an airfoil is taken and the computational domain is split into many blocks (up to 256). When a grid is split into several blocks, which are all smoothed simultaneously powerful line smoothers loose their line coupling over interior block boundaries.

This means that a typical multigrid property, the h-independent convergence rate is lost for problems in which anisotropies occur. In [14] this was observed and treatments to overcome this problem were given for incompressible Navier-Stokes equations. Here, we investigate whether these treatments are also satisfactory, when compressible equations are solved for first order accuracy and with defect correction for second order accuracy, and when a grid is partitioned into very many blocks.
The two-dimensional compressible Euler equations are discretized on a block-structured grid with a finite volume discretization. The discretization is a vertex-centered discretization based on Osher's flux difference splitting (as in [23]). With the defect correction iteration second order accuracy is obtained with van Leer's second order κ-scheme ([25]). Parallel multigrid is used to solve for first order accuracy and as inner iteration in the defect correction technique (see [8], [11]).
Furthermore, the multigrid code is tested on several parallel machines: on IBM SP1 with 10 nodes, on CM5 with 64 nodes and on a cluster of workstations containing 17 nodes. Solution times for grid partitionings up to 32 blocks are presented.

2 The discretization of the Euler equations

The 2D steady compressible Euler equations are commonly written in their differential form as follows

$$\frac{\partial}{\partial x}\begin{bmatrix} \rho u \\ \rho u^2 + p \\ \rho u v \\ (E+p)u \end{bmatrix} + \frac{\partial}{\partial y}\begin{bmatrix} \rho v \\ \rho u v \\ \rho v^2 + p \\ (E+p)v \end{bmatrix} = 0$$

$$p = (\gamma - 1)(E - \frac{1}{2}\rho(u^2 + v^2)), \tag{1}$$

ρ is the density, u and v the two Cartesian velocity components, E the total energy, p the pressure, and γ (assumed to be constant) is the ratio of the specific heats at constant pressure and constant volume.
The vertex-centered discretization adopted for the Euler equations is described briefly here. It is based on the cell-centered discretization described in [23] and [11]. The details on the discretization and on boundary conditions for LiSS are given in [4]; some trans- and supersonic channel flow test problems have been solved in [3]. For the finite volume discretization the domain Ω is divided into quadrilaterals $C_{i,j}$. For each quadrilateral (1) must hold in integral form.

$$\oint_{\partial C_{i,j}} (f(q)cos\phi + g(q)sin\phi)ds = 0 \tag{2}$$

for every i,j, where $cos\phi$ and $sin\phi$ are the x- and y- component of the outward normal vector on $\partial C_{i,j}$, and q is the state vector.
The rotational invariance of the Euler equations is used, and the discretization results in:

$$\sum_{(i,k)\in k(i)} F(q^L, q^R)|\partial C_{i,k}| = 0 \tag{3}$$

with $k(i)$ being the set of neighbouring cell indices of $C_{i,j}$; $|\partial C_{i,k}|$ is the length of the boundary between $C_{i,j}$ and $C_{i,k}$; $F(q^L, q^R)$ is an approximate Riemann solver, which depends on the left, q^L, and right state, q^R, along the cell boundary. The discretization requires a calculation of the convective flux at each cell face $\partial C_{i,j}$. With the Godunov upwind approach ([6]) the approximate solution $F(q^L, q^R)$ of the 1D Riemann problem is solved with an approximate Riemann solver proposed by Osher ([22]) in its P-variant ([23]):

$$F(q^L, q^R) = \frac{1}{2}(f(q^L) + f(q^R)) - \int_{q^L}^{q^R} |A(q)| dq \qquad (4)$$

where $|A(q)|(= A^+(q) - A^-(q))$ is a splitting of the Jacobian matrix A into matrices with positive and negative eigenvalues, and f is the one-dimensional flux along the normal vector. First order accuracy is obtained by substituting,

$$q_{i+\frac{1}{2}}^L = q_{i,j}$$
$$q_{i+\frac{1}{2}}^R = q_{i+1,j} .$$

As state vector $q = (u, v, c, z)^T$ is chosen, where $c \equiv \sqrt{\gamma p/\rho}$ is the speed of sound and $z \equiv ln(p\rho^{-\gamma})$ is unscaled entropy.

Second order accuracy is achieved with the defect correction technique ([8], [11]). In general it is favourable for iterative solution methods that from discretized equations so-called K-operators result. K-operators lead to M-matrices (see [26]) for which it is known that an efficient use of basic iterative methods is possible. However, often, as for steady compressible Euler equations, a second order discretization does not guarantee K-operator properties, while a first order upwind operator does. With defect correction second order accuracy can be obtained by iterating with a first order discretized operator. The right-hand-side is then corrected with a second order operator. It is shown in [8] that defect correction converges to the second order solution. Although the convergence of the second order residuals is slow, it is found (for example in [11] for Euler equations) that convergence to engineering accuracy is very fast. Lift and drag coefficients are converged in [11] after approximately 10 defect correction iterations. Multigrid with defect correction starts with nested iteration on the coarsest grid (FMG). After the finest grid has been reached and one additional FAS cycle is made, defect correction starts. Only on the finest grid the right-hand-side is corrected. The second order scheme used in defect correction is van Leer's κ-scheme ([25]). The states $q_{i+\frac{1}{2},j}^L$ and $q_{i+\frac{1}{2},j}^R$ in (4) are replaced by:

$$\begin{aligned} q_{i+\frac{1}{2},j}^L &= q_{i,j} + \frac{1+\kappa}{4}(q_{i+1,j} - q_{i,j}) + \frac{1-\kappa}{4}(q_{i,j} - q_{i-1,j}) \\ q_{i+\frac{1}{2},j}^R &= q_{i+1,j} + \frac{1+\kappa}{4}(q_{i,j} - q_{i+1,j}) + \frac{1-\kappa}{4}(q_{i+1,j} - q_{i+2,j}) . \end{aligned} \qquad (5)$$

All tests have been made with $\kappa = 0$, the Fromm scheme. In order to avoid wiggles, that may appear with these κ-schemes, the van Albada limiter as defined and proposed in [11] is implemented. With the cell-centered discretization and a multigrid solution algorithm very successful results are obtained for transonic flow problems by Koren ([11]).

3 Parallel multigrid

The parallel multigrid algorithm consists of a host and a node program. The host program takes care for the organization of in- and output. It creates node processes, mails initial data to node processes and receives calculated results, like residuals from nodes. These tasks are taken care for by PARMACS-based ([1]) routines of the 2-D Communications Library. With PARMACS-based routines portability is guaranteed for a large class of parallel computers. In a node program the calculation takes place, also communication among nodes is taken care for.

Grid partitioning, the technique to distribute parts of a domain to different processes, is explained in many papers ([13], [16]). The domain is split into blocks. Along the arisen block boundaries, called interior block boundaries, an overlap region is placed. Therefore all operations in multigrid, like restriction, prolongation and relaxation can be performed in parallel. Keeping the values in overlap regions up-to-date requires communication among the nodes. It should be noticed that a first order discretization requires an overlap region of one line of cells in order to achieve accuracy, while for the second order discretization of (4), (5) the stencil for evaluation of the right-hand-side grows and a straightforward implementation of the parallel algorithm would require an overlap region of two cells. Figure 1 shows an example of grid partitioning near a leading edge with an overlap region of two cells. In [14] a satisfactory strategy was found to get good multigrid

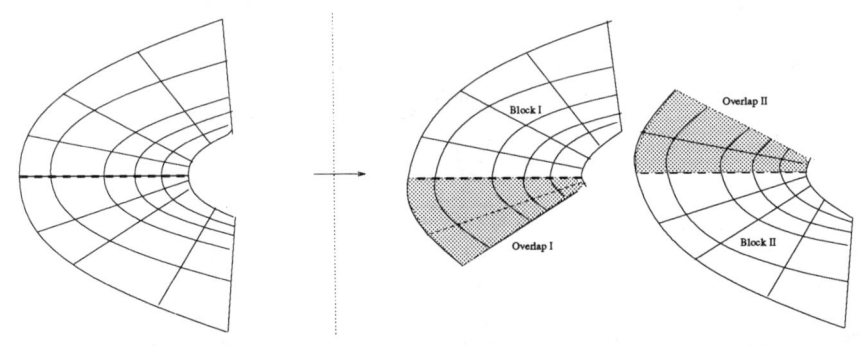

Figure 1: *An example of grid partitioning near a leading edge with an overlap region of two cells. One of the two blocks has the update right on the interior boundary.*

convergence, when many blocks are used. For an incompressible flow problem with 32 blocks similar convergence factors as for single block were obtained. The smoother used was a Collective Damped Alternating Line Gauß-Seidel (DALGS) relaxation method. Satisfactory multigrid convergence factors were achieved with an extra internal boundary relaxation after each smoothing iteration and with an extra update of the overlap region, when the smoothing along lines in the first direction of the alternating method was finished. This strategy is investigated for 2D steady compressible Euler equations. DALGS is a robust smoother. It is a necessary tool to tackle anisotropies, that may occur in coefficients of equations, or due to a large spatial discretization in one grid direction only. In [23] a Collective Symmetric Point Gauß-Seidel smoother leads to efficient results with the multigrid V-cycle. In our experiments with many blocks this combination was not as robust and efficient as Collective Damped Alternating *Line* Gauß-Seidel in a multigrid F-cycle, although smoothing analysis in [27] indicates that alternating line Gauß-Seidel is not robust for all convection angles of the convection-diffusion equation. A collective smoother updates all unknowns at each grid point simultaneously. It should be noted that a general block-structured grid split up into many blocks does not need to possess global directions. The lines along which the smoother updates are only local, i.e. per block. When a domain is partitioned into very many blocks, which are all updated simultaneously, the robustness of a line smoother with respect to anisotropies is impaired due to short line length.

Another problem exists in grid coarsening. The parallel inefficiency of coarse grid operations is a major problem for achieving good parallel performance on block-structured grids. At a certain stage on a coarse grid level every process contains the minimal number of grid points. Then all processors are busy, assuming that each processor contains one process. However, when a domain is split into many blocks this coarse level does not need to be the global coarsest grid possible. With an agglomeration strategy ([10], [18]) grids can be coarsened further, leaving some processors idle and re-mapping busy processors to an optimal configuration with little communication time. An agglomeration strategy for physically rectangular grids has, for example, been implemented successfully for parallel solution of time-dependent partial differential equations in [24]. Here, an agglomeration strategy is not implemented, due to enormous technical difficulties. Furthermore, a general block-structured grid does not need to contain a global coarsest grid, but re-mapping the partitioned grid to the user-defined grid should be a future implementation. In our investigations all processors on a coarsest level contain the minimal number of grid points per block. When a 257 × 65-grid is split into 256 equal sized 9 × 9 blocks only three multigrid levels remain for an overlap region of one cell.

4 Results

The computational domain around a NACA0012 airfoil is a C-grid consisting of 257 × 65 cells. It is partitioned into many blocks to investigate the convergence rate of multigrid for compressible Euler equations. The transonic testcase investigated is: Mach number $M_\infty = 0.85$ and angle of attack $\alpha = 1°$. This is a very popular airfoil testcase with a strong lee-side and a less strong wind-side shock. The Mach number distributions for the second order discretization is presented in Figure 2. The outer boundary of the domain is

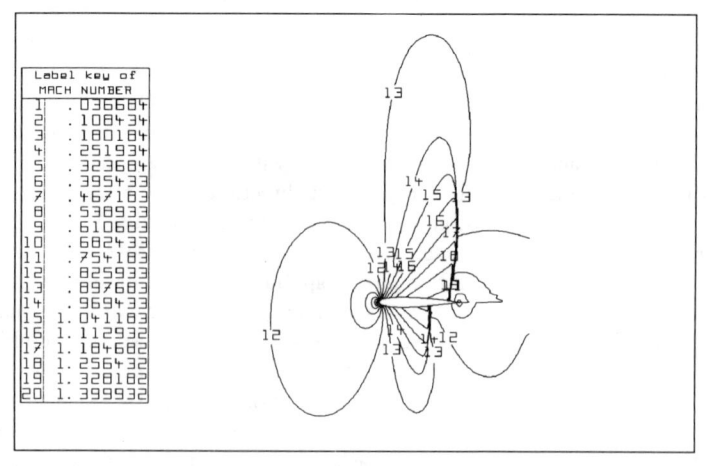

Figure 2: *Mach number distributions for flow around an airfoil, ; $M_\infty = 0.85, \alpha = 1°$.*

20 chord lengths away; the outer boundary condition is not corrected with a vortex, which is commonly done. For hypersonic problems it is not at all trivial to get a satisfactory multigrid algorithm for a first order discretization and to get convergence of defect correction. In ([12]) global upwind prolongations and local defect damping lead to satisfactory multigrid convergence for a first order accurate reentry flow around a blunt forebody at $M_\infty = 8.15, \alpha = 30°$. We did not implement these components in LiSS and therefore restrict ourselves to transonic test problems.

First order discretization. First, the investigations in [14] leading to satisfactory parallel multigrid convergence for incompressible Navier-Stokes equations are repeated here for the first order accurate discretization. The domain is split into 2 (129 × 65), 4 (65 × 65), 8 (65 × 33), 16 (33 × 33) and 64 (17 × 17) (see Figure 3) blocks is shown. Furthermore, the domain is also split into 256 (9 × 9) blocks. For these splittings the influence of an

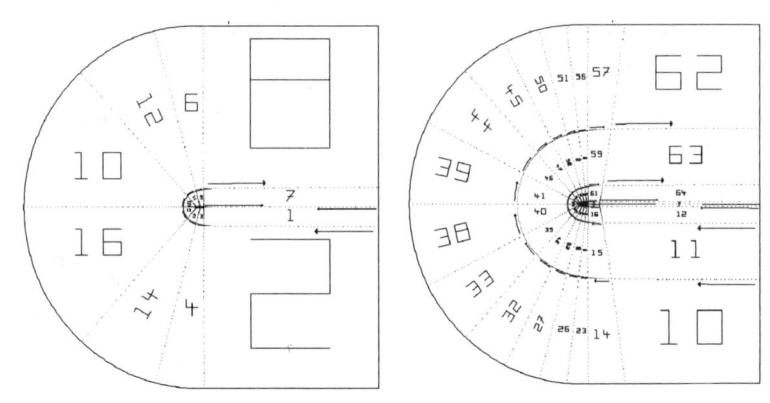

Figure 3: *The domain for a flow around an airfoil split into 16 and 64 equally sized blocks.*

additional update of the overlap region and/or an extra interior boundary relaxation on the convergence factor is investigated.

The details of the nonlinear multigrid algorithm for first order discretization are: The FAS F-cycle with 1 pre- and 1 post-smoothing iteration (=F(1,1)), starting on the coarsest grid (FMG) is used. Underrelaxation factor ω for DALGS resulting in best convergence factors depended on the number of blocks into which the grid was partitioned. This is probably due to the absence of an agglomeration strategy, because the coarsest level then visited is not always the global coarsest grid. Underrelaxation needs to be adapted in order to get convergence on this coarsest level, which leads to a reduction of the overall convergence rate. For splittings up to 16 blocks $\omega = 0.8$ resulted in satisfactory convergence, for 64 blocks $\omega = 0.7$; for 256 blocks $\omega = 0.45$ was best. The algorithm is fairly robust; with underrelaxation factors approximately 0.2 smaller convergence is still obtained. (A similar observation is also made for single block steady incompressible Navier-Stokes equations, amongst others in [20].) Figure 4 show a comparison between the convergence of single block and splittings into 16 blocks for the testcase with and without an additional update of the overlap region and with and without an extra interior boundary relaxation sweep. It appeared to be necessary to use both additions for robust convergence of F(1,1) cycles for the testcase, as can be seen in Figure 4. Convergence is not robust unless both additions are included. With only one smoothing iteration per grid visit the information from different blocks is not distributed well enough. With F(2,1)-cycles without additional boundary relaxations there is enough distribution, as we will see later. The convergence

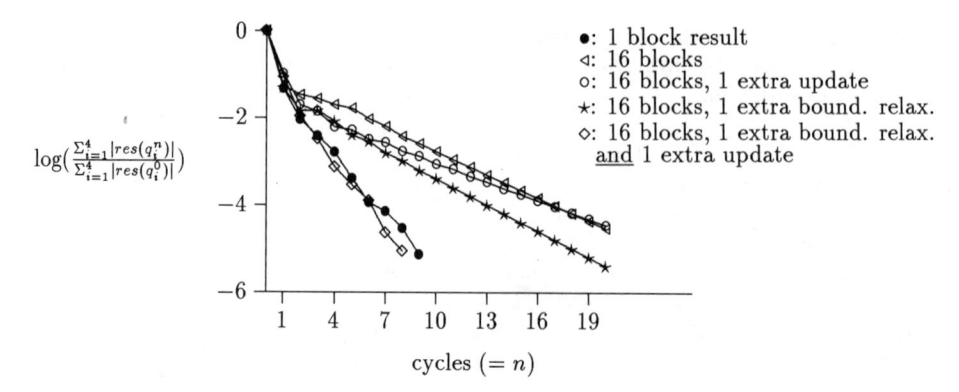

Figure 4: *F(1,1)-cycles with DALGS relaxations; the 1 block solution method compared to the 16 block method for* $\underline{M_\infty = 0.85, \alpha = 1°}$.

of several block splittings with both additional update and interior boundary relaxation is presented versus the single block case in one diagram in Figure 5. Up to 64 blocks (with

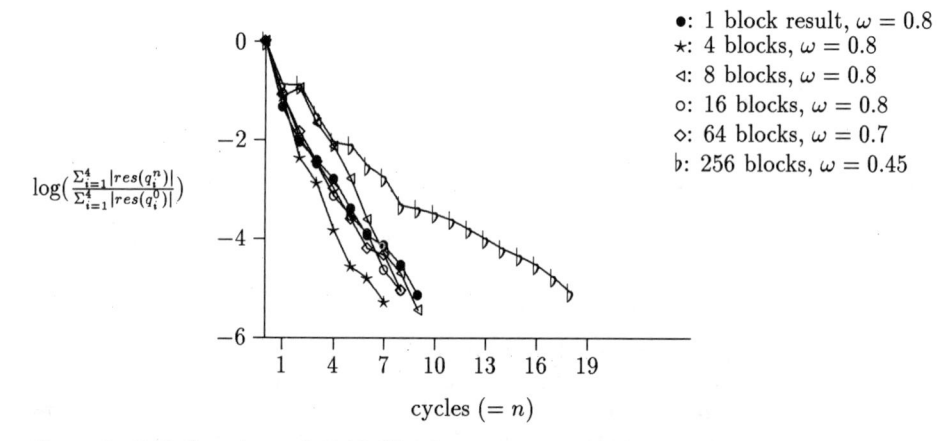

Figure 5: *F(1,1)-cycles with DALGS relaxations; several block results compared. All cases (except the 1 block case) consist of one additional interior boundary relaxation and an extra update of the overlap region during the alternating sweep.* $\underline{M_\infty = 0.85, \alpha = 1°}$.

17×17 points per block) the convergence behaviour is similar to single block convergence, as is underrelaxation factor ω. For some splittings the convergence behaviour with additions even improves compared to single block. For 256 blocks convergence slows down, due to the smaller underrelaxation factor that needs to be used to get convergence on the coarsest level. This splitting is investigated once more. The influence of two interior boundary relaxations and of two pre-smoothing iterations is presented in Figure 6. Also an influence of less coarse grid relaxations can be observed, because of the small number (=3) of multigrid levels. The F(2,1)-cycle gives an improvement towards F(1,1), while two interior boundary relaxations do not lead to faster convergence. Furthermore, the number of coarse grid relaxations must be sufficiently large, for example 20. For very many blocks

$$\log\left(\frac{\sum_{i=1}^{4}|res(q_i^n)|}{\sum_{i=1}^{4}|res(q_i^0)|}\right)$$

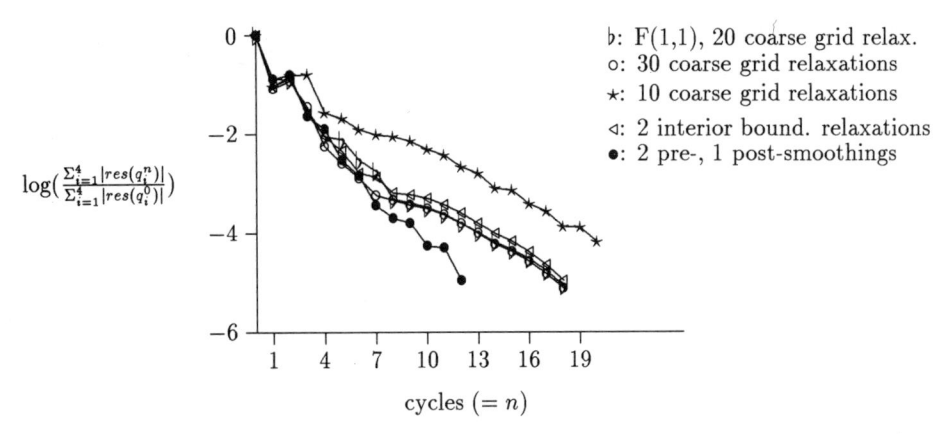

♭: F(1,1), 20 coarse grid relax.
○: 30 coarse grid relaxations
★: 10 coarse grid relaxations
◁: 2 interior bound. relaxations
●: 2 pre-, 1 post-smoothings

cycles ($= n$)

Figure 6: *F-cycles with DALGS relaxations for the 256 block case. The effect of more/less coarse grid iterations, of more pre-smoothing and of 2 extra interior boundary relaxations, $M_\infty = 0.85, \alpha = 1°$, first order discretization, grid $:= 257 \times 65$ points.*

with a small number of points per block satisfactory convergence can still be obtained, but level-independency is impaired without an agglomeration strategy, due to the smaller underrelaxation factor. For several block splittings the performance on parallel computers is investigated, like on a cluster of 17 workstations, on the IBM SP1 with 10 nodes and on the CM5 Connection Machine (vectorization is not employed) with 64 nodes. The cluster of workstations consists of 8 SP1 nodes and 9 additional IBM RS6000 350 workstations. For each partitioning an equal number of SP1 nodes and workstations are taken. Our implementation consists of a host and a node program, therefore it is necessary to use one node as host and for example 63 nodes for the partitioning on the CM5. However, a partitioning into 63 blocks would result in one (load balanced) multigrid level. Therefore, the maximal number of blocks used is 32. For all splittings the same algorithm is investigated: 5 F(1,1) multigrid cycles with two additions, started with nested iteration are performed for the first order discretization with one cell overlap. Figure 7 shows solution times in seconds for several splittings on different computers.

In Figure 8 results of two pre-smoothing iterations without additional boundary relaxation are shown for all partitionings. F(2,1) again leads to a robust algorithm, although convergence is worse than in Figure 5.

It is interesting to compare solution times of the different strategies. For this comparison the CM5 is used, because the range of splittings that can be investigated is largest. In Figure 9 the solution times of 5 F-cycles (with nested iteration) are compared of:

• F(1,1), every smoothing sweep an extra update and boundary relaxation are performed. (robust)
• F(1,1), every sweep an extra boundary relaxation is performed. (not robust)
• F(1,1), every sweep an extra update is performed. (not robust)
• F(2,1), every sweep an additional update is performed. (robust)

When the number of processors grows the difference in solution times between the two robust strategies reduces. Furthermore, the cost of an extra update is negligible, while an additional boundary relaxation is fairly expensive.

<u>Defect correction.</u> For defect correction similar investigations are performed. The accu-

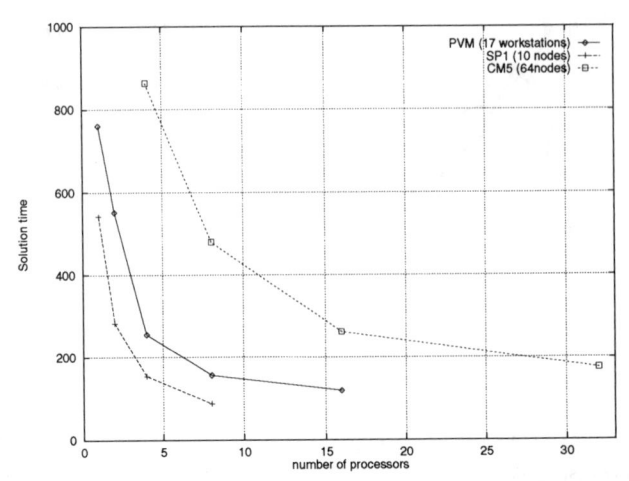

Figure 7: *Solution times for different partitionings on different machines.*

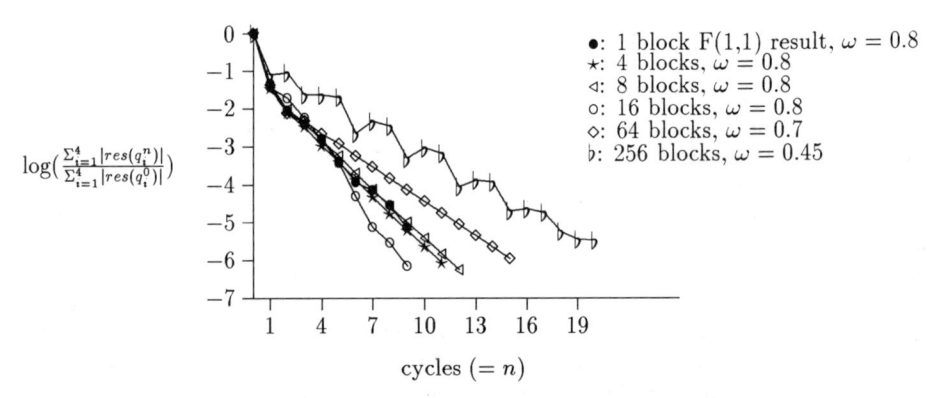

Figure 8: *F(2,1)-cycles with DALGS relaxations with extra update of the overlap compared to F(1,1) for single block,* $M_\infty = 0.85, \alpha = 1°$.

racy of the lift coefficient (c_L) is taken as second order convergence criterium. With an overlap region of two cells the finest partitioning is the 64 block case, resulting in three multigrid levels. F(2,1) is used for this splitting, while it resulted in best convergence for three multigrid levels (see Figure 6), for single block, 4, 8 and 16 block splittings F(1,1) is taken. Choosing underrelaxation 0.2 less than in the first order algorithm leads to robust algorithms. In Figure 10 it can be seen that the convergence of c_L is still fast when the domain is split into many blocks (again with one extra boundary relaxation and an extra update). We find $c_L = 0.3760, c_D = 0.055$.

5 Conclusions

In this paper the convergence behaviour of a parallel multigrid algorithm has been investigated for steady compressible Euler equations discretized on a block-structured grid,

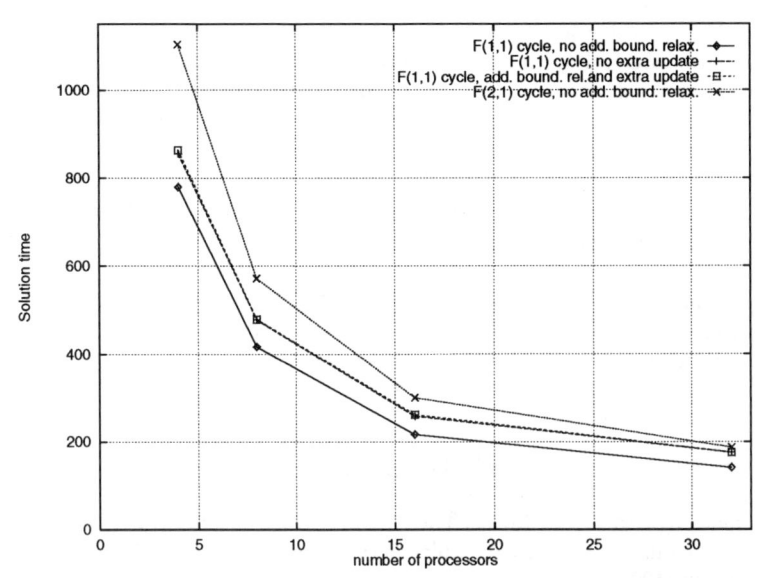

Figure 9: *Solution times for different strategies on the CM5.*

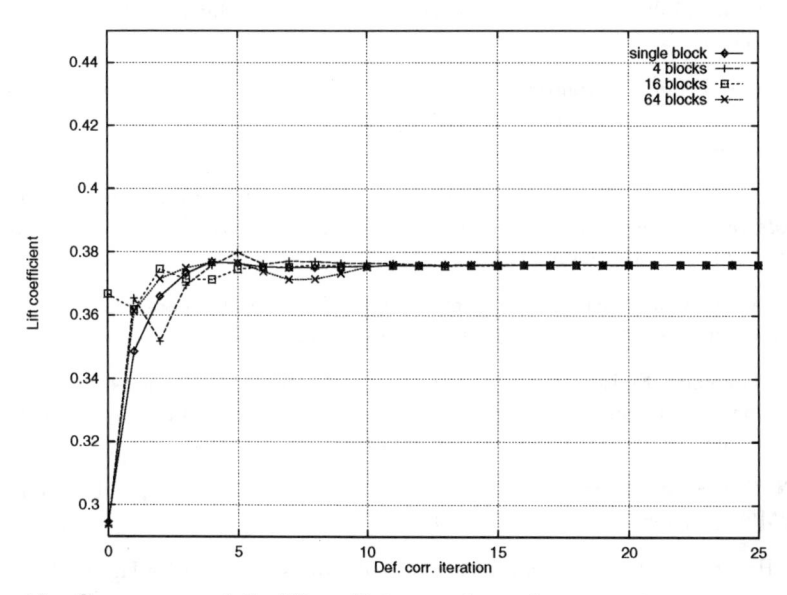

Figure 10: *Convergence of the lift coefficient c_L for a flow around an airfoil when the domain is split into 1, 4, 16 and 64 equal sized blocks.*

which is partitioned into many blocks. It is found that with one additional boundary relaxation and an extra update of the overlap region the multigrid convergence is not much affected by grid partitioning for many splittings investigated. However, the additional boundary relaxation appeared to be fairly expensive. The collective alternating line Gauß-Seidel smoother appeared to be a robust smoother for the test problems, but when many blocks are used a change in underrelaxation is necessary for convergence. For a splitting into very many blocks (256) satisfactory convergence can still be obtained with a small underrelaxation factor, but level-independency is impaired. Due to the fact that the algorithm does not start on the global coarsest grid without an agglomeration strategy the underrelaxation parameter needs to be adapted to get convergence on the starting level. The lower underrelaxation parameter reduces the multigrid convergence rate. For defect correction with a second order accurate κ-scheme an overlap of two cells is needed in the current LiSS implementation. The accuracy of the lift coefficient, the criterium for defect correction convergence, converges almost independently of the number of blocks, again including both additions. CPU time results on different parallel computers show a promising performance for the IBM SP1 architecture.

References

[1] L. Bomans, R. Hempel, D. Roose, The Argonne/GMD macros in Fortran for portable parallel programming and their implementation on the Intel iPSC/2. *Parallel Comp.* **15**, 119–132 (1990).

[2] A. Brandt, Multi–level adaptive solutions to boundary–value problems. *Math. Comp.* **31**, 333-390, (1977).

[3] J. Canu, J. Linden, *Multigrid solution of 2D Euler equations: A comparison of Osher's and Dick's flux difference splitting schemes.* GMD Arbeitspapiere 693, GMD St. Augustin, Germany (1992).

[4] C. Frohn-Schauf, *Flux-splitting-Methoden und Mehrgitterverfahren für hyperbolische Systeme mit Beispielen aus der Strömungmechanik.* Ph.D. Thesis, H. Heine University, Düsseldorf, Germany (1992).

[5] U. Gärtel, W. Joppich AND A. Schüller, First results with a parallelized 3D weather prediction code. *Parallel Comp.*, **19**, pp. 1427–1429 (1993).

[6] S.K. Godunov, Finite difference method for numerical computation of discontinuous solutions of the equations of fluid dynamics. Cornell Aeronautical Lab transl. from the Russian *Math. Sbornik* **47**, 271–306 (1959).

[7] S.N. Gupta, M. Zubair, C.E. Grosch, A multigrid algorithm for parallel computers: CPMG. *J. Scient. Comp.* **7**, 263–279 (1992).

[8] W. Hackbusch, *Multi-grid methods and applications.* Springer-Verlag, Berlin (1985).

[9] W. Hackbusch, The frequency domain decomposition multigrid method, part I: Application to anisotropic equations. *Num. Math.* **56**, 229–245 (1989).

[10] R. Hempel, A. Schüller, *Experiments with parallel multigrid algorithms using the SUPRENUM communications subroutine library.* GMD Arbeitspapiere 141, GMD St. Augustin, Germany (1988).

[11] B. Koren, Defect correction and multigrid for an efficient and accurate computation of airfoil flows. *J. Comp. Phys.* **77**, 183–206 (1988).

[12] B. Koren, PW. Hemker, Damped direction-dependent multigrid for hypersonic flow computations. *Appl. Num. Math.* **7**, 309–328 (1991).

[13] J. Linden, B. Steckel, K. Stüben, Parallel multigrid solution of the Navier-Stokes equations on general 2D-domains. *Parallel Comp.* **7**, 461-475 (1988).

[14] G. Lonsdale, A. Schüller, Multigrid efficiency for complex flow simulations on distributed memory machines. *Parallel Comp.* **19**, 23–32 (1993).

[15] G. Lonsdale, H. Ritzdorf, K. Stüben, *The LiSS package.* GMD Arbeitspapier 745, GMD St. Augustin, Germany (1993).

[16] O.A. McBryan, P.O. Frederickson, J. Linden, A. Schüller, K. Solchenbach, K. Stüben, C.A. Thole, U. Trottenberg, Multigrid methods on parallel computers - a survey of recent developments. *Impact Comp. Science and Eng.* **3**, 1–75 (1991).

[17] W.A. Mulder, A new multigrid approach to convection problems. *J. Comp. Phys.* **83**, 303–323 (1989).

[18] V.K. Naik, S. Ta'asan, Performances studies of the multigrid algorithms implemented on hypercube multiprocessor systems. In: M. Heath (ed.), *Hypercube multiprocessors 1987*, SIAM, Philadelphia (1987).

[19] N.H. Naik, V.K. Naik, M. Nicoules, Parallelization of a class of implicit finite difference schemes in computation fluid dynamics. *Int. J. High Speed Comp.* **5**, 1–50 (1993).

[20] C.W. Oosterlee, *Robust multigrid methods for the steady and unsteady incompressible Navier-Stokes equations in general coordinates.* Ph.D. Thesis, University of Technology Delft, Netherlands (1993).

[21] C.W. Oosterlee and P. Wesseling, *On the robustness of a Multiple Semi-coarsened Grid method.* TU Report 93-51, Delft Univ. Techn. (1993). To appear in ZAMM.

[22] S. Osher, F. Solomon, Upwind difference schemes for hyperbolic systems of conservation laws. *Math. Comp.* **38**, 339–374 (1982).

[23] S.P. Spekreijse, *Multigrid solution of the steady Euler equations*, CWI Tract 46 (1988).

[24] St. Vandewalle, *Parallel Multigrid Waveform Relaxation for Parabolic Problems.* B.G. Teubner, Stuttgart (1993).

[25] B. van Leer, Upwind-difference methods for aerodynamic problems governed by the Euler equations. In: B. Enquist, S. Osher, R. Somerville (eds.), *Large scale computations in fluid mechanics.* Lectures in Applied Mathematics, Vol. 22, II, Americ. Math. Soc., Providence, R.I., 327-336 (1985).

[26] R.S. Varga, *Matrix iterative analysis.* Prentice-Hall, Englewood Cliffs, N.J. (1962).

[27] P. Wesseling, *An introduction to multigrid methods.* John Wiley, Chichester (1992).

A Coupled Algebraic Multigrid Method for the 3D Navier-Stokes Equations

Michael Raw
Advanced Scientific Computing Ltd.
554 Parkside Drive, Unit 4, Waterloo
Ontario, N2L 5Z4. Canada

Summary

A new implementation of the Algebraic Multigrid method is presented. It is applied to the coupled, linearized, discrete equations arising from a finite volume discretization of the 3D Navier-Stokes equations. It employs a grid coarsening algorithm based on an evaluation of relative coefficient strengths. The restriction and prolongation operators are those implied by the Additive Correction Multigrid method. The relaxation scheme is a coupled Incomplete Lower Upper factorization and the W-Cycle is used. Verification of performance is made on a simple and a complex 3D viscous flow problem, and the method is shown to a very effective linear equation solver for the intended application.

1. Introduction

Finite Volume based Computational Fluid Dynamic (CFD) methods applied to the solution of elliptic flow problems result in very large, stiff, systems of discrete equations. Workstation class computers are expected to routinely solve systems with several hundred thousand degrees of freedom. This paper presents a method to solve the linearized discrete equations using a particular implementation of the Algebraic Multigrid Method (AMG).

The present work has evolved out of earlier efforts at the University of Waterloo, Refs. [1], [2], and previous research at ASC, Refs. [3], [4]. Many other Multigrid approaches have been developed; see, for example, Refs. [5], [6].

The linear Multigrid solver described herein is extensively tested and has been used in a commercial CFD code called **TASCflow**[1], Ref. [7], for nearly 2 years.

The system of equations that is solved can be written in the common form:

$$\sum_{nb \text{ of } i} a_i^{nb} \phi_{nb} = b_i \tag{1}$$

where ϕ is the solution, b the right-hand-side, a the coefficients of the equation, i is the number of the fine grid finite volume (or node) in question, and the nb means "neighbour" but also includes the central coefficient multiplying the solution at the i location. The set of these, for all finite volumes, constitutes the whole system. For a scalar equation (e.g. enthalpy or turbulent kinetic energy) each a_i^{nb}, ϕ_{nb} and b_i is a single number. For the coupled, 3D mass-momentum equation set they are a 4×4 matrix or a 4×1 vector, which can be expressed as:

$$a_i^{nb} = \begin{pmatrix} a_{uu} & a_{uv} & a_{uw} & a_{up} \\ a_{vu} & a_{vv} & a_{vw} & a_{vp} \\ a_{wu} & a_{wv} & a_{ww} & a_{wp} \\ a_{pu} & a_{pv} & a_{pw} & a_{pp} \end{pmatrix}_i^{nb} \tag{2}$$

and

$$\phi_i = \begin{pmatrix} u \\ v \\ w \\ p \end{pmatrix}_i \qquad b_i = \begin{pmatrix} b_u \\ b_v \\ b_w \\ b_p \end{pmatrix}_i \tag{3}$$

---
[1]TASCflow is a registered trademark of Advanced Scientific Computing Ltd.

where the subscripts within the matrix and vector follow from the traditional naming of the components of momentum and pressure as u, v, w, p. It is thus at the equation level that the coupling in question is retained and at no point are any of the rows of the matrix treated any differently (e.g. different solution algorithms for momentum versus mass). The advantages of such a coupled treatment over a non-coupled, or segregated approach (e.g. Ref. [10]) are several: robustness, efficiency, generality and simplicity. The principal drawback is the higher storage needed for all the coefficients.

In all the steps that follow, the Multigrid algorithm is applied to the full coupled equation set.

To highlight the ways in which the present Multigrid is similar and different from existing variants, the presentation focuses on 4 major aspects of any Multigrid method: the coarsening algorithm, the restriction and prolongation operators, the relaxation scheme and the cycle type.

2. Coarsening Algorithm

The original motivation for Geometric Multigrid was based on an important observation: iterative solvers tend to reduce short wavelength errors preferentially over long wavelength errors – short and long defined with respect to the grid spacing. The hierarchy of successively coarser grids is designed so that taken collectively, all fine grid error components are reduced — as a "short wavelength" error on a coarse grid is "long" on a relatively finer grid. A complete description of this kind of conception can be found in any basic Multigrid text, e.g. [11].

In multi-dimensions, however, this picture is incomplete. To illustrate this, consider a 2-D diffusion equation,

$$\frac{\partial^2 \phi}{\partial x^2} + \frac{\partial^2 \phi}{\partial y^2} = 0 \qquad (4)$$

which, using central differencing on a regular mesh, would have a corresponding finite difference equation as,

$$\frac{\phi_{i-1,j} - 2\phi_{i,j} + \phi_{i+1,j}}{\Delta x^2} + \frac{\phi_{i,j-1} - 2\phi_{i,j} + \phi_{i,j+1}}{\Delta y^2} = 0 \ . \qquad (5)$$

Using Gauss-Seidel as an example iterative solver, and sweeping in the $+i$ and $+j$ directions, the solution scheme would be

$$\phi_{i,j}^{n+1} = \frac{\Delta y^2 \left(\phi_{i-1,j}^{n+1} + \phi_{i+1,j}^n\right) + \Delta x^2 \left(\phi_{i,j-1}^{n+1} + \phi_{i,j+1}^n\right)}{2\Delta y^2 + 2\Delta x^2} \ . \qquad (6)$$

Notice how the coefficients in the i direction compare to the j direction. Their ratio scales as the grid cell aspect ratio squared. For instance, if $\Delta x = 10\Delta y$ then in the right-hand-side of Eq. (6) the j direction coefficients would be 100 times larger than the i direction coefficients. Consequently, a given change in $\phi_{i,j-1}$ would influence the new solution 100 times more strongly than the same change in $\phi_{i-1,j}$. This has the important effect that the error components in the large coefficient direction are reduced much faster than those in the small coefficient direction. So not only does Gauss-Seidel preferentially reduce short wavelength errors, in multi-dimensions it does so preferentially in "large coefficient directions". Thus the long wavelength error in the small coefficient direction is reduced least of all. The directional sensitivity of the iterative solver is caused by anisotropic coefficients in general, not just cell aspect ratio, so that, for instance, anisotropic material properties also cause the same behaviour. This poor performance in the small coefficient direction is a second observation that will lead to the Algebraic Multigrid Method (AMG). Unless specially designed iterative solvers are used (usually impractical in 3D), conventional Geometric Multigrid would not work well on a problem such as this. This is because error components of a short wavelength in the small coefficient direction will not be reduced on any grid level. As a final observation, note that there is always one or more large coefficient directions (as it is defined relatively) so that, in general, iterative solvers will be always be able to reduce *some* error component.

The objective, then, is to construct an alternative to Geometric Multigrid, based on these observations. To ensure that the Multigrid method can reduce all error components effectively, the relationship between the properties of the iterative solver, or "relaxation" scheme, and the various grids must be carefully examined.

A coarse grid, by virtue of being coarser, cannot reduce error components that have wavelengths shorter than its grid spacing. Therefore, the relatively finer grids must be able to handle these. A necessary coarse grid construction rule which, when applied between each grid level, ensures that the ability to reduce any specific error component is not lost can be described as follows. First identify the direction(s) for which the fine grid relaxation scheme is well able to reduce the error components. Then *create the coarse grid from the finer by only coarsening in this direction*. This is called Semi-Coarsening and forms the basis of this implementation of AMG. Thus the specific error component that has been lost by the coarsening process is exactly the one that can be handled on the relatively finer grid. The degree of coarsening possible is determined by how long an error component the iterative solver can effectively reduce.

To see how this principle can lead to a grid hierarchy, consider the very simple example shown in Fig. 1. Assume that the discrete equations on this grid are those arising from the previous 2D diffusion equation discretization. Starting with Grid 1, we now wish to generate the coarser meshes. On Grid 1, the

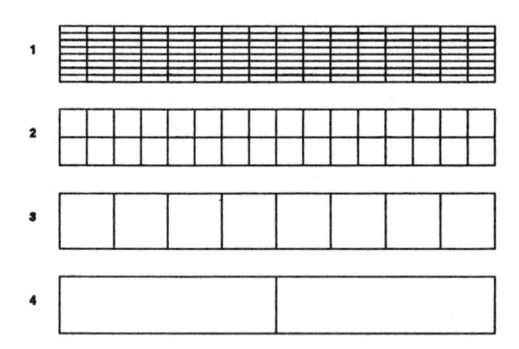

Figure 1: Repeated coarsening of a 2D grid

coefficients are largest in the vertical direction, implying that Gauss-Seidel can reduce errors in this direction well. Therefore, create a coarse grid by only coarsening in this direction. If the coarsening factor is 4, then the result is Grid 2. The degree of coarsening desirable is a trade-off: only coarsening by a factor of 2 will retain excellent relaxation scheme performance on each grid, but will require more storage; larger coarsening factors will diminish performance but require less storage (see Section 5 on the Relaxation Scheme for further discussion). Grid 2 has a unity aspect ratio and therefore leads to no preferential direction for the relaxation scheme. Coarsening in both direction, with a factor of 4 again, yields Grid 3. Now, as the system is 1D, there is only one choice for coarsening and the result is Grid 4.

Even though this example is trivial it illustrates several points. As the grid hierarchy is being created, the coarsening direction changes. Grids 1 and 3 were coarsened in opposite directions, for example, and Grid 2 used coarsening in both directions (like Geometric Multigrid). Even if the first grid is structured (i.e. logically Cartesian), the coarser grids need not be, which necessitates an unstructured-grid data structure. In this implementation, a linked-list data structure has been used, though others are possible. Consequently this Multigrid solver can solve equations on an unstructured grid, even though the examples presented herein use structured grids on the finest level. Lastly, and as will become clearer later, the cells in this implementation are, in fact, finite volumes, so that the coarse grid cells are larger blocks, physically covering the region of many fine grid finite volumes.

This semi-coarsening principle is not sufficient to prescribe a complete algorithm. An additional criterion used in the present implementation is designed to minimize the number of neighbouring connections on the coarser grids. The full algorithm, for example, is designed to recover the coarsening of Geometric Multigrid, which keeps the same number of neighbours on successively coarser grids, when applied to equations with isotropic coefficients.

Coarsening Algorithm Specification The full coarsening algorithm is presented with reference to Fig. 2. Two parameters control the blocking (i.e. the merging of fine grid finite volumes to coarsen the

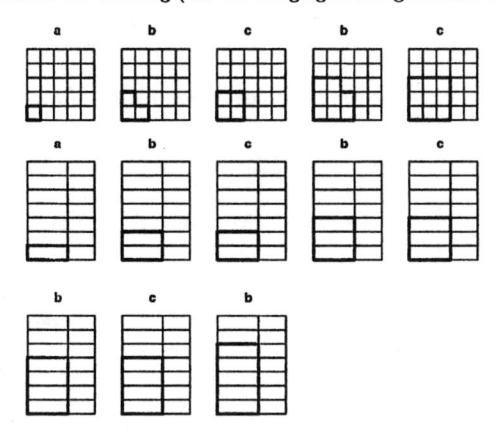

Figure 2: Examples of block formation

grid) of the finite volumes, **mgbmin** and **mgbmax**. The former is the minimum number of finite volumes to group together, the latter is the maximum. For the example in the figure, **mgbmin** is 5 and **mgbmax** is 9. The overall algorithm proceeds as follows:

1. Loop over all the nodes and for each equation, find its largest neighbouring coefficient, a_i^{max}, defined as,

$$a_i^{max} = \max(a_i^1, a_i^2, ..., a_i^{nb}) \tag{7}$$

   where the $a_i^1, a_i^2, ..., a_i^{nb}$ represent the list of all neighbouring coefficients for the equation at the given node i.

2. Loop over all the equations whose finite volume has not yet been assigned to a block; start a new block with the next un-assigned finite volume. This corresponds to step a in Fig. 2.

3. For each equation connected to the block in question, identify all the neighbouring coefficients that are "strong". A coefficient, a_i^n, is strong if,

$$a_i^n > \alpha \times a_i^{max} \tag{8}$$

   where α of $\frac{1}{3}$ has been found to work well.

4. Assign neighbouring finite volumes to the current block if they are connected with a strong coefficient, up to a maximum of **mgbmax** finite volumes. This corresponds to step b in Fig. 2.

5. Perform an accretion step. This is intended to make the blocks as "rounded" as possible, rather than extended with connections to many other blocks. The accretion step adds to the current block those finite volumes having 2 or more strong (as defined in step 3 above) connections to the block. Again this is up to a maximum of **mgbmax**. If any doubly (or more) strongly connected equations still exist, then perform the accretion step once more. This corresponds to step c in Fig. 2.

6. If the total number of nodes is less than **mgbmin**, then repeat steps 4 and 5 until at least **mgbmin** is reached.

The following are important factors about the coarsening algorithm necessary in the implementation:

- The steps given above did not distinguish between a strong connection from a block to a finite volume versus a strong connection from a finite volume to the block. Often one is strong, but the other is not. Whether such a case should result in the finite volume being added to the block is determined by requiring that the relaxation scheme is able to reduce all the error components within the resultant block.

- For the coupled momentum-mass set, the coarsening rule is based on the pressure coefficients in the mass equation, the a_{pp} coefficients in Eq. (2). This assumes that the behaviour of the relaxation scheme applied to the coupled set can be inferred from these coefficients. In practice, it appears that this is a reasonable assumption.

- In order to save storage, the minimum target block size is quite large. A value for mgbmin of 9 and for mgbmax of 13 is the default for the 3D coupled-set solution. This imposes a requirement that the relaxation scheme is able to reduce the error components in the large coefficient over several grid spacings, which is substantially longer than the "shortest wavelength" component.

Examples of other implementations of AMG methods can be found in Refs. [8] and [9].

One of the implicit premises of any Multigrid method is that the relatively simple iterative solver is retained, and a system built around it so that the whole functions well. An alternative approach, of course, is to devise more sophisticated iterative solvers, which do not suffer from the noted limitations. There are many such methods, outside the scope of this paper to review (e.g. conjugate gradient methods, Ref. [12]).

3. Restriction and Prolongation

Once a grid hierarchy is defined, the question of how the coarse grid equations are created and how their solution is used to reduce long wavelength errors on the finer grids must be answered.

The technique used is based on the Additive Correction Multigrid (ACM) method, Ref. [1]. This method was originally used within the context of Geometric Multigrid, but it can be applied identically once the grid hierarchy is established. Therefore the current implementation is sometimes referred to as the AMG/ACM method. The basic idea of ACM can be simply expressed: cause the coarse grid equations to be statements of conservation over a larger virtual finite volume, corresponding to the "blocks" described above. Then the job of the coarse grid equations is viewed as generating corrections to the fine grid solution so that conservation is enforced over the larger block. The advantages of this are several. Firstly, this is a physically meaningful principle, well suited to a finite volume method. Secondly, it results in coarse grid equations with well defined properties which is very useful in the debugging and validation process. Lastly, the implicit restriction and prolongation operators of the approach are exceedingly simple, making the method easy to implement and efficient to run.

Recall the form of the discrete equation:

$$\sum_{nb \text{ of } i} a_i^{nb} \phi_{nb} = b_i \ . \tag{9}$$

Now, define a correction to every fine grid solution variable,

$$\tilde{\phi}_i = \phi_i + \Phi_{I/i} \ . \tag{10}$$

where $\tilde{\phi}_i$ is the corrected solution and $\Phi_{I/i}$ is the correction for the block containing i, corresponding to finite volume I on the coarse grid. The requirement that the resultant residual sum is zero is,

$$\sum_{i \text{ in block } I} \tilde{r}_i = 0 \tag{11}$$

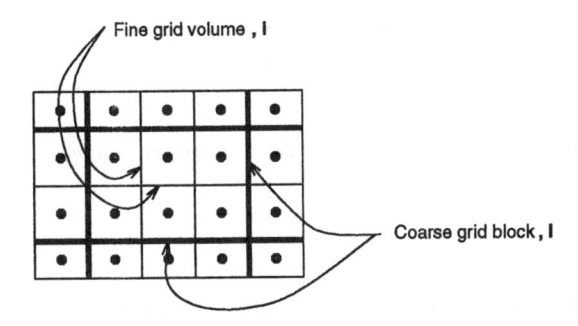

Figure 3: Coarse-to-Fine Grid Relationship

where $\tilde{r}_i$ is the new residual, and the summing is done over the fine grid finite volumes that are in block I. Refer to Fig. 3 for the relationship between equation i and block I.

Each new residual is,

$$\tilde{r}_i = b_i - \sum_{nb \text{ of } i} a_i^{nb} \tilde{\phi}_{nb} \tag{12}$$

so Eq. (10) can be substituted into Eq. (12) to yield,

$$\tilde{r}_i = r_i - \sum_{nb \text{ of } i} a_i^{nb} \Phi_{NB/nb} \tag{13}$$

where r_i is the original fine grid residual (defined as in Eq. (12) with ϕ instead of $\tilde{\phi}$), and $\Phi_{NB/nb}$ is the correction for the fine grid finite volume nb.

Next, Eq. (11) is imposed by summing Eq. (13) over the fine grid block and setting the result to zero, which gives, after minor re-arrangement,

$$\sum_{i \text{ in block } I} \sum_{nb \text{ of } i} a_i^{nb} \Phi_{NB/nb} = \sum_{i \text{ in block } I} r_i \;. \tag{14}$$

Grouping together coefficients that multiply the same Φ, this can be re-written as,

$$\sum_{NB \text{ of } I} A_I^{NB} \Phi_{NB} = B_I \tag{15}$$

where the NB now refers to the coarse grid neighbours of coarse grid finite volume I, including the I location. The values of A_I^{NB} and B_I follow directly from Eq. (14). The right-hand-side, for instance, is calculated as

$$B_I = \sum_{i \text{ in block } I} r_i \tag{16}$$

or in other words, the right-hand-side of a coarse grid equation is the sum of the fine grid residuals in the block that defines the coarse grid finite volume. The above definitions of A and B imply the restriction and prolongation operators. Equivalent operators are used by the Cell-Centered Multigrid method.

Equation (15) is written deliberately to look like Eq. (9), and indeed, the coarse grid equations are very similar to the fine grid equations in structure and connectivity, except that there are fewer of them. One important point to note is that in this form of Multigrid, the coarse grid equations are completely computable in terms of the finer grid equations. Thus the coarse grids themselves do not have to be actually created, only the coarse grid system of equations needs to be. With ACM, the coarse grid equations are fully determined from the blocking, or coarsening process, so this is the only aspect of the coarse grids that needs to be stored.

To summarize the information flow: Equation (15) is the coarse grid equation, with its coefficients and right-hand-side being a restriction of the fine grid residuals and coefficients respectively. Equation (10) represents the prolongation step in which the fine grid solution is updated by a correction from the coarse grid.

The next component of a complete Multigrid algorithm addresses how this information flow is managed for a hierarchy of grids.

3. Cycle Type

Flexible Cycle Multigrid One of the earliest Multigrid cycles used is called the "Flexible Cycle". After a restriction or prolongation step, the relaxation scheme iterates on the given grid level. Convergence rates are monitored and target residuals set to determine when control should pass up to a finer grid via another prolongation or down to a coarser grid via another restriction. The number of relaxation sweeps and grid sequence that results will thus vary. The main drawback to this cycle type is that basing control decisions on a residual norm (such are RMS residuals) is sometimes not an accurate measure of the relaxation scheme's performance and often the flexible cycle will end up doing too much or too little work on a given grid level.

V–Cycle One fixed cycle Multigrid sequence is the "V–Cycle". Starting with the finest grid and then moving down, a pre-determined number of relaxation scheme sweeps is made on all the grids. At the bottom, a different number of sweeps is made (or a direct solver is used). Control is passed all the way up, with another fixed number of sweeps on each grid. This is illustrated in Fig. 4. The number of sweeps going down is given by parameter "m", at the bottom "l", and going up "n". If the target reduction has

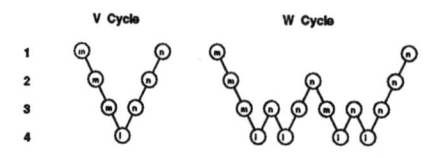

Figure 4: V and W fixed cycle Multigrid

not been met on the finest grid after one cycle, the whole algorithm is repeated again. This cycle type can work well for purely elliptic scalar systems.

W–Cycle For the coupled equation set under consideration a third cycle type has been found to be the most desirable, a modified fixed cycle called the W–Cycle. Basically, as each grid is visited on the way up, control is passed down again one time before it is allowed to continue up. This is illustrated also in Fig. 4. With more levels, the pattern is recursively repeated.

The main advantage of the W–Cycle is that each grid level gets to pass its residuals down to the coarser grids twice and receive corrections twice. In the current CFD code, due to memory restrictions, the block sizes have to be quite large, so this approach gives the Multigrid algorithm more opportunity to do work, and ensure that the error components are sufficiently reduced. Again, if the fine grid has not met its target reduction after one cycle, the whole cycle is repeated.

Default sweep counts are 1 relaxation sweep after restriction and 3 after prolongation. The unusually high number after prolongation is primarily required due to the large block sizes. On the coarsest grid (defined when fewer than 20 nodes are reached) a direct solver is used.

4. Relaxation Scheme

The relaxation scheme used must be able to reduce at least one error component on any given grid well. As discussed, the coarsening algorithm is designed to create a coarser grid from a finer, by removing the ability to represent precisely that component. It is in this way that these two aspects of the Multigrid method must be matched.

On a typical Navier-Stokes grid, there are very large aspect ratios in all dimensions. Therefore, it is usually the case that most applications of coarsening will result in blocks that are 1D stacks of plate-like finite volumes (e.g. the second example in Fig. 2). The larger these blocks can be made, in other words the larger the coarsening factor, then the less storage and CPU will be required for the coarser grids. A good relaxation scheme then, is one that is well able to reduce error components in the large coefficient direction of a wavelength extending as far as possible. Incomplete Lower Upper factorization is such a scheme.

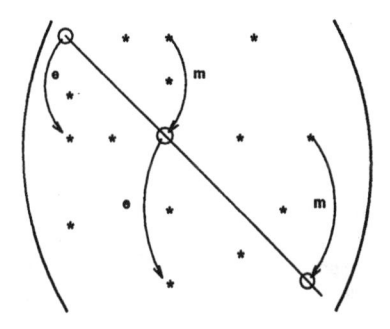

Figure 5: Incomplete Lower Upper Factorization TDMA Similarity

The current implementation is the simplest possible, with no additional fill-in being generated and only coefficients on the main diagonal being modified. This means that the storage is 16 words per node (for the modified 4×4 central coefficient. This scheme is sometimes referred to as ILU0 and works exactly as coupled Gaussian Elimination except that only the diagonal coefficient block is ever modified.

One of the advantages of this method is that in the limit of strong coefficients in one direction, it becomes an exact Tridiagonal Matrix solver (TDMA). In the general non-limiting case, it is still true of ILU that it can reduce quite long wavelength error components in the strong coefficient direction. This is explained with reference to Fig. 5. The main diagonal is a solid line, off-diagonal elements are marked as an asterisk, three selected modified central coefficients are marked with and empty circle. The "e" arrow denotes the elimination step and the "m" arrow denotes the modification of the diagonal coefficient. Independent of node numbering (i.e. equation ordering), if the coefficients connecting the three marked equations are relatively very large, that sub-system looks like a tri-diagonal system and is solved directly by the ILU algorithm. This aspect of ILU is one of the reasons it is well suited to the current implementation of AMG, with its large block sizes.

Another advantage of ILU over Gauss-Seidel is its implicitness. Much of the coupling between the momentum and continuity equations is through first derivatives which tend to create off-diagonal coefficients − coefficients that are at least partly involved in the factorization.

Finally, the coupled nature of the relaxation scheme is critical for one of the application areas of the CFD code. In rotating frames of reference, the Coriolis term in the momentum equations introduces extremely strong coupling between the components of momentum. If this were not handled implicitly, performance would degrade as rotation speeds increased.

5. Demonstration

Varying the Grid Size If a Multigrid method is working properly, then the work required to solve the equations to a given tolerance should scale linearly with the number of degrees of freedom. This property is demonstrated for the current solver on a turbulent, incompressible, internal flow problem. Figure 6

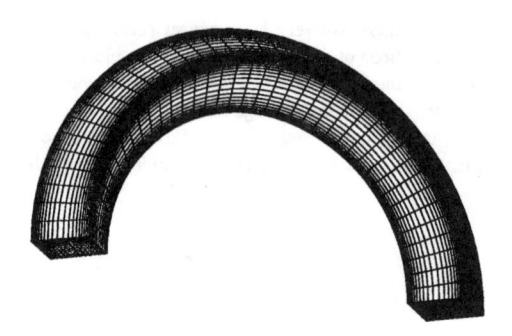

Figure 6: External Grid for Demonstration Problem 1

shows a $20 \times 20 \times 40$ grid of a curved duct. Velocity and turbulence levels are specified at the inlet, the outlet conditions are determined by the solution. Based on the width, the Reynolds number is about 100 000. An incompressible flow is chosen as this is the most elliptic. This particular flow has streamwise and radial pressure gradients and strong secondary flows, so even though the geometry is simple, the flow is not trivial.

Four grids were used with dimensions $(10 \times 10 \times 20)$, $(20 \times 20 \times 40)$, $(30 \times 30 \times 60)$ and $(40 \times 40 \times 80)$, giving grids of 2 000, 16 000, 54 000 and 128 000 nodes respectively. The grid expansion factors used to resolve the boundary layers result in severe cell aspect ratios in many parts of the grids.

The solver performance on the different grids was compared as follows. For each grid, the same initial guess was given and ten time steps, or coefficient re-linearizations with under-relaxation, were run. On the 11th time step, the solver was driven to a target reduction of about 1.E-5 (i.e. five orders of magnitude RMS residual reduction) and the convergence rates were plotted. Default control parameter values were used, except for the number of post-prolongation sweeps which was 2 (at the time of the runs, that was the default). The results for all four grids are shown in Fig. 7. The vertical axis is the logarithm to the base 10 of the root-mean-square continuity residual and the horizontal axis is the work units. A work unit is defined as a finest grid ILU sweep equivalent. For example, on the finest grid, one ILU sweep is 1 work unit; on a coarse grid with one eighth the number of nodes, one ILU sweep would be worth 0.125 work units. The slopes of these 4 curves are very similar, indicating near linear performance. This performance should also be evaluated in light of the block size used. Typical sizes vary between 7 and 10, thus the coarsening rate is very high — this gives the relaxation scheme a much tougher job than a smaller block size would. In practice, the target residual reduction that the solver needs to reach each non-linear iteration is only 0.1. This is typically reached in 6 to 12 work units, depending on the problem type. Since the work units accumulate in multiples of the total number of work units required by one W-Cycle, the target residual reduction is usually surpassed, so that linear performance (i.e. the same work unit count for differing grid sizes) is observed in practice.

Varying the Number of Grid Levels When all grid levels are present, and with a direct solver on the coarsest level, all error components ought to be effectively reduced by the Multigrid solver. If they are not all present, then some long wavelength error will not be reduced, which typically results in

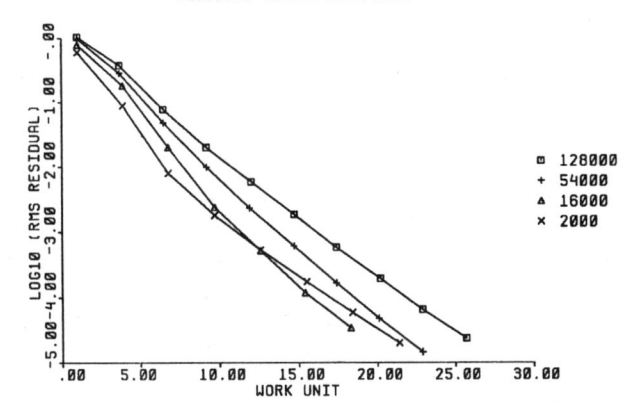

Figure 7: AMG/ACM solver performance for 4 grids

stalled or reduced convergence rates. These effects are demonstrated on a more complex and realistic demonstration problem.

The second demonstration problem is a twin bladed inducer, a typical rotating machinery problem. The flow is again incompressible and turbulent and is now in a rotating frame, which means Coriolis and Centripetal forces are present. A 4-grid multi-block grid is used, including grid-embedding (local refinement), non-parallel periodicity and resolved tip-clearance. The grid has severe aspect ratios and non-orthogonality and uses 130 000 nodes. This means 520 000 simultaneous degrees of freedom to solve. The problem was computed with TASCflow3D, Version 2.3, with default settings. The grid on the inducer surface is shown in Fig. 8. Again the problem was run for about 10 iterations, and then repeatedly run

Figure 8: Surface Grid for Demonstration Problem 2

from that point, while varying the number of grid levels. The Multigrid solver convergence rates are

213

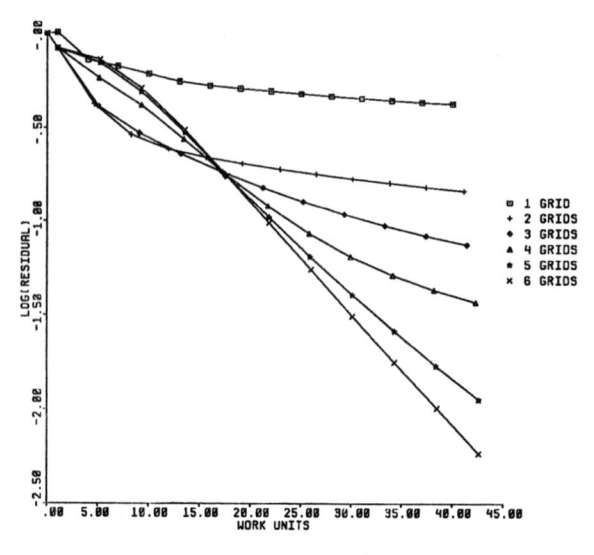

Figure 9: Convergence Rates for Varying Number of Grid Levels

shown in Fig. 9. For the cases with less than 6 grid levels, ILU was run with 5 sweeps on the coarsest grid level. Note for these cases how the convergence rates are slowing down, and how for the full case the rate is high and constant, as expected. Asymptotic convergence rate with all grids was 16 work units per factor of 10 residual reduction. In practice, with all error components present, a factor of ten reduction is achieved in 9.4 work units. On an IBM 37T RS/6000 Unix Workstation, this factor of ten reduction took 107 CPU seconds. (Note: total outer-loop iterations to convergence was 93, with 30% of the total CPU for Multigrid, for a total CPU time of 9.5 hours.) These results are typical for a very wide range of problems.

6. Limitations of the Method

1. Due to the requirements of ACM, all the equations on the fine grid must be conservative finite volume equations. This means that, for instance, Dirichlet boundary condition information must be substituted into the finite volume equations and eliminated before the Multigrid solver can be used.

2. The coupled ILU relaxation scheme requires a sufficiently diagonally dominant equation set to be convergent. This has required only minor adjustments of discretization as standard implementations of finite volume methods tend to be quite diagonally dominant to begin with.

3. It is possible to construct systems of equations where the a_{pp} coefficients are not good indications of the iterative behaviour of the coupled system. Some Stokes flow limits appear to exhibit this effect. Fortunately, the intended application for this software rarely requires the modelling of such flows. When this has been required, reducing the minimum block size restores the high performance level.

4. Storage is high. Fine grid coefficient storage is 112 words per node plus typical Multigrid storage overhead of about 40 words per node.

7. Conclusions and Acknowledgments

The Algebraic Multigrid implementation presented constitutes a viable choice for a coupled linear equation solver. The method is extremely robust, and has been validated over a period of 2 years on hundreds of flow problems. Its reliability rate, with default settings, is now virtually 100%, with most of this experience occuring in the hands of commercial users. Minor adjustment of control parameters (e.g. reducing the minimum block size) has always yielded convergence for the few cases that did not perform well initially.

The method is easy to implement. In a new unstructured grid platform currently under development, the whole linear solver, for scalar equations and the coupled mass-momentum set, takes about 1000 lines of Fortran. For the current class of CFD technology in use at ASC, the linear solver is no longer an area of active research or continued code development.

The author of this paper would like to express his appreciation and thanks to Brad Hutchinson, George Raithby, Paul Galpin and Jeff Van Doormaal for countless discussions and insights into Multigrid gained over many years. The support of Ontario Hydro and Aerojet during some of the early development phases is also gratefully acknowledged.

References

[1] B. R. Hutchinson and G. D. Raithby, "A Multigrid Method Based on the Additive Correction Strategy", *Numerical Heat Transfer*, Vol. 9, pp. 511-537, 1986.

[2] B. R. Hutchinson, P. F. Galpin and G. D. Raithby, "Application of Additive Correction Multigrid to the Coupled Fluid Flow Equations", *Numerical Heat Transfer*, Vol. 13, pp. 133-147, 1988.

[3] "TASCflow Theory Documentation", Advanced Scientific Computing Ltd., October, 1992.

[4] M. E. Thomas, N. R. Shimp, M. J. Raw, P. F. Galpin and G. D. Raithby, "The Development of an Efficient Turbomachinery CFD Analysis Procedure", AIAA/ASME/SAE/ASEE 25th Joint Propulsion Conference, Monterey, CA, July 10-12, 1989.

[5] A. Brandt, "Multi-Level Adaptive Solutions to Boundary-Value Problems", *Math. Comput.*, Vol. 31, pp. 333-390, 1977.

[6] W. Hackbusch, "The Fast Numerical Solution of Very Large Elliptic Differential Equations", *J. Inst. Maths. Applics.* 26 (1980), 119-132.

[7] M. J. Raw, P. F. Galpin. B. R. Hutchinson, G. D. Raithby and J. P. Van Doormaal, "An Element-Based Finite-Volume Method for Computing Viscous Flows", submitted for publication to the International Journal for Numerical Methods in Fluids.

[8] A. Brandt, S. McCormick, J. Ruge, "Algebraic multigrid for sparse matrix equations", *Sparsity and Its Applications* (D. J. Evans, Editor), Cambridge University Presss, 1984.

[9] J. W. Ruge, and K. Stüben, "Algebraic Multigrid", *Multigrid Methods* (S. F. McCormick, Editor), Frontiers in Applied Mathematics, SIAM, 1987.

[10] J. P. Van Doormaal and G. D. Raithby, "The Segregated Approach to Predicting Viscous Compressible Fluids Flows", *Journal of TurboMachinery*, vol. 109, pp. 268-277, 1987.

[11] W. L. Briggs, *A Multigrid Tutorial*, SIAM, Philadelphia 1987.

[12] P. Sonneveld, "CGS, A Fast Lanczos-Type Solver for Nonsymmetric Linear Systems", *SIAM J. Sci. Stat. Comput.*, Vol. 10, No. 1 pp. 36-52, 1989.

Robust Multigrid Methods for the Incompressible Navier-Stokes Equations

Henrik Reichert and Gabriel Wittum
Institut für Computeranwendungen (ICA/Numerik), Universität Stuttgart
Pfaffenwaldring 27, D-70 569 Stuttgart

Summary

We introduce and compare several smoothers for the stationary incompressible Navier-Stokes equations in primitive variables on unstructured and locally refined grids. Special emphasis is laid on the robustness of the linear multigrid solver in view of large convecting velocities and bad aspect ratios in the grid. We describe a new streamwise numbering algorithm for the unknowns with a special treatment of the cyclic dependencies due to vortices. Further we describe the implemented smoothers and show diagrams of the convergence rates (per grid level) versus the aspect ratios of the elements. Additionally we present a comparison of some theoretical results for the Stokes-solution in a corner with our calculations.

1 Introduction and the Notion of "Robustness"

The final goal of our recent work is a highly efficient solver for general CFD problems by using multigrid methods on locally refined and adapted unstructured meshes. The present paper is concerned with the first step on this way we have reached at. We discuss the following problem: Since we use a linear multigrid as inner solver we want it to be able to yield good convergence rates for all problems the discretization passes to it. Especially the cases of large convecting velocities (see section 2) and bad aspect ratios (see section 3) of the elements, which appear frequently in boundary layer fitted grids, should cause no severe problems for the smoother.

To achieve this we tried the following strategy: Take some variant of ILU_β as smoother and possibly decouple the equations by a transforming approach [Wi1]. Then the algorithm should prove to be robust in view of bad aspect ratios.

For the convection dominated case we will choose a special streamwise numbering of the unknowns.

2 Robustness in View of Large Convecting Velocities

2.1 Description of the numbering algorithm

Let us switch off the diffusion for the time being. Due to the quasi Newton linearization and to the upwind scheme a given node depends only on its upwind neighbours. If we can find a global ordering of the unknowns in a way that the stiffness matrix has nonzero entries only in the lower triangle, then of course we will be able to solve the system of equations in one step even by a Gauß-Seidel method. Unfortunately in most of the relevant cases there are vortices in the flow and therefore cyclic dependencies. But nevertheless we will obtain good results if we introduce arbitrary cuts through the vortices by removing just enough of the "cyclic" nodes to get rid of the cylic dependencies. We start the numbering at the inlet going in layers downstream but taking only nodes depending on the already numbered ones (those nodes will form the beginning of our new list). In a similar way we go upstream from the outlet (those nodes will make up the end of our new list). Finally we are left with nodes with cyclic dependencies. We cut it, appending those nodes to the beginning of our list. Then steps one to three are repeated until every node is processed.

After this rough description we introduce the following algorithm that does the job (the basic ideas can be found in [BW]):

```
while (some nodes are not numbered)
{
     /* find FIRST set */
     do {
          Find all nodes having at most such UPWIND
               neighbour nodes that are already numbered
          Number them starting with the least number not used yet.
     } while (no further nodes are found).

     /* find LAST set */
     do {
          Find all nodes having at most such DOWNWIND
               neighbour nodes that are already numbered
          Number them starting with the greatest number not used yet.
     } while (no further nodes are found).

     /* find CUT set (only cyclic dependencies are left) */
     Cut one vortex transverse to the streamlines.
     Number the nodes on this cut starting with the least number
          not used yet.
}
```

Algorithm 1: Streamwise numbering

An example of the resulting sparsity pattern of the stiffness matrix could look like this:

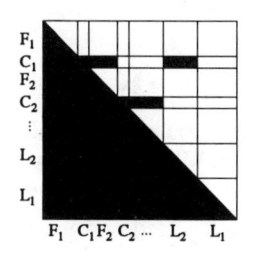

For the Backward Facing Step this could look like:

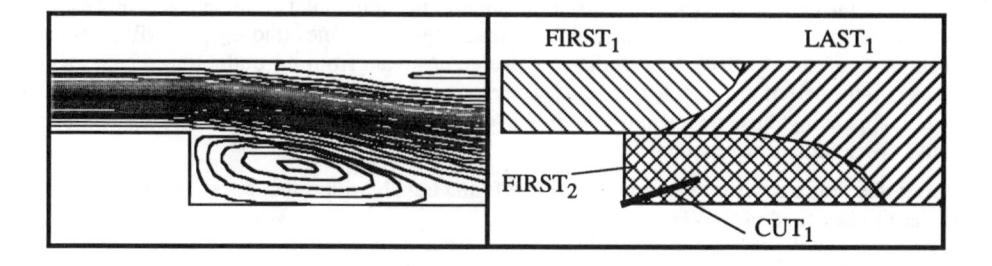

Fig. 1: Streamwise numbering of the Backward Facing Step.

For the Driven Cavity at a Reynolds number of 500 as a more complicated example the FIRST, CUT and LAST sets look like:

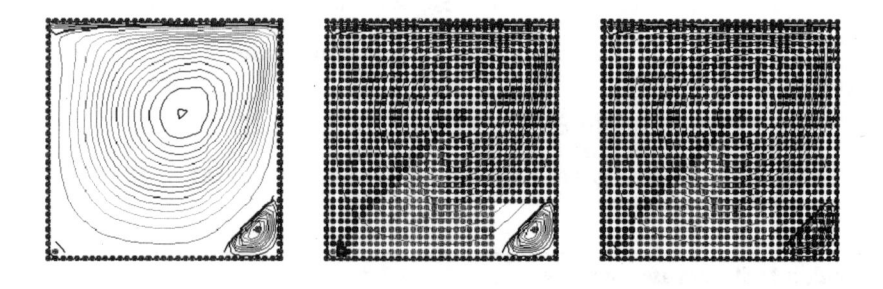

Fig. 2: Left side: $FIRST_1$ (marked nodes): Dirichlet boundary nodes ($LAST_1$ is empty)
middle: CUT_1 (black nodes), $FIRST_2$ (medium gray nodes) and $LAST_2$ (gray nodes)
right side: CUT_2 (black), $FIRST_3$ (gray) ($LAST_3$ is empty).

2.2 Results

As test examples for the efficiency of this numbering strategy we chose a simple Pipe flow and the Backward Facing Step from above. We calculated a velocity field for Re=100 (called $\vec{u}_{old}$) and treated the equations linearized in $\lambda \vec{u}_{old}$ with our smoother:

$$-\Delta \vec{u} + Re\,(\lambda \vec{u}_{old} \cdot \nabla)\, \vec{u} + \nabla p = 0$$
$$\nabla \vec{u} = 0 \qquad\qquad (2.2.1)$$

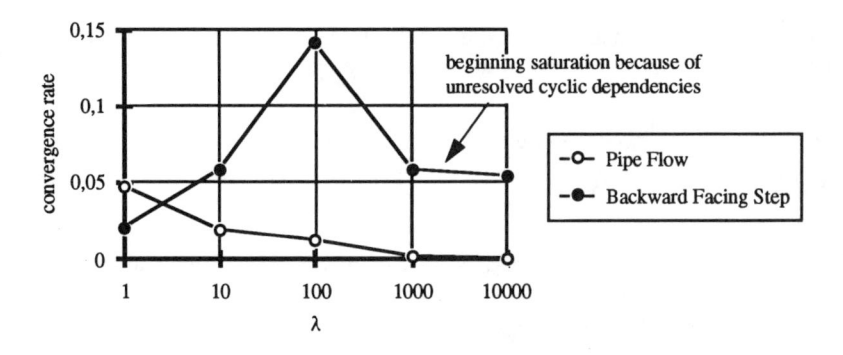

Fig. 3: Convergence rate over λ for streamwise numbering.

3 Robustness in View of Large Aspect Ratios

For this section we want to introduce the following simplifications: a) we stick to the Stokes equation (pure diffusion) to avoid the mixing of various effects and b) we use a rectangular equidistant grid with meshsizes h_x, h_y in x- and y-direction resp. for the calculations since we want to have only elements of one type with the same aspect ratio $s = h_x/h_y$.

3.1 Description of the implemented smoothers

For the classification and description of the implemented smoothers we need to introduce some notation:

We have to solve the linear system of equations

$$Kx = b \tag{3.1.1}$$

$$x = \begin{bmatrix} u \\ v \\ p \end{bmatrix} \qquad b = \begin{bmatrix} f_x \\ f_y \\ 0 \end{bmatrix} \tag{3.1.2}$$

with the stiffness matrix

$$K = \begin{bmatrix} -\Delta & 0 & \frac{\partial}{\partial x} \\ 0 & -\Delta & \frac{\partial}{\partial y} \\ \frac{\partial}{\partial x} & \frac{\partial}{\partial y} & -c_0 \Delta \end{bmatrix} \tag{3.1.3}$$

where c_0 is the stability parameter introduced in a natural way by the discretization procedure. It behaves like $O(h^2)$ and is necessary to supress artificial pressure oscillations.

Once the ordering of the nodes in the grid is given there are two obvious ways to remove the remaining arbitraryness in the order of the $3N$ unknowns:

- $u_1, \ldots, u_N, v_1, \ldots, v_N, p_1, \ldots, p_N$
- $u_1, v_1, p_1, \ldots, u_N, v_N, p_N$

with their associated block structures, the first of which we will refer to as *equationwise* ($3\ N \times N$ blocks), the ladder as *nodewise* ($N\ 3 \times 3$ blocks).

Classification of implemented smoothers:

variant / ordering	equationwise	nodewise
scalar	ILU	–
block	Gauß-Seidel (inner solver ILU)	ILU, Gauß-Seidel (inner solver exact)

Additionally we implemented two (right) transforming smoothers.

The transforming iteration step reads:

$$x^{i+1} = x^i + \overline{K}M^{-1}(b - Kx^i) \qquad (3.1.4)$$

where M is a regular decomposition of

$$K\overline{K} = M - N \qquad (3.1.5)$$

with some rest matrix N.
Now choose for $\overline{K}$

- **distributive ILU:**

$$\overline{K} = \begin{bmatrix} 1 & 0 & \frac{\partial}{\partial x} \\ 0 & 1 & \frac{\partial}{\partial y} \\ 0 & 0 & \Delta \end{bmatrix} \Rightarrow K\overline{K} = \begin{bmatrix} -\Delta & 0 & \{\frac{\partial}{\partial x}\Delta - \Delta\frac{\partial}{\partial x}\} \\ 0 & -\Delta & \{\frac{\partial}{\partial y}\Delta - \Delta\frac{\partial}{\partial y}\} \\ \frac{\partial}{\partial x} & \frac{\partial}{\partial y} & \{\Delta - O(h^2)\} \end{bmatrix}$$

if we neglect the commutators (in curly brackets) coupling the momentum equations to the pressure and if we then perform an ILU on the diagonal blocks we yield the TILU (transforming ILU) by Wittum [Wi3], which essentially is a DGS-like algorithm (Distributive Gauß-Seidel, introduced by Brandt/Dinar [BD]).

Or take

- **SIMPLE-ILU:**

$$\bar{K} = \begin{bmatrix} 1 & 0 & D^{-1}\frac{\partial}{\partial x} \\ 0 & 1 & D^{-1}\frac{\partial}{\partial y} \\ 0 & 0 & 1 \end{bmatrix} \Rightarrow K\bar{K} = \begin{bmatrix} -\Delta & 0 & 0 \\ 0 & -\Delta & 0 \\ \frac{\partial}{\partial x} & \frac{\partial}{\partial y} & \{-c_0\Delta - \nabla^T D^{-1}\nabla\} \end{bmatrix}$$

this holds exactly if we take D to be the Laplacian. Now we leave the zero entries in the upper triangle but replace D by some easily invertible approximation.

We took as a very crude approximation $D = \text{diag}(\Delta)$. This leads to the well known SIMPLE-method by Patankar/Spalding [PS]. The resulting system we treated with ILU (instead of the original Gauß-Seidel).

As a much more elaborate version we treated the Schur complement with the Frequency-Filtering method by Wittum [Wi3]. There an approximate matrix is defined by its sparsity pattern and by requiring that it acts on a certain subspace exactly like the original matrix. We want to emphasize that our SIMPLE smoothers have nothing in common with the original SIMPLE but the very basic idea of choosing an approximation for D.

3.2 Results

To test the performance of our smoothers we measured the dependence of the convergence rate (mean value over 10 V-cycles, denoted as κ_{10}) on a rectangular grid for the Driven Cavity on the aspect ratio (the y-coordinates of the grid are scaled by powers of 2)

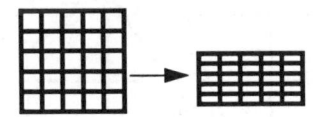

as well as the dependence of the mean convergence rate on the meshsize of the finest grid (h-independence expected).

We did the tests with various parameters as there are damping factors ω and the β-parameters of the ILU$_\beta$ each chosen independently for the velocity and the pressure.

All calculations where done with V-type multigrid cycles and one pre- and two post-smoothing steps.

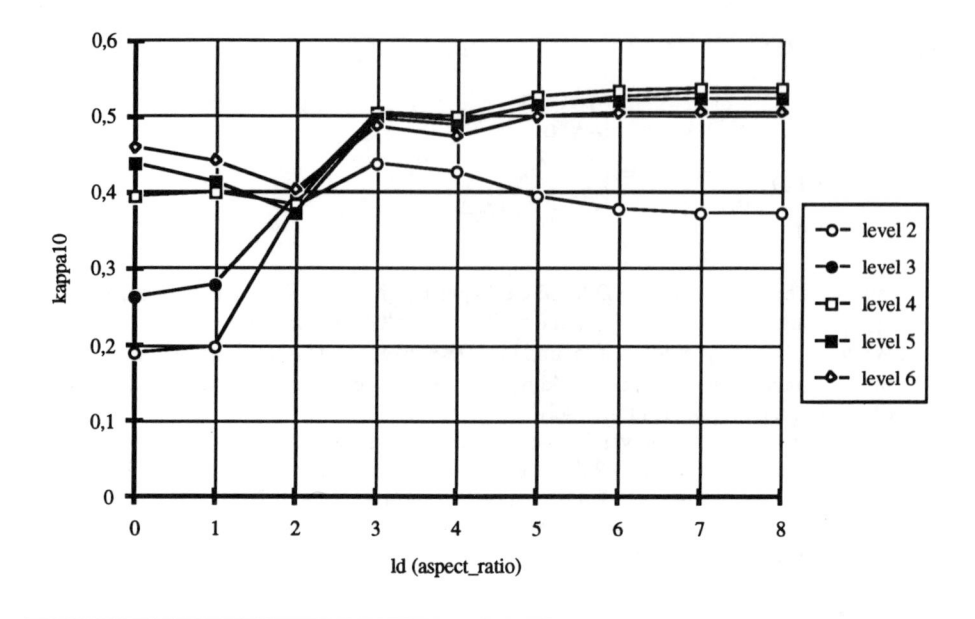

Fig. 4: Nodewise block Gauß-Seidel (inner solver exact), $\omega_u=1.4$, $\omega_p=0.5$.

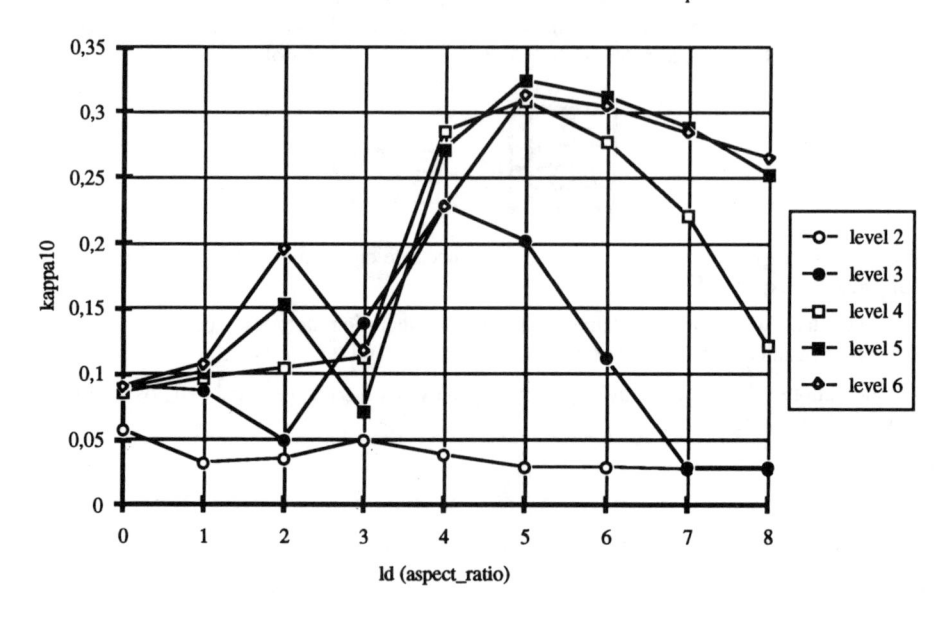

Fig. 5: Nodewise block ILU (inner solver exact), $\beta_u=0$, $\beta_p=0$, $\omega_u=1$, $\omega_p=1$.

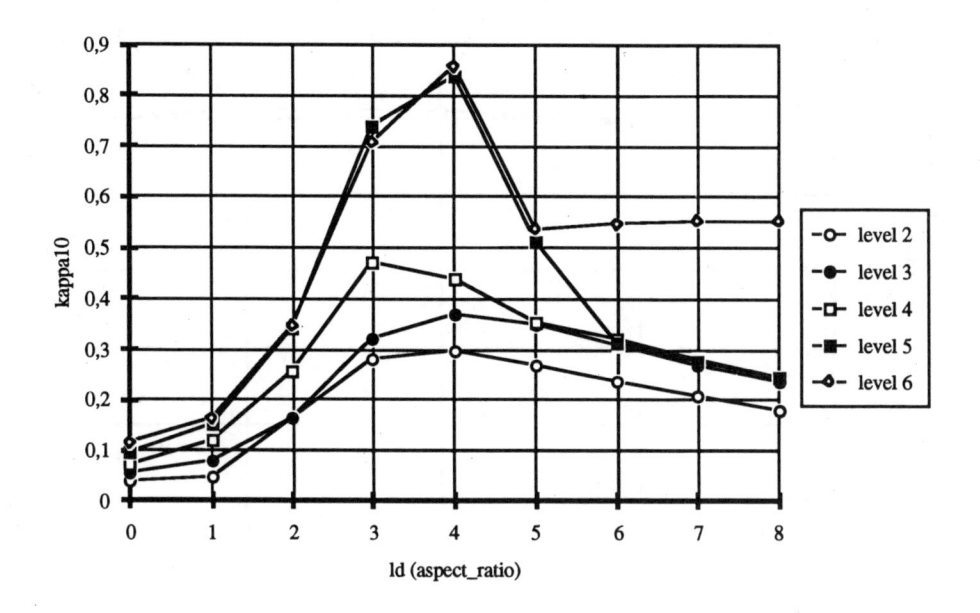

Fig. 6: Scalar equationwise ILU, $\beta_u=0$, $\beta_p=0$, $\omega_u=1$, $\omega_p=0{,}8$.

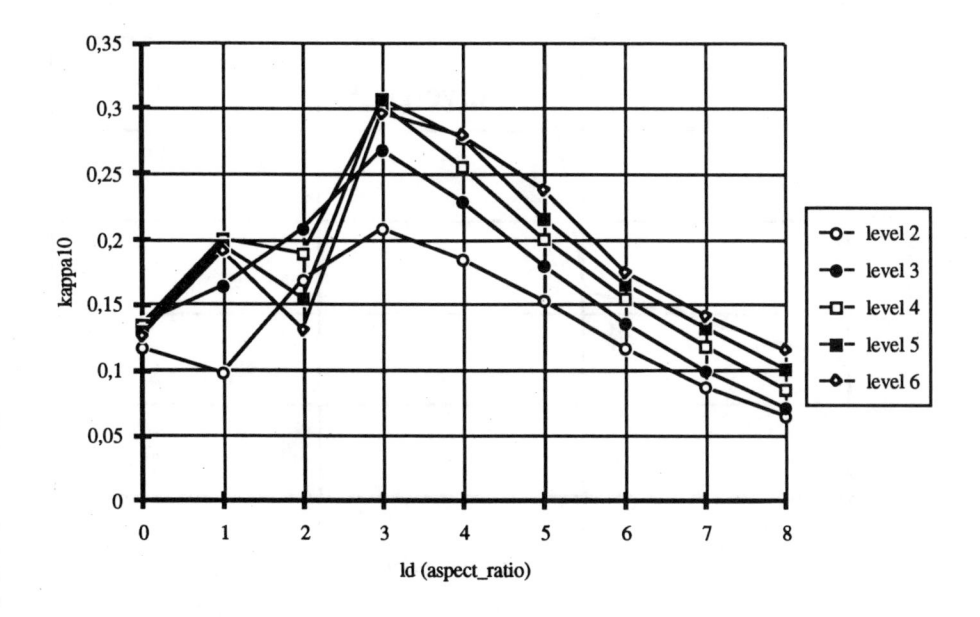

Fig. 7: Equationwise block Gauß-Seidel (inner solver ILU), $\beta_u=0$, $\beta_p=10$, $\omega_u=1$, $\omega_p=0.6$.

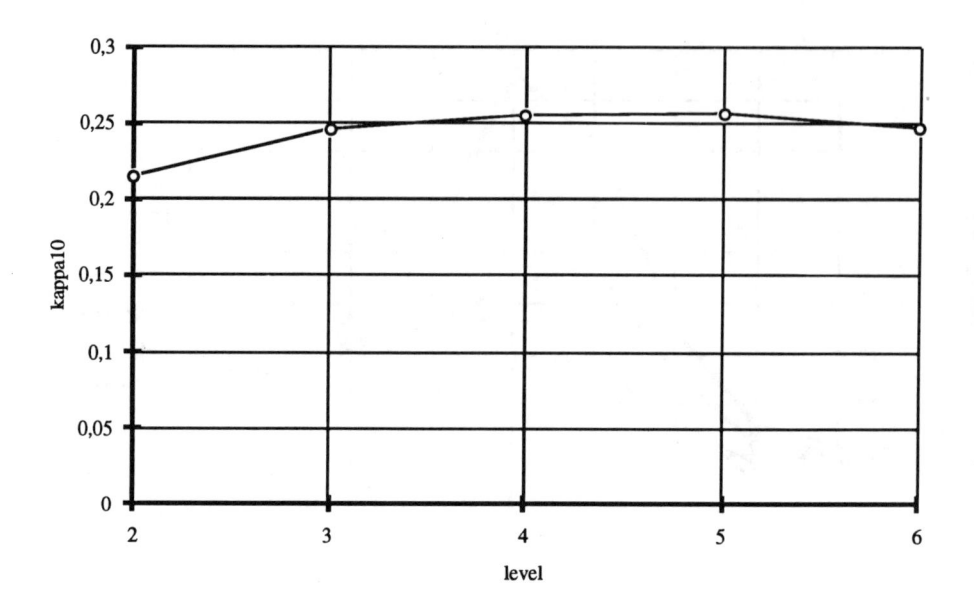

Fig. 8: Distributive ILU, $\beta_u=0$, $\beta_p=0$, $\omega_u=1$, $\omega_p=1$.

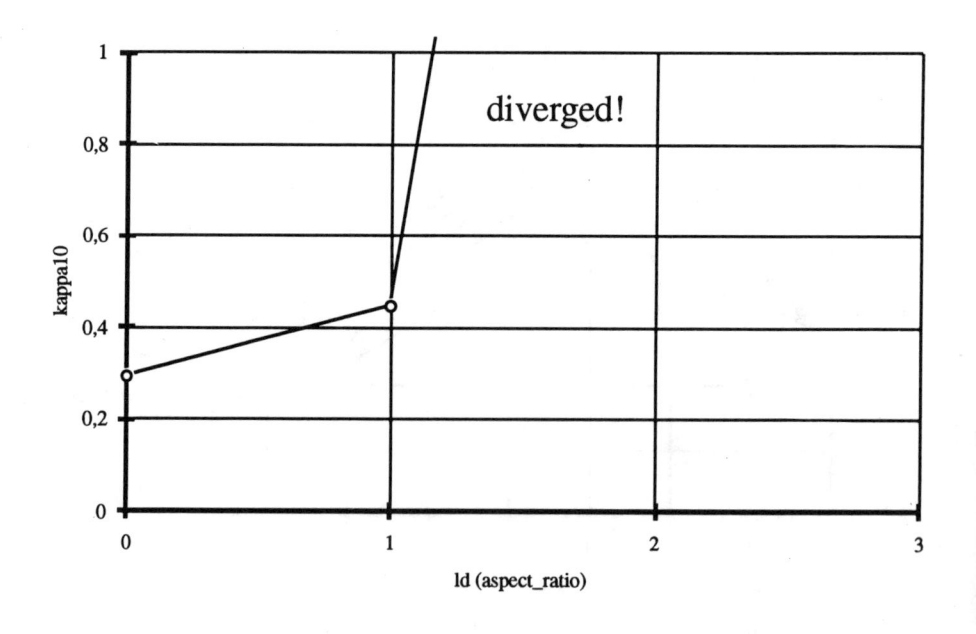

Fig. 9: Distributive ILU, $\beta_u=0$, $\beta_p=0$, $\omega_u=1$, $\omega_p=1$.

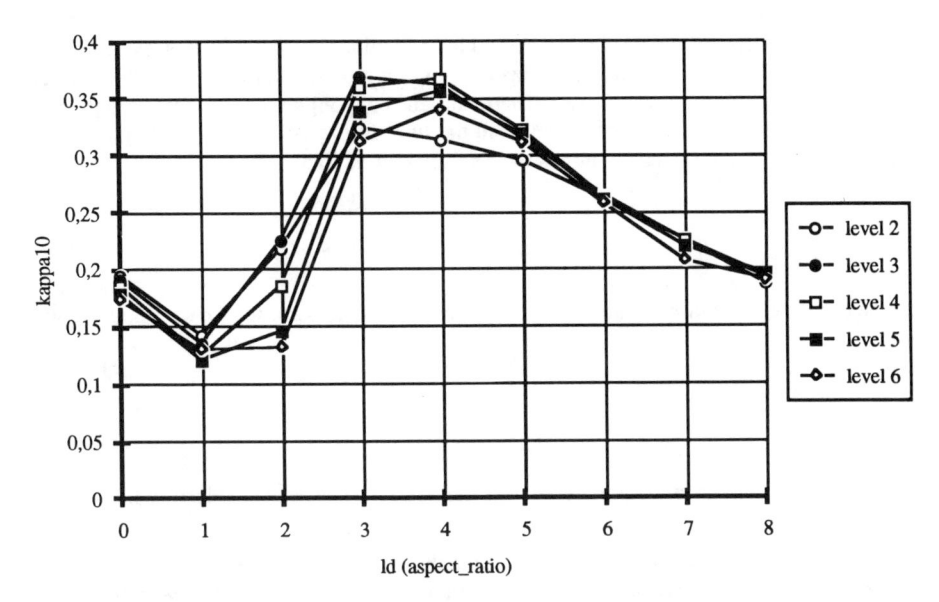

SIMPLE-ILU, $\beta_u=0$, $\beta_p=10$, $\omega_u=1$, $\omega_p=1$.

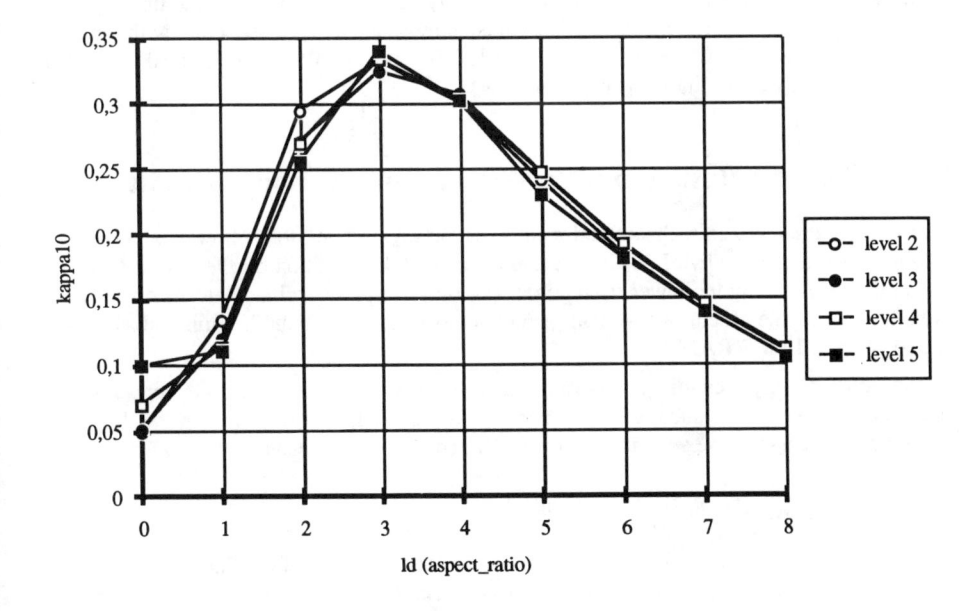

Fig. 10: SIMPLE-Frequency-Filter, $\omega_u=1$, $\omega_p=1$.

For the interpretation of the results one should have in mind that the finite volume discretization – similar to the finite element discretization – will not be singularly perturbed in the sense the finite difference discretizations are. There only the coupling to the two next neihgbours remains finite whereas the other ones vanish in the limit of the aspect ratio tending to ∞. That leads to an asymptotic stencil for minus the Laplacian of the type

$$\frac{1}{h_y^2}\begin{bmatrix} 0 & -1 & 0 \\ 0 & 2 & 0 \\ 0 & -1 & 0 \end{bmatrix}. \tag{3.2.1}$$

On the other hand our discretization yields (no factor of $1/h^2$ because of the volume integration)

$$\frac{h_x}{8h_y}\begin{bmatrix} -1 & -6 & -1 \\ 2 & 12 & 2 \\ -1 & -6 & -1 \end{bmatrix}. \tag{3.2.2}$$

Note the "wrong" signs in the horizontal neighbours of the center which destroy the M-matrix property!

This behavior of the stencils explains why the Gauß-Seidel convergence rates are still bounded below 0.6 while in the scalar case with finite differences the convergence rates approach 1 for aspect_ratio → ∞.

Unfortunately, the effects are rather complex, mainly due to the coupling of the three differential equations. But experiment shows that the equationwise block Gauß-Seidel with inner solver ILU, the SIMPLE-ILU as well as the SIMPLE-Frequency-Filter prove to tend to solve the equations exactly in the limit aspect_ratio → ∞.

4 Stokes-Solution in a Corner – a Comparison with Theory

For a long time there exist theoretical results for the asymptotic Stokes flow in a corner with arbitrary opening angle [Mo]. We have calculated the Stokes solution for the Driven Cavity on 19 grid levels refining the lower right corner (*ug* only stores the refined patches on higher levels). On an Apple Macintosh we had a computational time of about 2.3 min and storage requirements of 14 MB RAM.

If the Driven Cavity is cut diagonally from the lower right (refined) corner to the upper left one and if we denote the distance on this line from the lower left corner by r, then the theoretical formula for the transverse velocity component (direction (-1,-1)) reads

$$u_{\text{transverse}} = u_0 \left(\frac{r}{r_0}\right)^p \sin\left(q \ln \frac{r}{r_0}\right) \tag{4.0.1}$$

where the exponents p and q are only depending on the opening angle of the corner and which are given by the asymptotic theory. For our case (opening angle = $\pi/2$) we have $p = 2.74$, $q = 1.13$. u_0 and r_0 are a velocity and a length scale resp. which are to be fitted. We found them to be u_0 = 9.71e-11 and r_0 = 1.70e-4.

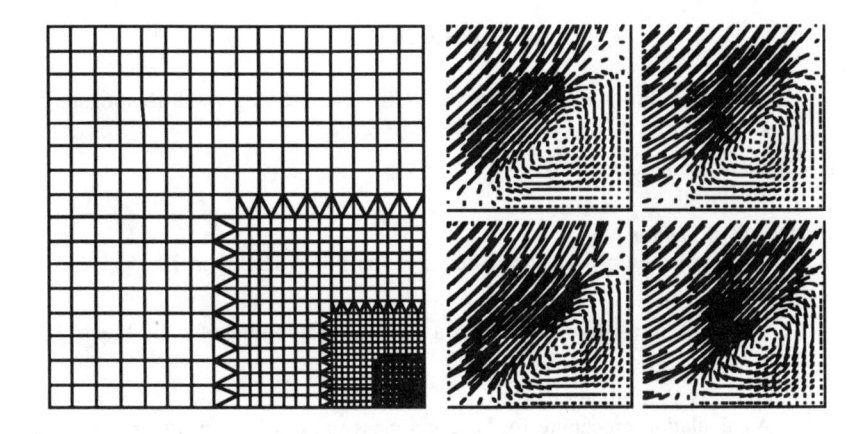

Fig. 11: Driven Cavity with locally refined grid and zooms into the lower left corner showing the four first of an infinite series of secondary vortices (zoom factors are 10, 180, 2800, 4.5e4 resp.).

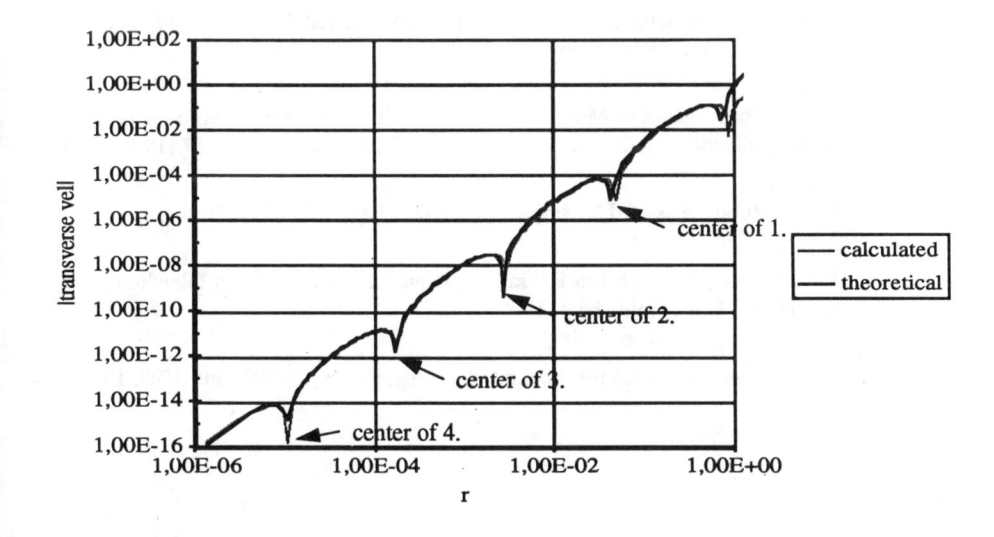

Fig. 12: Comparison with the results in [Mo].

References

[Ba] *Bastian, P.:*
ug 2.0: Ein Programmbaukasten zur effizienten Lösung von Strömungsproblemen. IWR Preprint 92-14 (Univ. Heidelberg), 1992.

[BW] *Bey, J., Wittum, G.:*
To appear.

[BD] *Brandt, A., Dinar, N.:*
Multigrid solutions to elliptic flow problems. ICASE Report 79-15 (1979).

[Mo] *Moffatt, K.H.:*
Viscous and resistive eddies near a sharp corner. Journal of Fluid Mech. 18 (1964) 1-18.

[PS] *Patankar, S.V., Spalding, D.B.:*
A calculation procedure for heat and mass transfer in threedimensional parabolic flows. Int. J. Heat Mass Transfer 15 (1972), 1787-1806.

[SR] *Schneider, G.E., Raw, M.J.:*
Control volume finite-element method for heat transfer and fluid-flow using colocated variables. Numer.Heat.Transf. 11 (1988) 363.

[Re] *Reichert, H., Wittum, G.:*
Solving the Navier-Stokes-Equations on Unstructured Grids. NNFM, Vol. 39, p. 321-333, Vieweg, Braunschweig 1993.

[Wi1] *Wittum, G.:*
On the Convergence of Multi-Grid Methods with Transforming Smoothers. Theory with Applications to the Navier-Stokes Equations. Numer. Math. 57 (1990) 15-38.

[Wi2] *Wittum, G.:*
On the Robustness of ILU-Smoothing. SISSC 10 (1989) 699-717.

[Wi3] *Wittum, G.:*
Filternde Zerlegungen: Ein Beitrag zur schnellen Lösung großer Gleichungssysteme. Habil., Univ. Heidelberg, 1990.

[Wi4] *Wittum, G.:*
Distributive Iterationen für indefinite Systeme. Ph. D. Thesis, Univ. Kiel, 1986.

A new robust multigrid method for 2D convection–diffusion problems

ARNOLD REUSKEN

Department of Mathematics and Computing Science
Eindhoven University of Technology
P.O. Box 513, 5600 MB Eindhoven, The Netherlands
e-mail: wsanar@win.tue.nl

Summary

In this paper we introduce a new multigrid approach for solving 2D convection–diffusion problems. As in the hierarchical basis multigrid method we use a transformed matrix with a relatively low generalized condition number. The transformation we use is not based on the nesting of the coarse and fine grid only, but uses more information from the matrix. After transforming back to the original matrix this results in a robust multigrid algorithm with matrix–dependent prolongations and restrictions.

1. Introduction

Multigrid methods are very fast methods for the solution of the large systems of equations arising from the discretization of partial differential equations. Today multigrid methods are used in nearly every field where partial differential equations are solved by numerical methods. Concerning the theoretical analysis of multigrid methods different fields of application have to be distinguished. For selfadjoint and coercive linear elliptic boundary value problems the convergence theory has reached a mature, if not its final state, cf. [14]. In other areas the state of the art is (far) less advanced. For example, for convection–dominated problems the development of a satisfactory theoretic analysis is still in its infancy. In this field only a few theoretical results are known in the literature, e.g. [2], [7], [9].
In this paper we consider multigrid for 2D convection–diffusion problems. We introduce a new multigrid approach for solving these problems, which hopefully contributes to further theoretical insights in this field of multigrid application.
The basic two–grid approach is as follows. We assume two nested grids ("coarse" and "fine") and on the finest grid the new mesh points are ordered first and then the old ones. Now, as in the hierarchical basis multigrid method (HBMG; [4]), we use a transformation from the original matrix to another matrix that has a significantly lower generalized condition number. To this transformed matrix we apply a block Jacobi type of method. The actual algorithm results after transformation back to the original matrix and is formulated in the classical multigrid framework. A fundamental difference between the approach of this paper and the HBMG approach is that the transformation we use is not based on the nesting of the coarse and fine grid only, but uses more information from the matrix. This transformation is based on the block LU- factorization with characterizes block Gaussian elimination. Thus nonsymmetry in the matrix influences the transformation. To preserve

sparsity we use the block Gaussian factorization of a modified matrix in which the equations in the new grid points (i.e. points in the fine grid which are not in the coarse grid) can be decoupled from the equations in the coarse grid points without fill–in. A proper modification is crucial for the method and will be explained for the case with square grids and standard $h - 2h$ coarsening. In the multigrid setting the matrix–dependent transformation results in a matrix–dependent restriction and prolongation and in a coarse grid matrix that has a Galerkin property (cf. [8]). The grid transfer operators and coarse grid matrices are computed in a preprocessing phase. In the two–grid method a correction coming from the coarse grid points (i.e. old points) is combined with a correction coming from the new points. This structure is as in the HBMG, however, we use matrix dependent grid transfers and another coarse grid matrix.

The remainder of this paper is organized as follows. In Section 2 we discuss connections between two–grid solvers and block factorizations. This yields a motivation for the particular method we propose. This method is then explained in Section 3. In Section 4 we discuss how the resulting method can be put in the classical multigrid framework. In Section 5 we present results of numerical experiments which indicate the robustness of the method.

2. Two–grid solvers based on block factorization

We consider a second order linear elliptic boundary value problem on a plane polygonal domain Ω. Let Ω_H be a "coarse" mesh on Ω. For ease of exposition we assume that Ω_H is a uniform mesh that consists of triangles or rectangles. By Ω_h we denote the "fine" mesh that results after a standard uniform refinement of Ω_H. The space of grid functions on Ω_H (Ω_h) is denoted by U_H (U_h). In U_H (and U_h) we use the standard nodal basis functions. The ordering of the basis functions in U_h is chosen such that the basis functions corresponding to nodes in $\Omega_h \backslash \Omega_H$ are taken first. This induces a partitioning for $u \in U_h$ as $u = (u_1 \; u_2)^T$, and $u \in U_H$ can be considered as an element of U_h through the injection $u \to (0 \; u)^T$.

We assume a given (finite element or finite difference) discretization method on Ω_h. This results in a linear system $Ax = b$, with $A : U_h \to U_h$. We consider the block partitioning

$$A = \begin{bmatrix} A_{11} & A_{12} \\ A_{21} & A_{22} \end{bmatrix}, \tag{2.1}$$

where A_{11} corresponds to the nodal basis functions in $U_h \backslash U_H$ and A_{22} corresponds to the (fine grid) nodal basis functions in U_H. Note that we do not assume symmetry of A.

Below we briefly discuss some known two–grid approaches which are based on LU-type factorizations of the form

$$A = \begin{bmatrix} I & \emptyset \\ R_{21} & I \end{bmatrix} B \begin{bmatrix} I & R_{12} \\ \emptyset & I \end{bmatrix}, \tag{2.2a}$$

i.e.

$$B = \begin{bmatrix} B_{11} & B_{12} \\ B_{21} & B_{22} \end{bmatrix} = \begin{bmatrix} I & \emptyset \\ -R_{21} & I \end{bmatrix} A \begin{bmatrix} I & -R_{12} \\ 0 & I \end{bmatrix} =: S_L A S_R \tag{2.2b}$$

with

$$B_{11} = A_{11}, \; B_{12} = -A_{11}R_{12} + A_{12}, \; B_{21} = -R_{21}A_{11} + A_{21},$$
(2.2c)
$$B_{22} = R_{21}A_{11}R_{12} - A_{21}R_{12} - R_{21}A_{12} + A_{22}.$$

Below, for a given block–matrix $C = \begin{bmatrix} C_{11} & C_{12} \\ C_{21} & C_{22} \end{bmatrix}$ the Schur complement of C_{11}, i.e. $C_{22} - C_{21}C_{11}^{-1}C_{12}$ is denoted by $\mathcal{S}_C$.

Clearly, the classical block Gaussian elimination is of the form (2.2) and is characterized by

resulting in
$$R_{21} = A_{21}A_{11}^{-1}, \quad R_{12} = A_{11}^{-1}A_{12},$$
(2.3a)
$$B_{12} = B_{21} = 0, \quad B_{22} = \mathcal{S}_A.$$
(2.3b)

Another approach based on a factorization as in (2.2) is the basic form of the hierarchical basis multigrid method (HBMG; cf. [4], [5]). We assume that A is the stiffness matrix corresponding to the standard finite element space of continuous piecewise linear functions on the triangulation Ω_h. By A_H we denote the stiffness matrix resulting from the coarse triangulation Ω_H. In HBMG a basis transformation is used which maps the nodal basis onto the hierarchical basis. This transformation is characterized (in the two–grid case) by a matrix $R: U_H \to U_h \backslash U_H$ that is related to the nested triangulations. Every row of R contains only zeros except for two entries which are equal to $\frac{1}{2}$. In this setting the factorization in (2.2) is given by

and
$$R_{21} = -R^T, \quad R_{12} = -R,$$
(2.4a)
$$B_{12} = A_{11}R + A_{12}, \quad B_{21} = R^T A_{11} + A_{21}, \quad B_{22} = A_H.$$
(2.4b)

The classical hierarchical two–grid method is mathematically equivalent to a symmetric block Gauss–Seidel iteration applied to the transformed matrix B. The off–diagonal blocks B_{12}, B_{21} are not zero (as in the case of Gaussian elimination) but they are small in the following sense. If we assume *symmetry*, i.e. $A = A^T$ and thus $B_{21} = B_{12}^T$, then a strengthened Cauchy inequality holds (cf. [3], [1], [5]) with a constant $\gamma < 1$, which can be formulated in this setting as

$$|x^T B_{21} y| \le \gamma (x^T A_H x)^{\frac{1}{2}} (y^T A_{11} y)^{\frac{1}{2}}.$$
(2.5)

For (nearly) symmetric problems, in a finite element setting, there are other approaches based on hierarchical type of block factorizations. For example, polynomial multilevel preconditioners based on incomplete block factorizations are treated in [1].

In the 2D case the approach in (2.3) is not feasible because then the matrices $R_{21}, R_{12}, \mathcal{S}_A$ will not be sparse. However, in the 1D case there is no fill–in and recursive application of the factorization yields the odd–even cyclic reduction method (i.e. a direct solver). In the 1D case a multigrid method can be derived which is based on the block Gaussian factorization (2.3). This approach is as follows. Based on the transformed matrix $B = \begin{bmatrix} A_{11} & \emptyset \\ \emptyset & \mathcal{S}_A \end{bmatrix}$ we use the iterative method with iteration matrix

$$M_B = I - \begin{bmatrix} \emptyset & \emptyset \\ \emptyset & \mathcal{S}_A^{-1} \end{bmatrix} B \quad \left(= \begin{bmatrix} I & \emptyset \\ \emptyset & \emptyset \end{bmatrix}\right).$$
(2.6)

We transform this back to a method for the original matrix A characterized by the iteration matrix $M_A := S_R M_B S_R^{-1}$. This yields

$$M_A = I - \begin{bmatrix} -A_{11}^{-1} A_{12} \\ I \end{bmatrix} \mathcal{S}_A^{-1} [-A_{21} A_{11}^{-1} \ I] A \ . \tag{2.7}$$

If A corresponds to a 3-point stencil then A_{11} is diagonal and (2.7) results, in the multi-grid setting, in a feasible coarse grid correction with matrix dependent prolongation and restriction and coarse grid matrix $\mathcal{S}_A$. This coarse grid correction is analyzed in [9]. It is shown there that this coarse grid correction is much more robust than the standard coarse grid correction, for example with respect to nonsymmetry in the problem. This robustness is closely related to the algebraic background of (2.7).

We summarize some important observations from above:

- Block Gaussian elimination yields a block diagonal matrix B with unacceptable fill–in.
- The HB two–grid method is based on a matrix B which is, in the symmetric case, "close to block diagonal" and sparse.
- In the block Gaussian factorization (2.2), (2.3) in general we have $S_L \neq S_R^T$. In the HB two–grid method we always have a symmetric factorization: $S_L = S_R^T$.
- In the 1D case one can derive a very robust multigrid coarse grid correction that is based on the factorization (2.2), (2.3).

In view of these observations we propose the following approach that will be discussed in detail in the next section. We use a nonsymmetric factorization that is similar to the block Gaussian factorization (2.2), (2.3). After transformation we introduce a suitable block iterative method based on B. The actual algorithm results after transformation back to the original system with matrix A, and is formulated in a multigrid framework. A main topic is a systematic procedure to avoid fill-in. This will be explained in detail for the situation with square grids and standard $h \to 2h$ coarsening.

3. A two–grid approach based on approximate LU- factorization

In this section we derive a two–grid method based on an LU-type of factorization as in (2.2), (2.3). To avoid fill-in we replace $A_{11}^{-1} A_{12}$ in S_R by a suitable approximation $\tilde{A}_{11}^{-1} \tilde{A}_{12}$. This approximation is made using a so called lumping approach (also used in [10]). In this lumping method we use information about the underlying differential equation.

We will explain the approach for the following class of boundary value problems:

$$\begin{cases} -\varepsilon \Delta u + a(x,y) u_x + b(x,y) u_y = f & \text{in } \Omega =]0,1[^2 \quad (\varepsilon > 0) \\ u = g & \text{on } \partial\Omega \ . \end{cases} \tag{3.1}$$

We use a standard square mesh Ω_H with mesh size $H = 2^{-k}$ and a uniform refinement Ω_h of Ω_H with mesh size $h = \frac{1}{2} H$. For Δ we use the standard 5-point star and for u_x, u_y upwind differences are used. This results in a linear system that is represented with the block partitioning as in §2:

$$\begin{bmatrix} A_{11} & A_{12} \\ A_{21} & A_{22} \end{bmatrix} x = b . \tag{3.2}$$

In the lumping procedure that we discuss below, the equations in the grid points of $\Omega_h\backslash\Omega_H$ are modified, resulting in an approximation

$$\tilde{A} = \begin{bmatrix} \tilde{A}_{11} & \tilde{A}_{22} \\ A_{21} & A_{22} \end{bmatrix} \text{ of } A .$$

Consider a grid point P of $\Omega_h\backslash\Omega_H$ (cf. Fig. 1). For the coefficients a, b in (3.1) we assume

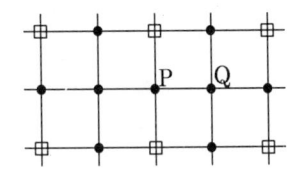

$\{\square\}$: Ω_H
$\{\bullet\}$: $\Omega_h\backslash\Omega_H$

Fig. 1

$a(P) \geq 0$, $b(P) \geq 0$. Then the equation in P consists of a linear combination of the difference stars

$$\begin{bmatrix} 0 & -1 & 0 \\ -1 & 4 & -1 \\ 0 & -1 & 0 \end{bmatrix} , \quad \begin{bmatrix} 0 & 0 & 0 \\ -1 & 1 & 0 \\ 0 & 0 & 0 \end{bmatrix} \text{ and } \begin{bmatrix} 0 & 0 & 0 \\ 0 & 1 & 0 \\ 0 & -1 & 0 \end{bmatrix} .$$

We get a modified equation, represented in $[\tilde{A}_{11} \ \tilde{A}_{12}]$, by the following substitution:

$$h^{-2} \begin{bmatrix} 0 & -1 & 0 \\ -1 & 4 & -1 \\ 0 & -1 & 0 \end{bmatrix} \longrightarrow h^{-2} \begin{bmatrix} -1/8 & 0 & -3/4 & 0 & -1/8 \\ 0 & 0 & 2 & 0 & 0 \\ -1/8 & 0 & -3/4 & 0 & -1/8 \end{bmatrix} \tag{3.3a}$$

$$h^{-1} \begin{bmatrix} 0 & 0 & 0 \\ -1 & 1 & 0 \\ 0 & 0 & 0 \end{bmatrix} \longrightarrow h^{-1} \begin{bmatrix} -1/4 & 0 & -1/4 & 0 & 0 \\ 0 & 0 & 1 & 0 & 0 \\ -1/4 & 0 & -1/4 & 0 & 0 \end{bmatrix} \tag{3.3b}$$

Taylor expansion shows that for smooth functions the difference between the results of the two stars in (3.3a), (3.3b) is $\mathcal{O}(h^2)$, $\mathcal{O}(h)$ respectively. In a point Q (cf. Fig. 1) we make the following substitution (again $a(Q) \geq 0$, $b(Q) \geq 0$):

$$h^{-2} \begin{bmatrix} 0 & -1 & 0 \\ -1 & 4 & -1 \\ 0 & -1 & 0 \end{bmatrix} \longrightarrow h^{-2} \begin{bmatrix} -1/2 & 0 & -1/2 \\ 0 & 2 & 0 \\ -1/2 & 0 & -1/2 \end{bmatrix} \quad (\mathcal{O}(h^2) \text{ accurate}) \tag{3.4a}$$

$$h^{-1} \begin{bmatrix} 0 & 0 & 0 \\ -1 & 1 & 0 \\ 0 & 0 & 0 \end{bmatrix} \longrightarrow h^{-1} \begin{bmatrix} -1/2 & 0 & 0 \\ 0 & 1 & 0 \\ -1/2 & 0 & 0 \end{bmatrix} \quad (\mathcal{O}(h) \text{ accurate}) \tag{3.4b}$$

$$h^{-1} \begin{bmatrix} 0 & 0 & 0 \\ 0 & 1 & 0 \\ 0 & -1 & 0 \end{bmatrix} \longrightarrow h^{-1} \begin{bmatrix} 0 & 0 & 0 \\ 0 & 1 & 0 \\ -1/2 & 0 & -1/2 \end{bmatrix} \quad (\mathcal{O}(h) \text{ accurate}) \tag{3.4c}$$

Clearly this approach uses information from the underlying differential equation. We may combine this with a more algebraic approach in which only the structure of the grid is used. In the latter approach a relation between unknowns in $\Omega_h \backslash \Omega_H$ is "eliminated" by a (linear) interpolation procedure (as in HBMG). For example, if in the point $P = (x, y)$ the unknown $u(x - h, y - h)$ is replaced by $\frac{1}{2}(u(x - 2h, y - h) + u(x, y - h))$ (cf. Fig. 1) then the star $\begin{bmatrix} 0 & 0 & 0 \\ 0 & 1 & 0 \\ -1 & 0 & 0 \end{bmatrix}_P$ changes into $\begin{bmatrix} 0 & 0 & 0 & 0 & 0 \\ 0 & 0 & 1 & 0 & 0 \\ -1/2 & 0 & -1/2 & 0 & 0 \end{bmatrix}_P$. In the implementations we always used a lumping in two steps. In the first step, in a given grid point in $\Omega_h \backslash \Omega_H$, we modify the star by replacing certain finite differences, that are selected using the BVP, by other finite differences (as in (3.3), (3.4)). In the second step we eliminate remaining relations between grid points in $U_h \backslash U_H$ by an algebraic elimination process (e.g. linear interpolation).

The lumping procedure described above yields $\tilde{A}_{11}$, $\tilde{A}_{12}$ (note: $\tilde{A}_{11}$ diagonal) that will be used in an approximate LU-factorization. We introduce the following notation:

$$p = \begin{bmatrix} -A_{11}^{-1} A_{12} \\ I \end{bmatrix}, \quad \tilde{p} = \begin{bmatrix} -\tilde{A}_{11}^{-1} \tilde{A}_{12} \\ I \end{bmatrix}, \quad r_{inj} = [0 \; I], \quad r = [-A_{21} A_{11}^{-1} \; I]. \qquad (3.5)$$

Note that $\mathcal{S}_A$, $\mathcal{S}_{\tilde{A}}$ are the Schur complements of A, $\tilde{A}$ respectively.

Lemma 3.1. The following holds:
a) $\mathcal{S}_A = r_{inj} A p$, $\mathcal{S}_{\tilde{A}} = r_{inj} \tilde{A} \tilde{p} = r_{inj} A \tilde{p}$.

b) $\mathcal{S}_{\tilde{A}}$ has a 9-point stencil.

c) If $\tilde{A}$ is an M-matrix then $\mathcal{S}_{\tilde{A}}$ is an M-matrix.

Proof. The results in a) follow directly from the definitions. The lumping procedure is such that, in a point $x_j \in \Omega_H$, $\tilde{p}$ has a 17-point star in which the points $\{(x_j \pm (kh, \ell h)) \cap \Omega_h \backslash \Omega_H \mid 1 \leq k, \ell \leq 2\}$ are used. The coefficient $(\mathcal{S}_{\tilde{A}})_{i,j}$ $(x_i, x_j \in \Omega_H)$ is given by

$$< \mathcal{S}_{\tilde{A}} e_j^H, e_i^H > = < r_{inj} A \tilde{p} e_j^H, e_i^H > = < \tilde{p} e_j^H, A^T e_i^h > .$$

Note that $\tilde{p} e_j^H$ is just the star of $\tilde{p}$ in x_j and $A^T e_i^h$ is just the star of A in x_i. From the geometry of Ω_h it is now obvious that $\mathcal{S}_{\tilde{A}}$ has a 9-point star. The result in c) is classical. □

Corollary 3.2. The lumping procedure is defined in such a way that stability is preserved. So if A is stable then $\tilde{A}$ is stable too. A precise statement concerning this stability is given in [10]. From the results in Lemma 3.1 we then conclude that $\mathcal{S}_{\tilde{A}}$ is a stable 9-point operator on the coarse grid Ω_H. The same approach as on the fine grid, i.e. lumping and Schur complement computation, can be applied to the coarse grid operator $\mathcal{S}_{\tilde{A}}$, too. This results in stable 9-point operators on all coarser grids.

Remark 3.3. The results in Lemma 3.1 show that $\mathcal{S}_{\tilde{A}}$ has some interesting properties. A further very important claim is that $\mathcal{S}_{\tilde{A}}$ is a "good" approximation of $\mathcal{S}_A$ in the sense that for a suitable constant α $\rho(I - \alpha \mathcal{S}_{\tilde{A}}^{-1} \mathcal{S}_A)$ has an upper bound smaller than one that is uniform for a large class of problems (with varying h and ε). For problems where the diffusion is dominant such a result is expected e.g. from the analysis in [1], [3], [6] and mathematically founded by the strengthened Cauchy inequality. It appears that the preconditioning of S_A by $S_{\tilde{A}}^{-1}$ is also good for many strongly nonsymmetric problems. This is a subject of current research. Here we only give three typical results which can be obtained using Fourier analysis (assuming constant coefficients and periodic boundary conditions).

- diffusion problem: $A \sim h^{-2} \begin{bmatrix} 0 & -1 & 0 \\ -1 & 4 & -1 \\ 0 & -1 & 0 \end{bmatrix}$, then $\rho(I - 0.8\, S_{\tilde{A}}^{-1} S_A) \leq 0.2$.

- convection problem, no alignment: $A \sim h^{-1} \begin{bmatrix} 0 & 0 & 0 \\ -1 & 2 & 0 \\ 0 & -1 & 0 \end{bmatrix}$,

  then $\rho(I - 0.6\, S_{\tilde{A}}^{-1} S_A) \leq 0.5$.

- convection problem with alignment: $A \sim h^{-1} \begin{bmatrix} 0 & 0 & 0 \\ -1 & 1 & 0 \\ 0 & 0 & 0 \end{bmatrix}$, then $I - S_{\tilde{A}}^{-1} S_A = 0$.

Within the framework of Section 2 we now use $\tilde{A}_{11}$, $\tilde{A}_{12}$, $S_{\tilde{A}}$ to derive a two-grid method based on approximate LU-factorization. Let S_L, S_R be as in (2.2) with

$$R_{21} = A_{21} A_{11}^{-1}, \quad R_{12} = \tilde{A}_{11}^{-1} \tilde{A}_{12}. \tag{3.6}$$

Then for $B = S_L A S_R$ we get

$$B = \begin{bmatrix} A_{11} & -A_{11} \tilde{A}_{11}^{-1} \tilde{A}_{12} + A_{12} \\ \emptyset & S_A \end{bmatrix}.$$

Based on the results of Lemma 3.1 and the observations in Remark 3.3 we introduce a block iterative method for $Bz = d$ with iteration matrix

$$M_B = I - W_B B, \quad W_B = \begin{bmatrix} A_{11}^{-1} & \emptyset \\ \emptyset & \alpha S_{\tilde{A}}^{-1} \end{bmatrix}. \tag{3.7}$$

This is a block Jacobi method with an inner iteration. The rate of convergence depends on the rate of convergence of the inner iteration. Two estimates are given in the lemma below.

Lemma 3.4. The following holds:

a) $\rho(M_B) = \rho(I - \alpha S_{\tilde{A}}^{-1} S_A)$

b) $\|M_B^2\|_\infty \leq \|I - \alpha S_{\tilde{A}}^{-1} S_A\|_\infty \max(\|\tilde{A}_{11}^{-1} \tilde{A}_{12} - A_{11}^{-1} A_{12}\|_\infty, \|I - \alpha S_{\tilde{A}}^{-1} S_A\|_\infty)$.

Proof. Note that

$$M_B = \begin{bmatrix} \emptyset & \tilde{A}_{11}^{-1} \tilde{A}_{12} - A_{11}^{-1} A_{12} \\ \emptyset & I - \alpha S_{\tilde{A}}^{-1} S_A \end{bmatrix} =: \begin{bmatrix} \emptyset & M_{12} \\ \emptyset & M_{22} \end{bmatrix}. \text{ So}$$

$$M_B^2 = \begin{bmatrix} \emptyset & M_{12} M_{22} \\ \emptyset & M_{22}^2 \end{bmatrix}, \text{ and thus,}$$

$$\|M_B^2\|_\infty \leq \max(\|M_{12} M_{22}\|_\infty, \|M_{22}^2\|_\infty) \leq \|M_{22}\|_\infty \max(\|M_{12}\|_\infty, \|M_{22}\|_\infty). \quad \square$$

As in the HBMG we transform the block iterative method based on B back to the original A. The result is stated below.

Lemma 3.5. With $\tilde{p}$ as in (3.5) and $\tilde{r} := [-A_{21}\tilde{A}_{11}^{-1} \ I]$ the following holds:

$$M_A := S_R M_B S_R^{-1} = I - W_A A \text{, with}$$

$$W_A = \left(\alpha \tilde{p} S_{\tilde{A}}^{-1} \tilde{r} + \begin{bmatrix} \tilde{A}_{11}^{-1} & \emptyset \\ \emptyset & \emptyset \end{bmatrix}\right) \begin{bmatrix} \tilde{A}_{11} A_{11}^{-1} & \emptyset \\ \emptyset & I \end{bmatrix}. \tag{3.8}$$

Also the following equality holds:

$$W_A = \alpha \tilde{A}^{-1} \begin{bmatrix} \tilde{A}_{11} A_{11}^{-1} & \emptyset \\ \emptyset & I \end{bmatrix} + (1-\alpha) \begin{bmatrix} A_{11}^{-1} & \emptyset \\ \emptyset & \emptyset \end{bmatrix}. \tag{3.9}$$

Proof. $S_R M_B S_R^{-1} = I - S_R W_B S_L A$ holds and thus $W_A = S_R W_B S_L$. This yields

$$\begin{aligned} W_A &= \begin{bmatrix} I & -\tilde{A}_{11}^{-1}\tilde{A}_{12} \\ \emptyset & I \end{bmatrix} \begin{bmatrix} A_{11}^{-1} & \emptyset \\ \emptyset & \alpha S_{\tilde{A}}^{-1} \end{bmatrix} \begin{bmatrix} I & \emptyset \\ -A_{21}A_{11}^{-1} & I \end{bmatrix} \\ &= \alpha \tilde{p} S_{\tilde{A}}^{-1}[-A_{21}A_{11}^{-1} \ I] + \begin{bmatrix} A_{11}^{-1} & \emptyset \\ \emptyset & \emptyset \end{bmatrix} \\ &= \alpha \tilde{p} S_{\tilde{A}}^{-1} \tilde{r} \begin{bmatrix} \tilde{A}_{11} A_{11}^{-1} & \emptyset \\ \emptyset & I \end{bmatrix} + \begin{bmatrix} A_{11}^{-1} & \emptyset \\ \emptyset & \emptyset \end{bmatrix}. \end{aligned}$$

From this the result in (3.8) follows. The result in (3.9) is a direct consequence of inverting the block Gaussian factorization of $\tilde{A}$ (i.e. (2.2), (2.3) applied to $\tilde{A}$):

$$\tilde{A}^{-1} = \tilde{p} S_{\tilde{A}}^{-1} \tilde{r} + \begin{bmatrix} \tilde{A}_{11}^{-1} & \emptyset \\ \emptyset & \emptyset \end{bmatrix}. \qquad \square$$

Note that $\sigma(M_A) = \sigma(M_B)$ and also that the similarity transformation with S_R is well-conditioned if $\tilde{A}_{11}$ is strongly diagonally dominant, which is the case if there is no alignment.

Finally, to get a feasible method we have to replace A_{11}^{-1} in (3.6) by $(I - M^\mu)A_{11}^{-1}$ where M is the iteration matrix of a basic iterative method for solving $A_{11}z = d$. Note that the latter system is only on the grid points in $\Omega_h \backslash \Omega_H$. In the HBMG a system with matrix A_{11} also occurs and is solved approximately by a few Gauss–Seidel iterations. In general (also for strongly nonsymmetric problems) the matrix A_{11} has condition number $\mathcal{O}(1)$ and thus, in principle, any basic iterative methode will work. However, if we have strong alignment (e.g. third example in Remark 3.3) then $\text{cond}(A_{11})$ deteriorates. So, to get a robust method it is better to take, for example, a line Jacobi method using the "odd" horizontal and vertical lines which together form the pattern of $\Omega_h \backslash \Omega_H$. An alternative, which may be attractive if one wants to use only local operations, is a (robust) multigrid method for solving systems with matrix A_{11}. This then in a natural way introduces a shifted coarse grid $\Omega_H^{\text{shift}} = \{(h,h) + H(k,\ell) \mid k, \ell \in \mathbb{N}\} \cap \Omega_h$, that is complementary to the original coarse grid $\Omega_H = \{H(k,\ell) \mid k, \ell \in \mathbb{N}\} \cap \Omega_h$.

4. Two–grid and multigrid algorithm

In this section we describe the multigrid algorithm based on the preconditioner W_A as in Lemma 3.5.

We assume a sequence of grids Ω_h with $h = 2^{-\ell}$, $k_{\min} \le \ell \le k_{\max}$. In a preprocessing phase the matrix dependent transfer operators $\tilde{p}$, $\tilde{r}$ and the coarse grid operators are constructed. We start with the matrix on the finest grid Ω_h ($h = 2^{-k_{\max}}$) and apply the lumping procedure in the points of $\Omega_h\backslash\Omega_H$ to construct $[\tilde{A}_{11}\ \tilde{A}_{12}]$. From this the stars of $\tilde{p}_{h\leftarrow H}$ and $\tilde{r}_{H\leftarrow h}$ (cf. Lemma 3.5) are computed. Then the coarse grid operator is formed using the Galerkin property: $A_H := S_{\tilde{A}} = r_{inj}A\tilde{p}$. Now the lumping is repeated on the coarse grid, etc. In the algorithm below we also need the diagonal matrices $\tilde{A}_{11}$ on $\Omega_h\backslash\Omega_H$ for all $h = 2^{-\ell}$.

A two–grid iteration on grid Ω_h ($h = 2^{-\ell}$) for solving $A_h x_h = b_h$ consists of the following steps:

1. $x_h := S^\nu(x_h; b_h)$; presmoothing (optional).

2. $d := A_h x_h - b_h$.

3. $z_h := J^\mu((A_h)_{11}; 0; d_{|\Omega_h\backslash\Omega_H})$; apply μ iterations of an iterative method for solving $(A_h)_{11} z_h = d_{|\Omega_h\backslash\Omega_H}$ with start 0.

4. $d_H := \tilde{r}_{H\leftarrow h} \begin{bmatrix} (\tilde{A}_h)_{11} z_h \\ d_{|\Omega_H} \end{bmatrix}$; compute coarse grid defect.

5. solve $A_H v_H = d_H$.

6. $x_h := x_h - \alpha \tilde{p} v_H - \begin{bmatrix} z_h \\ 0 \end{bmatrix}$; add corrections.

We briefly discuss these steps. The presmoothing step 1 is optional and may (for a diffusion dominated problem) accelerate the convergence. As in HBMG, W_A from Lemma 3.5 yields the standard choice $\nu = 0$ (i.e. no presmoothing!). Step 3 corresponds to the A_{11}^{-1} factor in W_A; we use an approximate solve for A_{11}^{-1}. As discussed in §3, in view of robustness we take for one iteration of J a horizontal line Jacobi followed by a vertical line Jacobi; both only using the "odd" lines (i.e. the lines forming the pattern of $\Omega_h\backslash\Omega_H$). Step 4 corresponds to $\tilde{r}\begin{bmatrix} \tilde{A}_{11}A_{11}^{-1} & 0 \\ 0 & I \end{bmatrix} d$ in W_A. Step 5 is clear; as usual, in a multigrid approach this step is replaced by a γ-fold recursive call ($\gamma = 1, 2$). In step 6 the corrections from the coarse grid Ω_H and from $\Omega_h\backslash\Omega_H$ are added to the current approximation. The parameter α is used for acceleration; in our experiments we take $\alpha \in [0.6, 1]$.

5. Numerical experiments

As stated in §3, we consider the following class of convection–diffusion problems ($\varepsilon > 0$):

$$\begin{cases} -\varepsilon \Delta u + a(x,y)u_x + b(x,y)u_y = f & \text{in } \Omega =]0,1[^2 \\ u = g & \text{on } \partial\Omega. \end{cases}$$

We use standard square meshes. The finest mesh always has $h = 2^{-6}$, the coarsest mesh size is $h = 2^{-2}$. We only need a discretization on the finest grid; there we use the standard 5-point stencil for Δ and upwind differences for u_x, u_y.

Standard in our experiments is $\nu = 0$ (no presmoothing), $\gamma = 2$ (W-cycle), $\mu = 2$ iterations of the line Jacobi method as discussed in §4. Based on numerical experiments, in step 6 we take $\alpha = 1$ on coarse grids and $\alpha = 0.8$ on the finest grid.

Our main interest in this paper is a *robust* multigrid approach based on approximate LU-factorization. The first three experiments below are meant to test the robustness of our method w.r.t. the amount of convection and its direction.

In our experiments we always take the data such that the exact solution is equal to zero and we take an arbitrary starting vector. As a measure for the error reduction we computed the harmonic average of the error reduction (w.r.t. the Euclidean norm) in the first ten iterations. Between brackets we give the error reduction in the tenth iteration.

Experiment 1. (standard test problem as in [13]). We take $a(x,y) = \cos\varphi$, $b(x,y) = \sin\varphi$. In Table 1 the results are given for different values of φ and ε.

Table 1

ε \ φ	0	$\pi/8$	$2\pi/8$	$11\pi/8$
10^0	0.13 (0.21)	0.13 (0.21)	0.13 (0.21)	0.13 (0.21)
10^{-2}	0.14 (0.24)	0.13 (0.23)	0.13 (0.25)	0.14 (0.26)
10^{-4}	0.20 (0.18)	0.31 (0.21)	0.27 (0.35)	0.35 (0.32)
10^{-6}	0.22 (0.20)	0.32 (0.22)	0.28 (0.36)	0.36 (0.32)

Experiment 2. (rotating flow). We define $\Omega_R := \{(x,y) \mid (x-\frac{1}{3})^2 + (y-\frac{1}{3})^2 \leq \frac{1}{16}\}$. $a(x,y) = \sin(\pi(y-\frac{1}{3}))\cos(\pi(x-\frac{1}{3}))$ if $(x,y) \in \Omega_R$, and zero otherwise; $b(x,y) = -\cos(\pi(y-\frac{1}{3}))\sin(\pi(x-\frac{1}{3}))$ if $(x,y) \in \Omega_R$, and zero otherwise. The results for different values of ε are given in Table 2.

Table 2

ε	10^0	10^{-2}	10^{-4}	10^{-6}
	0.13	0.13	0.30	0.30
	(0.21)	(0.21)	(0.33)	(0.27)

Experiment 3. (as in [11], [15]). We take $a(x,y) = (2y-1)(1-x^2)$, $b(x,y) = 2xy(y-1)$. For $\varepsilon = 0$, the characteristics are shown in Fig. 2. The results for different values of ε are given in Table 3.

Table 3

ε	10^0	10^{-2}	10^{-4}	10^{-6}
	0.13	0.13	0.36	0.37
	(0.21)	(0.22)	(0.33)	(0.35)

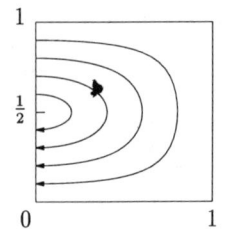

Fig. 2

The experiments above seem to indicate that the method is robust with respect to the amount of convection and its direction. Note that the unknowns in the coarse grid points are not updated by the line Jacobi method (which only uses the "odd" lines) and a reasonable update there has to come from the coarse grid correction. The line Jacobi method should not be called a "robust smoother" (cf. [8], [13]); the line Jacobi method is introduced in §3 because it is a robust method for solving systems with A_{11} and arguments based on possible smoothing properties are not used.

The two experiments below show that in certain situations special properties of the underlying problem can be used to make another choice for the components in our method, resulting in faster convergence.

Experiment 4. (pure diffusion problem). We take $a(x,y) = b(x,y) = 0$, i.e. a pure diffusion problem. The line Jacobi method is not needed for solving systems with A_{11}. Based on Lemma 3.5 we replace A_{11}^{-1} by $\tilde{A}_{11}^{-1}$; in step 3 of the algorithm we now perform a solve with the diagonal matrix $\tilde{A}_{11}$. We take $\alpha = 1$, then (cf. (3.9)) the iteration matrix is $I - \tilde{A}^{-1}A$ which will yield a large error reduction for smooth errors. We use the optional step 1 to smooth the error by ν iterations of a damped Jacobi iteration (damping 0.65). The results are given in Table 4 and show the usual multigrid convergence rates for Poisson-type of equations.

Table 4

	$\nu = 2$	$\nu = 4$
V-cycle	0.14 (0.13)	0.081 (0.089)
W-cycle	0.15 (0.21)	0.062 (0.095)

Experiment 5. (alignment). We take $a(x,y) = 1$, $b(x,y) = 0$. Based on Remark 3.3 $(I - S_{\tilde{A}}^{-1}S_A = 0)$ we take $\alpha = 1$ on all levels. The results are shown in Table 5. As expected from the analysis in §3 we get fast convergence for $\varepsilon \downarrow 0$.

Table 5

ε	10^0	10^{-2}	10^{-4}	10^{-6}
	0.27	0.39	0.14	0.0024
	(0.27)	(0.37)	(0.062)	(0.00064)

Finally we note that an approach similar to the one presented in this paper resulted in a robust solver for a class of diffusion problems with strongly varying coefficients in [12].

References

[1] O. AXELSSON and P. VASSILEVSKI, *Algebraic multilevel preconditioning methods*, II, SIAM J. Numer. Anal., 27 (1990), pp. 1569-1590.

[2] R.E. BANK and M. BENBOURENANE, *The hierarchical basis multigrid method for convection-diffusion equations*, Numer. Math., 61 (1992), pp. 7-37.

[3] R.E. BANK and T. DUPONT, *Analysis of a two-level scheme for solving finite element equations*, Techn. Report CNA-159, Center for Numerical Analysis, University of Texas, Austin, TX, 1980.

[4] R.E. BANK, T. DUPONT and H. YSERENTANT, *The hierarchical basis multigrid method*, Numer. Math., 52 (1988), pp. 427-458.

[5] R.E. BANK and J. XU, *The hierarchical basis multigrid method and incomplete LU decomposition*, to appear in the proceedings of the Sixth International Symposium on Domain Decomposition Methods for Partial Differential Equations 1993.

[6] V. EIJKHOUT and P. VASSILEVSKI, *The role of the strengthened Cauchy-Buniakowskii-Schwarz inequality in multilevel methods*, SIAM Review, 33 (1991), pp. 405-419.

[7] W. HACKBUSCH, *Multigrid convergence for a singular perturbation problem*, Linear Algebra Appl., 58 (1984), pp. 125-145.

[8] W. HACKBUSCH, *Multigrid Methods and Applications*, Springer, Berlin, 1985.

[9] A. REUSKEN, *Multigrid with matrix-dependent transfer operators for a singular perturbation problem*, Computing, 50 (1993), pp. 199-211.

[10] A. REUSKEN, *Multigrid with matrix- dependent transfer operators for convection-diffusion problems*, to appear in the Proceedings of the European Multigrid Conference '93.

[11] J. RUGE and K. STÜBEN, *Efficient solution of finite difference and finite element equations by algebraic multigrid*, in: D.J. Paddon and H. Holstein, Eds.: Multigrid methods for Integral and Differential Equations, Inst. Math. Appl. Conf. Ser. New Ser. (Oxford Univ. Press, New York, 1985), pp. 169-212.

[12] C. WAGNER, *Ein robustes Mehrgitterverfahren für Diffusions-Transport-Probleme in der Bodenphysik*, Diplomarbeit Universität Heidelberg, IWR Preprint 93-70 (1993).

[13] P. WESSELING, *An Introduction to Multigrid Methods*, Wiley, Chichester, 1992.

[14] H. YSERENTANT, *Old and new convergence proofs for multigrid methods*, Acta Numerica (1993), pp. 285-326.

[15] P.M. de ZEEUW, *Matrix-dependent prolongations and restrictions in a blackbox multigrid solver*, J. Comput. Appl. Math., 33 (1990), pp. 1-27.

Adapting Meshes by Deformation
Numerical Examples and Applications

M. Rumpf
Institut für Angewandte Mathematik
Hermann–Herder–Str. 10, 79104 Freiburg
Germany

Summary

The goal of this paper is to give a strictly derived definition for an optimal deformation of a given simplicial grid and to discuss in detail some properties and useful applications. Many grid handling approaches generate or adapt a mesh using combinatorial techniques such as refining or coarsening of simplices or continuously adding new points and reconnecting the vertices. Different to those methods we keep the connectivity of the vertices fixed solely changing their location and thereby deforming the mesh. It can rigorously be demonstrated that under a few reasonable assumptions the functional describing optimality of a mesh is of a type wellknown from elasticity theory. This approach is also suitable for numerical usage. It is possible to improve and adapt 3D grids by a stable minimization algorithm with useful properties. Interesting applications are the reconstruction of a mesh for a domain with a free boundary, the concentration of a mesh at singularities, or its compression near an arbitrary given boundary to resolve a layer structure.

1. Introduction

We will give a definition for optimal deformations of simplicial grids in two and three dimensions. Our definition for "optimal" will be with respect to a given mesh and we will retain its connectivity. It essentially differs from other mesh optimizing and adaption algorithms based on Delaunay triangulations or refinement and coarsening [2, 4, 3, 10], but can be combined with them. The details of the arguments can be found in [7]. This paper will extensively discuss in its second part properties and possible applications. The key point is to search for an admissible set of deformations φ on a triangulation $\mathcal{T}$ and an appropriate functional $\mathcal{F}$ such that this functional achieves its minimum over the latter set at a deformation φ^*. We will call this deformation the optimal one. The resulting triangulation $\varphi^*(\mathcal{T})$ will be denoted by $\mathcal{T}^*$.

$$\mathcal{F}(\varphi^*) = \min_{\varphi \text{ admissible}} \mathcal{F}(\varphi).$$

For these considerations we mainly focus on the three dimensional case. For planar domains the same conclusions hold and could be derived by analogy.

The outline of the paper is as follows: First we will define a subset of all possible deformations which is appropriate according to the resulting triangulations and useful for the actual numerical minimization. Then starting from a very basic ansatz and applying some principles from rational mechanics we will end up with a small class of possible functionals. Finally the properties of this approach will be underlined by some numerical examples and several useful applications will be discussed.

2. Admissible Deformations

An initial simplicial triangulation $\mathcal{T} = \{T_i\}_{i \in I^3}$ of a domain Ω in $\mathbb{R}^3$ with piecewise smooth boundary should be given. By an uppercase I with an index n we mean a specific finite index set $I^n \subset \mathbb{N}_0$. For simplicity we denote by $\mathcal{T}$ the simplicial complex <u>and</u> the corresponding polyhedral domain, which will be the domain for the mappings we are interested in. Let $\mathcal{N}(T)$, $\mathcal{N}(\mathcal{T})$ be the set of vertices of T respectively $\mathcal{T}$. Let us assume that vertices at $\partial \mathcal{T}$ are boundary points of Ω.

An admissible deformation has to be continuous and piecewise linear to retain the mesh connectivity. To avoid local overlapping we assume that φ restricted to a simplex T is orientation preserving: $\nabla \varphi \mid_T \in SL_3$. Furthermore the boundary shape $\partial \Omega$ should be respected by φ. If we solely expect that vertices at $\partial \mathcal{T}$ stay on $\partial \Omega$ under deformation by φ, it is possible that vertices on $\partial \mathcal{T}$ may switch over an edge between different smooth sections of $\partial \Omega$, for example from one boundary surface to another over the common edge. Checking which of these configurations decreases the functional is a combinatorial task and does not fit well into a variational setting and is therefore numerically difficult to realize. Moreover vertices on edges or at singular points of $\partial \Omega$ are allowed to move away. But that would possibly destroy the approximation of $\partial \Omega$. These considerations lead us to a smaller set of admissible deformations. Since there is no need for combinatorial testing we will call them variations.

Definition (Optimal Variation)

Let Ω be bounded by piecewise smooth surfaces. Denote the smooth surface segments by $\{\Gamma_i^2\}_{i \in I^2}$, smooth curves by $\{\Gamma_i^1\}_{i \in I^1}$ and singular points on $\partial \Omega$ by the sets $\{\Gamma_i^0\}_{i \in I^0}$ of isolated points. It always holds that $\Gamma_i^1 = \Gamma_j^2 \cap \Gamma_k^2$ for fixed $j, k \in I^2$. Then φ^* is an optimal variation of the triangulation $\mathcal{T}$ of Ω with respect to a functional $\mathcal{F}$, iff

$$\mathcal{F}(\varphi^*) = \min_{\varphi \in V} \mathcal{F}(\varphi)$$
$$V = \{\varphi : \mathcal{T} \to \mathbb{R}^3 \mid \varphi \in C^0(\mathcal{T}) \, ; \, \nabla \varphi \mid_T \in SL_3 \, ; \, \bigvee_{x \in \mathcal{N}(\mathcal{T})} \bigvee_{j \in I^k} x \in \Gamma_j^k \Rightarrow \varphi(x) \in \Gamma_j^k \}.$$

3. Appropriate Functionals

Having fixed the set of admissible deformations we shall now derive a class of well suited functionals $\mathcal{F}$. For notational brevity and if no misunderstanding is possible we will use the same symbol for functions and functionals if only the parameter structure varies. It turns out under the following basic and reasonable ansatz that $\mathcal{F}$ is of a specific and handsome type.

Ansatz

The functional $\mathcal{F}$ on a deformation φ should be the weighed sum over local functionals $\mathcal{F}(T, \varphi)$ depending solely on the deformation restricted to one single simplex T.

$$\mathcal{F}(\varphi) = \sum_{T \in \mathcal{T}} \mu_T \mathcal{F}(T, \varphi).$$

Here $\mu_T \in \mathbb{R}^+$ is a positiv weight, fixed for each simplex. We assume that $\sum_{T \in \mathcal{T}} \mu_T = 1$. For each $T \in \mathcal{T}$ there exists a fixed optimal shaped $\hat{T}$ of the same orientation, such that an orientation preserving deformation φ mapping T onto $\hat{T}$ would minimize $\mathcal{F}(T, \cdot)$ over the admissible deformations restricted to T.

In [7] we consider several well-grounded assumptions and principles, analogous to those well known in the theory of rational mechanics [5, 6, 8, 9]. Our local functional on the piecewise linear φ should be *translation invariant*, therefore there exists a function F such that $\mathcal{F}(T, \varphi) = F(T, \nabla \varphi)$. If we furthermore strictly rule out any other dependencies except those sketched in our ansatz and denote by R_T the reference mapping from $\hat{T}$ onto T, then the set of local functionals for each T can be replaced by one distinct function $\mathcal{F}(T, \varphi) = F(\nabla R_T(\varphi))$ where $R_T(\varphi) = \varphi \circ R_T$. Now a fundamental principle is that this function F should not depend on the frame or, in other words, on the observer's position while weighing the shape of $\varphi(T)$. And in addition it should be isotropic or more precisely independent of rotations of the coordinate system for the reference element $\hat{T}$. Then due to the Rivlin–Erikson Representation Theorem [5, 9] the function F can be expressed in terms of the principle invariants or equivalently

$$F(A) = F(\|A\|^2, \|\mathrm{Cof} A\|^2, \det A).$$

The arguments of F also have a geometrical meaning. $\|\nabla R_T(\varphi)\|$ measures the length change of the edges with respect to the reference simplex. $\|\mathrm{Cof} \nabla R_T(\varphi)\|$ controls the deformation of the boundary surfaces of an element. Finally $\det \nabla R_T(\varphi)$ could be understood as a change of the spatial angle or volume of a simplex. Therefore it seems natural that the functional depends on these quantities. Above it is pointed out that they are also sufficient to model a reasonable functional.

In our ansatz we have asked for a local functional which should be minimized by a deformation φ mapping T onto $\hat{T}$. With respect to $(\|\mathrm{Id}\|^2, \|\mathrm{Cof\,Id}\|^2, \det \mathrm{Id}) = (3, 3, 1)$ this leads us to the property $F(3, 3, 1) = \min_{a, c, d \in \mathbb{R}^+} F(a, c, d)$. Furthermore it is an additional and well grounded *regularity* assumption that a deformation $R_T(\varphi)$ with vanishing determinant should blow up the local functional value:

$$\lim_{\det A \to 0} F(A) = \infty.$$

It is a direct consequence that this condition rules out convexity of $F(\cdot)$ [5]. The class of *polyconvex* functionals seems to be simpliest class of functionals allowing to model all the necessary properties. Polyconvexity of $\mathcal{F}$ means that $F(\cdot, \cdot, \cdot)$ is convex in each argument.

$$F(a, c, d) = C(a - 3)^2 + h(d)$$
$$h(d) = g(\frac{1}{d}) + G(d)$$

with monotone $g, G : \mathbb{R}^+ \to \mathbb{R}$, $g'(1) = G'(1)$ and $\lim_{t \to \infty} g(t) = \infty$ is an example for such a function. Based on these considerations we state and proof in [7] the following theorem

Theorem
Suppose Ω is a bounded domain in $\mathbb{R}^3$ with a piecewise smooth boundary and $\mathcal{T}$ is a triangulation of Ω. Let us restrict ourselves to functionals being consistent with our ansatz

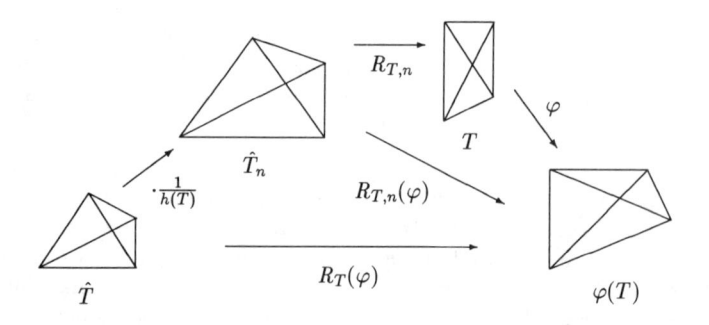

Fig. 1: The mapping $R_{T,n}(\varphi)$

and claim the principles of frame indifference and isotropy. Then the functional describing optimal variations is of the type

$$\mathcal{F}(\varphi) = \sum_{T \in \mathcal{T}} \mu_T F(\|\nabla R_T(\varphi)\|^2, \|Cof \nabla R_T(\varphi)\|^2, \det \nabla R_T(\varphi))$$

for $\mu_T \geq 0$ and $\sum_{T \in \mathcal{T}} \mu_T = 1$. If we claim in addition that F is polyconvex, $\mathcal{F}(Id) < \infty$, and that the regularity condition $\lim_{d \to 0} F(a, c, d) = \infty$ holds, then there exists an optimal variation φ^*. If φ^* has no self intersections at the boundary then φ^* is globally injective. For example $\Omega = \mathcal{T}$ rules out these self intersections. A subsequence of any minimization sequence converges to an optimal variation. The optimal variation is in general not unique.

We should mention here that depending on a very bad choice of the reference elements $\hat{T}$ local self intersections at the boundary of a triangulated, strictly convex 3D domain Ω could occur. Furthermore there are two major reasons which in general rule out uniqueness of an optimal variation. On the one hand depending on the set $\{\Gamma_i^0\}_{i \in I^0}$ the group of *rigid body motions* might not be ruled out. On the other hand the nonlinearity of the functionals under consideration may cause the optimal variation to be not unique.

In the case of a concrete application we choose a functional in the frame we have discussed up to now. What is left is to define an optimal simplex $\hat{T}$ for every $T \in \mathcal{T}$. Let us suppose the local functional to be in addition *independent of permutations* of the vertex indices of T then it can easily be demonstrated that up to scaling, rotation and translation the optimal simplex $\hat{T}$ is the normalized standard simplex $\hat{T}_n$

$$\hat{T} = h(T) \, \hat{T}_n \, .$$

Slightly differing from R_T and $R_T(\varphi)$ let us introduce the normalized reference mappings $R_{T,n}$ and $R_{T,n}(\varphi)$ which denote the unique affine mappings from $\hat{T}_n$ onto T, respectively $\varphi(T)$. Without any restriction we may assume that $R_T(\varphi) = \frac{R_{T,n}(\varphi)}{h(T)}$ (Fig. 1). What is finally left is to prescribe the optimal scale of $\hat{T}$ for each $T \in \mathcal{T}$. If we take into account the preservation of the volume of Ω and the minimum property of the function F we can evaluate the scale $h(T)$ of the reference element $\hat{T}$. Prescribing a concentration $c(T)$, proportional to the expected size of the optimal simplex, we define $\lambda(T)$

$$\lambda(T) = \frac{c(T)}{\sum_{T \in \mathcal{T}} c(T)}$$

and obtain [7] in terms of $\lambda(T)$ for $n = 2, 3$

$$h(\mathcal{T}) = \sqrt[n]{\lambda(T) \sum_{T \in \mathcal{T}} \det \nabla R_{T,n}(\text{Id})} \ .$$

The concentration $c(T)$ depends on the specific application we have in mind.

4. Numerical Examples and Applications

Let us now discuss some properties and applications of this mesh optimization strategy. Many effects are much easier to study in 2D than in 3D, therefore some of the following figures are two dimensional. In three dimensions the results are similar. The numerical algorithm is based on a nonlinear conjugate gradient approach with restart and a line search algorithm including an efficient step width control.

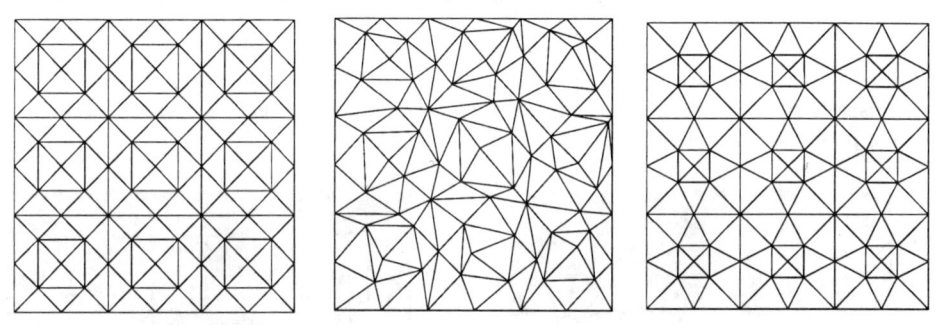

Fig. 2: *A triangulation of the domain $[0,1]^2$ (left), a random deformation (middle), the optimal triangulation (right)*

Let us first examine how the proposed optimization behaves on a simple mesh. In Fig. 2 we demonstrate that a standard triangulation is only slightly changed by the algorithm. If we then randomly disturb this mesh it will return by optimization to the starting grid.

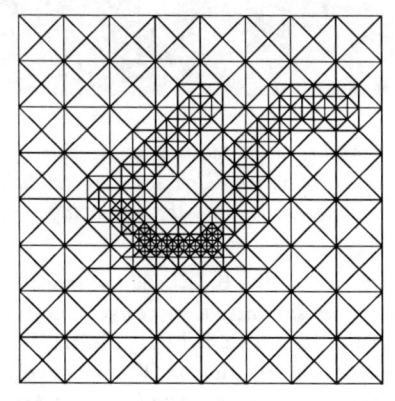

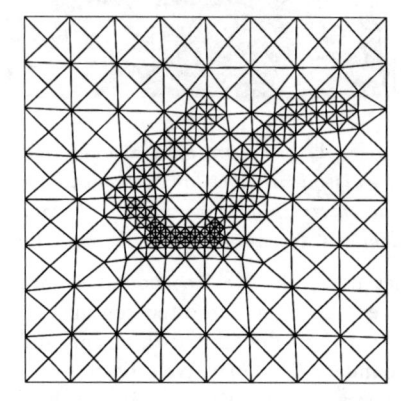

Fig. 3: *A locally refined 2D grid (left) and the corresponding optimal grid (right). The boundary of the domain is fixed. The refinement zone is not destroyed after the optimal deformation.*

If we deform a domain and want to conserve the level of refinement on an adaptively refined mesh we introduce the concentration $c(T) = b^{l(T)}$, where $l(T)$ is the level of refinement and b the refinement base. For example $b = \frac{1}{2}$ for simplices divided into two on the next level [2] or $b = \frac{1}{8}$ for the refinement based on 8 new simplices for each simplex [4] (Fig. 3).

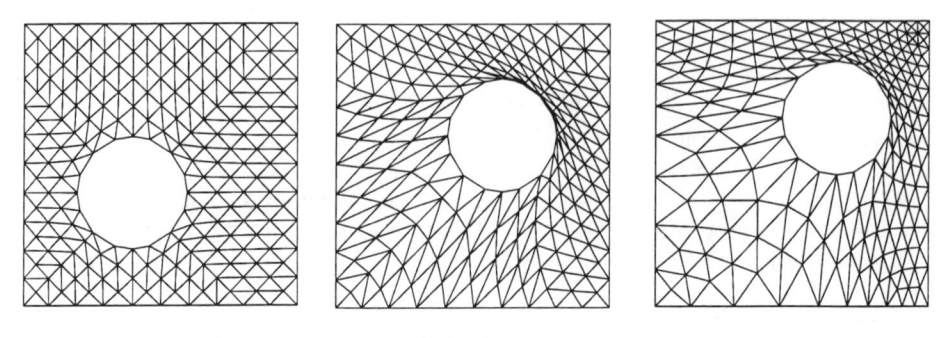

Fig. 4: *Starting with the 2D Domain Ω^0 (left) we apply the two methods. The result for the discrete Poisson problem (middle) and the optimal mesh (right) corresponding to one of the functionals discussed here.*

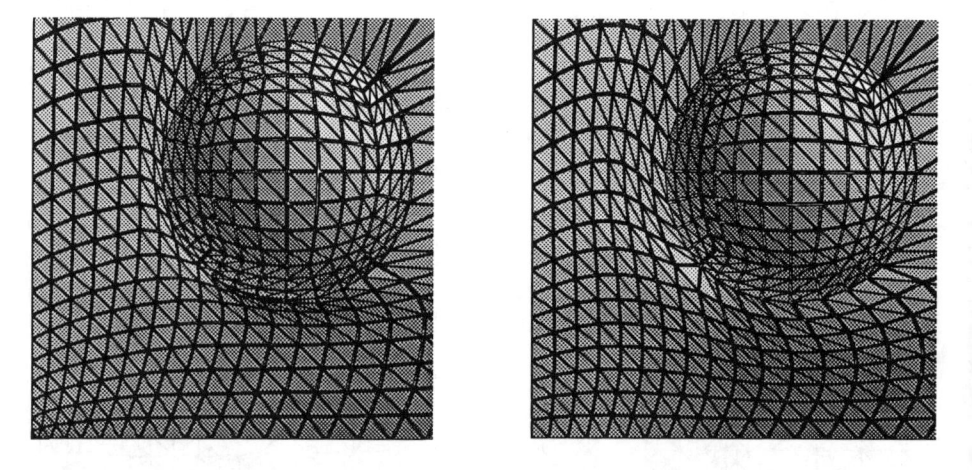

Fig. 5: *A significant slice through the two simplicial meshes in 3D gained by the solution of Poisson problem for the coordinates (left) and by the approach discussed here (right). The interesting zone arround the boundary of the ball is shown. On the left a layer of elements degenerates near the boundary.*

Let us think of a free boundary problem which we would like to attack numerically. Then the algorithm often contains a fixpoint iteration or similarly a succesive timestep calculation. We start with an initial domain $\Omega^0 = \Omega$ partially bounded by an initial free boundary $\Gamma^0 \subset \partial\Omega^0$. We solve for n starting at 0 a differential equation on Ω^n successively. Then we deform Γ^n according to the calculated solution and receive a new free boundary location Γ^{n+1} and an updated domain Ω^{n+1} and return to the previous step. In general we only know the deformation on Γ^n and have to remesh the deformed domain Ω^{n+1}. A first approach to do this is to solve a discrete Poisson problem for the coordinates of the mesh points on the old domain. Therefore we choose Dirichlet boundary conditions prescribing the point locations on $\partial\Omega^n$ including those on Γ^n computed just before [1]. If the deformation on Γ^n is small this leads to a smooth and useful new mesh on Ω^{n+1}. But in the case of large deformations on Γ^n this procedure will fail. As a model problem we inspect the domain $\Omega^0 = [-1,1]^{2(3)} \setminus \overline{B_r(x)}$ and $\Gamma^0 = \partial B_r(x)$ for $r < 1$. The prescribed deformation on Γ^0 should be a constant translation $x \to x - c\,(1,1,1)$ defining Γ^1. Let us think of an initial triangulation $\mathcal{T}$ of Ω^0. By the above sketched standard approach we would obtain for large c a very irregular triangulation especially in the zone near the translated boundary Γ^1. If c increases, the new mesh will even overlap itself. Fig. 4, 5 show a comparison of the latter two approaches in two respectively three dimensions.

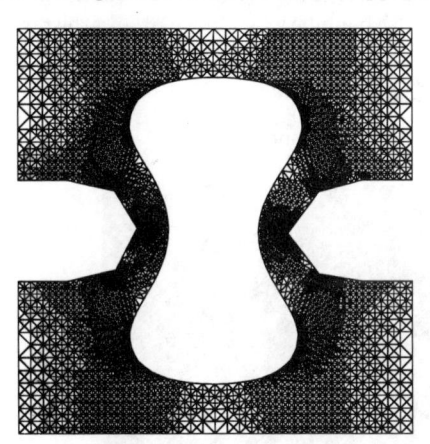

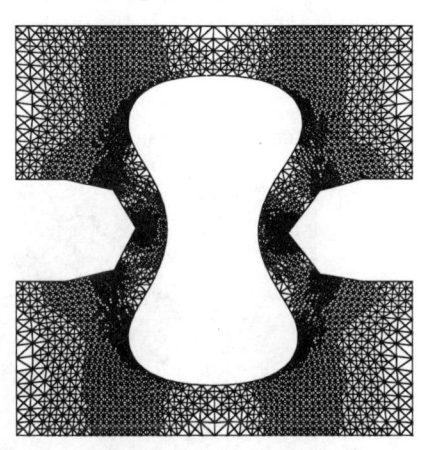

Fig. 6: *The grids generated by the adaptive solver for the Bernoulli free boundary problem, again based on grid deformation by solving discrete Poisson equations for the coordinates (left) and by our mesh optimization method (right)*

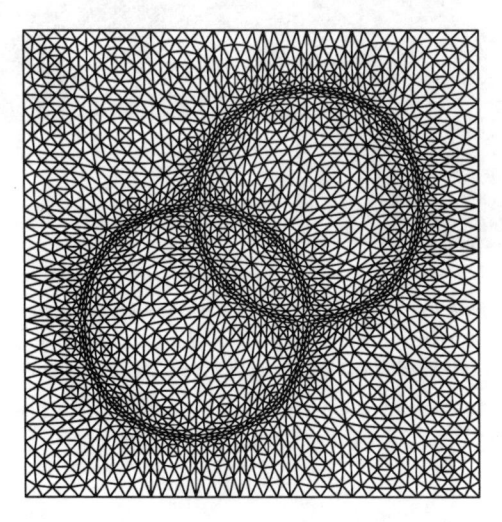

Fig. 7: *A uniform mesh is concentrated in a surrounding of two circles*

Now we turn to a typical example for a free boundary problem, which has been formulated by Bernoulli. The task is to find a set of constant area and minimal capacity. Starting with a coarse grid and an initial coarse free boundary an adaptive refinement strategy based on a residual error estimator and bisection refinement is applied to get an improved approximation of the free boundary (the inner boundary in Fig. 6) Requesting for the same error bound the strategy to solve discrete Poisson equations for this application needs 12 percent more elements than the optimization discussed in this paper. The underlying numerical algorithm has been developed in cooperation with Martin Flucher from the ETH at Zürich and will be discussed in a forthcoming paper.

It is also possible to condense a mesh in certain areas. Therfore choose a concentration of the following type to contract the mesh near the boundary, a surface or an arbitrary set S (Fig. 7).

$$c(T) = \Pi(\text{dist}(s_{\varphi^*(T)}, S)) .$$

Here s_T is the center of the simplex T and Π is a monotone function.

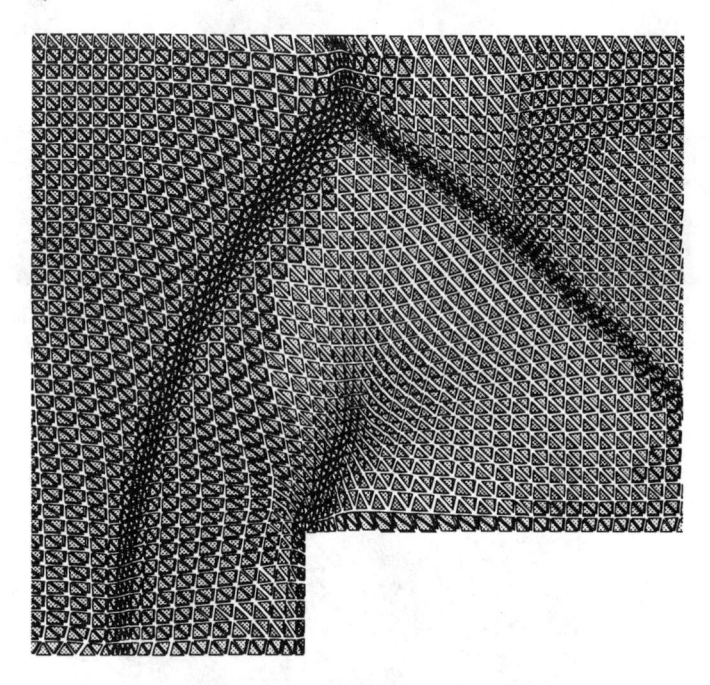

Fig. 8: *An optimal grid for a timestep of the numerical solution of the forward facing step problem in 3D. A slice of tetrahedra from the optimal triangulation is shown. The algorithm started on an equidistant mesh with given piecewise constant density.*

A similar problem is to concentrate a mesh in areas where an a posteriori error estimator indicates large errors. If $\eta(T)$ ist the estimator at an element T we define

$$c(T) = \pi(\eta(T))$$

for a monotonously decreasing function π with

$$\pi(\max_{T\in\mathcal{T}}\{\eta(T)\}) = c_{\min} \ll 1$$
$$\pi(\min_{T\in\mathcal{T}}\{\eta(T)\}) = 1.$$

As an example we will pick up here the well known *forward facing step* problem from Computational Fluid Dynamics. The underlying compressible Euler equations describe the supersonic flow over a step in a 3D channel. Let us suppose the numerical data to be piecewise constant on the simplices. Now we take an $\eta(T)$ which is related to the jumps of the data over the faces of the simplices (Fig. 8). Here we do not discuss how to use this method in a timedependent moving grid algorithm. We only point out that it is possible to adapt meshes according to numerical error information by our approach.

In all of the above examples and for dimensions $n = 2, 3$ we have choosen

$$F(a,c,d) = C_1 (a-n)^2 + \frac{2}{d} + C_2 (d-C_3)^2$$

except in the examples with an interior ball shaped boundary where the starting triangulation is overlapping itself. Here negative determinants occur. Therefore we have applied a homotopy of Functionals with an increasing penalty for small or negative determinants.

The pictures are produced with GRAPE, the graphic environment developed at the SFB 256 at Bonn University and the Institute for Applied Mathematics at Freiburg University. I would like to thank Monika Geiben, who has kindly supported me and provided the underlying numerical data, and Annette Schneider, who has ported the sources from 3D to 2D.

References

[1] J. E. Akin : *Application and Implementation of Finite Element Methods*, Academic Press, London, New York, 1982.

[2] E. Bänsch : *Local Mesh Refinenment in 2 and 3 Dimensions*, IMPACT of Computing in Science and Engineering, No. 3, 181-191, 1991.

[3] T. J. Barth : *Aspects of Unstructured Grids and Finite-Volume Solvers for the Euler and Navier–Stokes Equations*. AGARD Report 787, Special Course on Unstructured Grid Methods for Advection Dominated Flows, 1992.

[4] F. A. Bornemann, B. Erdmann, R. Kornhuber : *Adaptive Multilevel-Methods in Three Space Dimensions*, Preprint SC 92-14, Konrad–Zuse–Institut, Berlin, 1992.

[5] P. G. Ciarlet : *Mathematical Elasticity , Volume I : Three-Dimensional Elasticity*, North-Holland, Amsterdam (1988).

[6] Jerrold E. Marsden, Thomas J.R. Hughes : *Mathematical Foundation of Elasticity* Prentice-Hall, Englewood Cliffs, New Jersey (1983).

[7] M. Rumpf: *A Variational Approach to Optimal Meshes*, Preprint 331, SFB256, Bonn, 1993.

[8] C. Truesdell: *A First Course in Rational Continuum Mechanics*, Volume 1, Academic Press, London (1977).

[9] C. Truesdell, W. Noll: *The Non-Linear Field Theories of Mechanics, Handbuch der Physik*, Vol III.3, Springer, Berlin (1965).

[10] *Numerical Grid Generation*, VKI for Fluid Dynamics, Lecture Series 1990-06, 1990.

MULTIGRID CONVERGENCE RATES OF A SEQUENTIAL AND A PARALLEL NAVIER–STOKES SOLVER

Friedhelm Schieweck

Institute for Analysis and Numerical Mathematics, University Magdeburg

PSF 4120, D-39016 Magdeburg, Germany

Abstract

We consider a sequential and a parallel version of a multigrid method for a finite element discretization of the stationary incompressible Navier–Stokes equations. The sequential algorithm is based on a blockwise Gauss–Seidel smoother which passes successively through all elements of the domain using the already updated values from the previous elements. In the parallel version this strategy is only possible within each subdomain of some domain decomposition. We study how this algorithmical difference affects the convergence rate of the parallel multigrid method compared with the sequential one and investigate its effect on the numerical efficiency of our parallel solver. Especially we are interested in the case of a large number of subdomains.

1 Introduction

If we today consider the field of Computational Fluid Dynamics (CDF) and its requirements for the future we see that we always have to do with an increasing problem size. One reason is that we wish to have more and more grid points in order to get a higher accuracy of the solution. That means, for solving the corresponding problems, we need more and more storage capacity and CPU power. In order to satisfy these increasing requirements we need for the future a parallel computer with more and more processors. Therefore, we should construct our parallel CFD solver in such a way that it will be also reasonable for larger problems with a larger number p of employed processors. That means we want the *parallel efficiency* E to behave acceptable if we increase both the problem size and the number p of processors. This efficiency E is defined by

$$E = \frac{T_1^{seq}}{pT_p^{par}} \quad (1)$$

where T_1^{seq} denotes the CPU time of the sequential algorithm for solving the actual problem on one processor and T_p^{par} the CPU time of the parallel algorithm on p processors. For the sequential algorithm we should take a fast one (for instance a multigrid solver) in order to get a realistic impression about the quality of the parallel solver. The efficiency E can be regarded as the product of two other efficiencies

$$E = E^{num} E^{par} \tag{2}$$

with

$$E^{num} = \frac{T_1^{seq}}{T_1^{par}}, \qquad E^{par} = \frac{T_1^{par}}{pT_p^{par}}, \tag{3}$$

where T_1^{par} denotes the CPU time of the parallel algorithm on one processor (i.e. there are no communication costs). E^{num} is called the *numerical efficiency*. It shows us how fast the parallel algorithm converges compared with the sequential one. The quantity E^{par} is the *efficiency of the parallel implementation* which tells us how large the losses are that come from communication and idle times in the implementation of the parallel algorithm on p processors.

Our parallel solver is based on a decomposition of the domain into several subdomains each of which is assigned to another processor. In order to get a reasonable efficiency of the parallel implementation we have to avoid recursive data dependencies between the subdomain processors. That means, in the parallel version we have to modify the Gauss–Seidel type smoother of our sequential multigrid solver such that the unknowns are successively updated only within a subdomain. This leads to a difference in the convergence rates of the parallel and sequential algorithm. Hence the numerical efficiency E^{num} depends on the subdomain decomposition and therefore on the number p of processors. Thus, if we want to achieve a reasonable efficiency E also for large p, we have to ensure at least the condition that (because of $E^{par} \leq 1$) the numerical efficiency E^{num} does not become small for increasing p. The aim of this paper is to check this condition by means of a model problem. This leads to the investigation of the multigrid convergence rates for the sequential and the parallel algorithm. Of course, we have to guarantee also that E^{par} does not tend to zero for increasing p which is a nontrivial problem that has to be solved for the future (see [3] for a discussion). Therefore, the above condition for E^{num} is only a necessary condition for our parallel solver to be reasonable for the future where we want to solve larger problems on a larger number of processors.

In this paper we consider as a CFD model the stationary incompressible Navier–Stokes equations in primitive variables

$$-\nu\Delta u + (u\cdot\nabla)u + \nabla p = f, \quad \nabla\cdot u = 0 \quad \text{in } \Omega, \qquad u = g \quad \text{on } \partial\Omega \tag{4}$$

where u and p denote the unknown velocity and pressure field, respectively, in a bounded domain $\Omega \subset \mathbf{R}^2$, f a given body force, g the prescribed Dirichlet boundary data and $\nu=1/Re$ with the *Reynolds number Re*.

2 Grids and Discretization

We use a finite element method based on quadrilateral elements. For our multigrid algorithm we need a sequence of different fine grids. We start with a decomposition $\Omega = \bigcup_i M_i$ of the domain Ω into several quadrilateral elements M_i which are called *macroelements*. These macroelements define the finite element mesh on grid level $l = 1$. All finer grids on a level $l>1$ can be defined separately (i.e. also in parallel) on each macroelement by uniformly dividing the coarse grid cells of the previous grid level $l-1$ into four finer cells. At curved boundary parts we make the usual modification that we shift the bisection point of the corresponding coarse grid edge to the boundary. Figure 1 shows an example of such a grid generation. Thus we have defined some kind of blockstructured grids

grid level 1   grid level 2   grid level 3

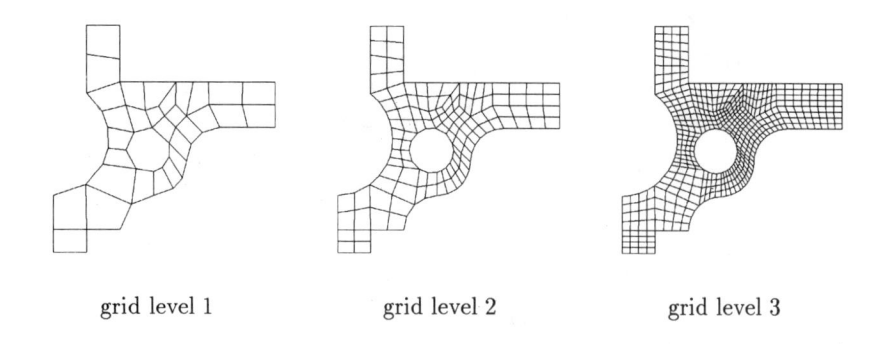

Figure 1: *Example of a multilevel grid sequence*

where the blocks correspond to the macroelements M_i. Such grids on the one hand give a relatively high flexibility for complex geometries and on the other hand they guarantee the local regularity of the grid on each macroelement which is important for the stability and convergence of our finite element discretization. Moreover, we get a natural and nearly optimal load balancing for a parallelization since we have the same number of elements on all macroelements. Therefore, it is reasonable to assign one processor to each macroelement.

For the finite element approximation of the Navier–Stokes problem (4) we use nonconforming rotated bilinear Q_1^{rot} elements for the velocity and piecewise constant Q_0 elements for the pressure. The degrees of freedom are the velocity values at the midpoints of the element edges and the pressure values at the centres of the elements (see Figure 2). This

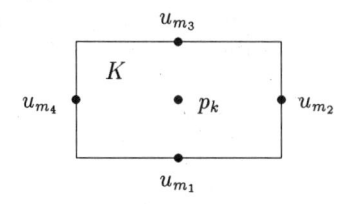

Figure 2: *Unknowns belonging to an element K*

element pair was introduced and analyzed by Rannacher and Turek [2] for the Stokes problem. They showed the uniform Babuška-Brezzi stability and optimal error estimates under the assumption that the mesh is sufficiently regular which is satisfied for our blockstructured grids. For the Navier–Stokes problem we use an upwind discretization of the additional convective term in order to stabilize the numerical scheme for high Reynolds numbers. We will omit here the details of the discretization (see for instance [5],[3],[4]) and give only the structure of the corresponding discrete Navier–Stokes problem. On some finest grid level $l=k$ we want to find a solution vector $y=(u,p)^\mathsf{T}$, consisting of the velocity vector u and the pressure vector p, which satisfies the following nonlinear system of equations

$$L(y)\,y = F \tag{5}$$

with

$$L(y) := \begin{pmatrix} A(u) & B \\ B^\mathsf{T} & 0 \end{pmatrix}, \quad y := \begin{pmatrix} u \\ p \end{pmatrix}, \quad F := \begin{pmatrix} f \\ g \end{pmatrix} \tag{6}$$

where the matrix block $A(u)$ contains the discretization of the nonlinear convection term.

3 Sequential Multigrid Solver

We use a linear multigrid method within an outer nonlinear iteration. On the finest grid level $l=k$ we solve our nonlinear problem (5) by means of the *simple iteration* where the idea is to linearize in the n-th step the nonlinear convection term by

$$(u \cdot \nabla)\,u \approx (u^{n-1} \cdot \nabla)\,u^n\,.$$

That means, in the n-th step of the outer iteration we first solve the linear system of equations

$$L(y^{n-1})\,\bar{y}^n = F \tag{7}$$

and get the new iterate $y^n = (u^n, p^n)^\mathsf{T}$ after an underrelaxation step

$$y^n = y^{n-1} + \omega(\bar{y}^n - y^{n-1})$$

with some parameter ω (we have used $\omega=0.9$).

For solving the linear problems (7) we apply a multigrid algorithm which is characterized as follows (for details see [3]). We use an F–cycle since it is a good compromise between the cheap V–cycle and the more accurate but expensive W–cycle. The coarse grid correction is damped by a factor $\alpha=0.8$. For the grid transfer we use standard finite element prolongation and restriction with a natural modification at points, where the finite element function is discontinuous (note that we have nonconforming elements). The coarsest grid problems are solved by an augmented Lagrange algorithm due to Fortin/Glowinski [1]. We use a blockwise Gauss–Seidel smoother similar to that proposed by Vanka [6].

In the following we want to explain the smoother in a little more detail. Let us indicate by

$$Ly = F \tag{8}$$

with $L=L(y^{n-1})$ the global system of equations which has to be solved on some actual grid level within the multigrid algorithm. With any finite element K on the current mesh we can associate the following modified local system

$$\tilde{L}_K y_K = \tilde{F}_K \tag{9}$$

with
$$\tilde{L}_K := \begin{pmatrix} D_K & b_K \\ b_K^T & 0 \end{pmatrix}, \quad y_K := \begin{pmatrix} u_K \\ p_K \end{pmatrix}, \quad \tilde{F}_K := \begin{pmatrix} \tilde{f}_K \\ \tilde{g}_K \end{pmatrix} \quad (10)$$

which corresponds to the local block y_K of all 9 unknowns belonging to the element K. These unknowns are the 8 velocity values at the 4 midpoints of the element edges and the one constant pressure value p_K on element K (see Figure 2). D_K denotes the diagonal matrix consisting of the diagonal entries of the corresponding 8×8 block A_K of the matrix block A of L in (8) and b_K the block of the column vector of matrix B which is associated with the unknown p_K. $\tilde{F}_K$ is derived from the corresponding block vector F_K of F in (8) by taking the parts of Ly, which are not associated with $\tilde{L}_K$, from the left to the right hand side. Therefore, $\tilde{F}_K$ depends on the vector y which was used for its generation. The local 9×9 system (9) has a very simple structure and can be solved by a few arithmetical operations. Now, our smoother in the sequential multigrid version is defined as follows.

Sequential Smoothing Procedure: $y^{\text{old}} \to y^{\text{new}}$

(a) initialize $y := y^{\text{old}}$;
  for all elements $K \subset \Omega$ **do**
    solve the local system $\tilde{L}_K y_K = \tilde{F}_K$ generated with y;
    update $y_K \to y$;
  enddo

(b) $y^{\text{new}} := y^{\text{old}} + \omega_S(y - y^{\text{old}})$;

In the underrelaxation step (b) we have used the value $\omega_S = 0.9$.

4 Parallel Multigrid Solver

We assign to each macroelement M_i a task T_i which is responsible for all operations of our algorithm with data belonging to $\overline{M}_i := M_i \cup \partial M_i$. Then we map the set of all tasks $\{T_i\}$ to a given set of processors $\{P_j\}$. Let us assume that we have enough processors such that we can assign to each task another processor (otherwise we would have to solve a load balancing problem to get a suitable mapping). Thus, the whole work and data of the multigrid algorithm is distributed to several tasks T_i which are executed in parallel on different processors. In order to perform operations with global Ω-data or with data belonging to the common edge of two macroelements the tasks have to exchange data by communication. However, the communication and also idle times for the synchronization within the algorithm are a matter of the parallel implementation and do not have any influence on the numerical algorithm.

From the algorithmical point of view, which is our primary interest in this paper, the parallel and sequential multigrid solver are different only in the smoothing procedure. All other parts are equivalent and differ only by their implementation. In the parallel smoother, we split up the one loop of the sequential version over all elements of the whole domain Ω into several loops over the elements within the macroelements which are executed in parallel. That means, all tasks T_i update at the same time their own unknowns on $\overline{M}_i$. Consequently, at velocity nodes on a common edge of two macroelements M_i and M_j the tasks T_i and T_j produce different values. In an additional step, these different values are equalized by taking the mean value. Let us denote by $\Theta(i)$ the set of all indices k of the global solution vector $y=(u,p)^\mathsf{T}$ for which the component y_k belongs to $\overline{M}_i$. Then, our parallel smoother can be described as follows.

Parallel Smoothing Procedure: $y^{\text{old}} \to y^{\text{new}}$

    for all macroelements M_i **do parallel**

    (a)   initialize $y^{M_i} = (u^{M_i}, p^{M_i})^\mathsf{T}$ by $y^{M_i} := y^{\text{old}}$;
          for all elements $K \subset M_i$ **do**
             solve the local system $\tilde{L}_K y_K = \tilde{F}_K$ generated with y^{M_i};
             update $y_K \to y^{M_i}$;
          enddo

    (b)   { update u at the edges of M_i }
          for all macroedges $\Gamma_{ij} = \partial M_i \cap \partial M_j$, $j \neq i$, of M_i **do**
             for all u-nodes $x_m \in \Gamma_{ij}$ **do**
$$u_m^{M_i} := \tfrac{1}{2}\left(u_m^{M_i} + u_m^{M_j}\right);$$
             enddo
          enddo

    (c)   **for** all $k \in \Theta(i)$ **do**
$$y_k^{\text{new}} := y_k^{\text{old}} + \omega_S \left(y_k^{M_i} - y_k^{\text{old}}\right);$$
          enddo
    enddo parallel

In contrast with the sequential smoother there is no spreading of new updated information over the macroelement boundaries within one iteration of the parallel version. This difference in general will affect the convergence rate of the parallel multigrid algorithm compared with the sequential one. We investigate this effect by means of numerical experiments in Section 6. Figure 3 shows for the sequential and the parallel smoother the order for passing through the elements in the case of 3 macroelements and grid level $l=4$. We use a multilevel order where we start with the elements located at the vertices of the level-1-elements. Then we take the elements at the vertices of the level-2-elements (in counterclockwise sense) followed by those at the vertices of the level-3-elements and so on. In some numerical tests, where we have compared different ordering strategies (red-black-like, zebra-like, row-wise within each macroelement), we obtained the best convergence results for the multilevel order. However, the differences in the convergence rates for the various ordering strategies have been relatively small.

5 A Jacobi Type Smoother

For the purpose of a comparison we consider a further multigrid method with a blockwise Jacobi type smoother. This smoother is defined as follows.

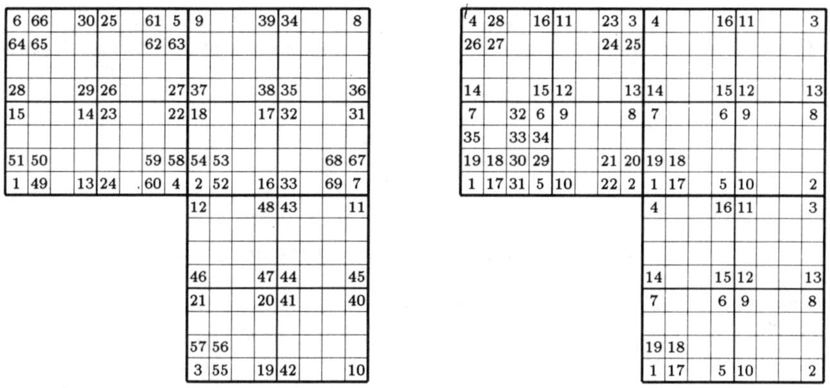

sequential version   *parallel version*

Figure 3: *Multilevel order for passing through the elements within the smoother*

Jacobi Type Smoothing Procedure: $y^{\text{old}} \to y^{\text{new}}$

(a)   **for** all elements $K \subset \Omega$ **do**
       solve the local system $\tilde{L}_K y_K = \tilde{F}_K$ generated with y^{old};
       save u_K;   update $y_K \to y$;
   enddo

(b)   { update u at inner nodes }
     for all inner u-nodes $x_m \in \Omega$ **do**
$$u_m := \tfrac{1}{2}\left(u_{K_1}(m) + u_{K_2}(m)\right);$$
     { where K_1, K_2 denote the two elements with $x_m \in \partial K_1 \cap \partial K_2$
     and $u_{K_i}(m)$, $i=1,2$, the velocity values in the local block vectors u_{K_i}
     corresponding to the node x_m }
   enddo

(c) $y^{\text{new}} := y^{\text{old}} + \omega_S(y - y^{\text{old}})$;

In the above defined smoother there is no spreading of new updated information from one element to another within one iteration step. The sequential version of the corresponding Jacobi type multigrid method and the parallelized one (according to the macroelement decomposition) would be algorithmically equivalent. Thus, we would expect that the convergence rate of this multigrid method is an upper bound for the convergence rates of our sequential and parallel multigrid algorithm for any macroelement decomposition of the domain, i.e. also for a large number of subdomains and processors, respectively.

6   Numerical Results

For our investigations we choose as a test model the well-known driven cavity problem with $\Omega = (0,1)^2$, $f = 0$, $g(x,y)_{|\partial\Omega} = 1$ for $y = 1$ and $g(x,y)_{|\partial\Omega} = 0$ otherwise. We consider uniform $N \times N$-decompositions of Ω into macroelements (see Figure 4) with $N \in \{2,4,5,6,8,10,12,20,24\}$, i.e. the number of macroelements $N^{mac} = N^2$ ranges be-

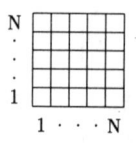

Figure 4: *Macroelement decomposition for the driven cavity problem*

tween 4 and 576. The numerical multigrid convergence rate ρ which we will compare for our different multigrid methods is defined as follows. Let y^i be the approximation of the solution y of (5) on the finest grid level after the i-th nonlinear iteration step and let $y^{i,0}$ and y^{i,m_i} be the approximations of y^i at the beginning and at the end of the linear multigrid iteration within the i-th nonlinear step. Then we define ρ by

$$\rho = \frac{1}{n}\sum_{i=1}^{n} \rho_i \quad \text{with} \quad \rho_i = \left(\frac{||L(y^{i-1})\, y^{i,m_i} - F||}{||L(y^{i-1})\, y^{i,0} - F||}\right)^{1/m_i} \qquad (11)$$

where $||\cdot||$ denotes the Euclidean norm and n the number of the nonlinear iteration steps. Figure 5 shows for the fixed finest grid level $l=4$ and the different Reynolds numbers $Re=20,200,2000$ the behaviour of the numerical convergence rates ρ of the sequential, parallel and Jacobi type multigrid algorithm for an increasing number N^{mac} of macroelements. Figure 6 shows the same for the fixed Reynolds number $Re=2000$ and the different finest grid levels $l=4,5,6$. Because of memory restrictions of our computer the highest possible value for N^{mac} is decreasing for an increasing finest grid level l. We see that our sequential multigrid solver works quite robust and efficient for the various Reynolds numbers and grid levels. The differences between the numerical convergence rates of the parallel and sequential version are in general very small. Only for $Re=2000$ and moderate values of N^{mac} they are somewhat larger. However, the differences decrease if the number N^{mac} of macroelements is increasing which is important for massive parallel applications. The differences to the convergence rate of the Jacobi type multigrid method are essentially larger especially for the high Reynolds number case. However, they decrease too for an increasing value of N^{mac} which is an interesting effect.

In the following we want to present the results for the numerical efficiency E^{num} of our parallel method compared with the sequential one. The definition (3) of E^{num} contains a problem from the practical point of view. On the one hand the CPU times T_1^{seq} and T_1^{par} have to be measured on a processor with the same performance as one node of the parallel computer. On the other hand such a processor in general has not enough memory to compute the very large global problems. Therefore, it is necessary to measure corresponding CPU times $\tilde{T}_1^{seq}$ and $\tilde{T}_1^{par}$ on a powerful serial computer with enough memory and to take $\tilde{E}^{num} = \tilde{T}_1^{seq}/\tilde{T}_1^{par}$ instead of E^{num}. We have implemented our multigrid methods on a Cray Y-MP2E/132. Since we have not used any special features of the Cray compiler we suppose that $\tilde{E}^{num}$ gives a relatively good approximation for E^{num}. Figure 7 shows for the fixed finest grid level $l=4$ with different Reynolds numbers and for the fixed Reynolds number $Re=2000$ with different finest grid levels the behaviour of the numerical efficiency E^{num} for an increasing number N^{mac} of macroelements. We obtain a good numerical efficiency for the various Reynolds numbers and grid levels

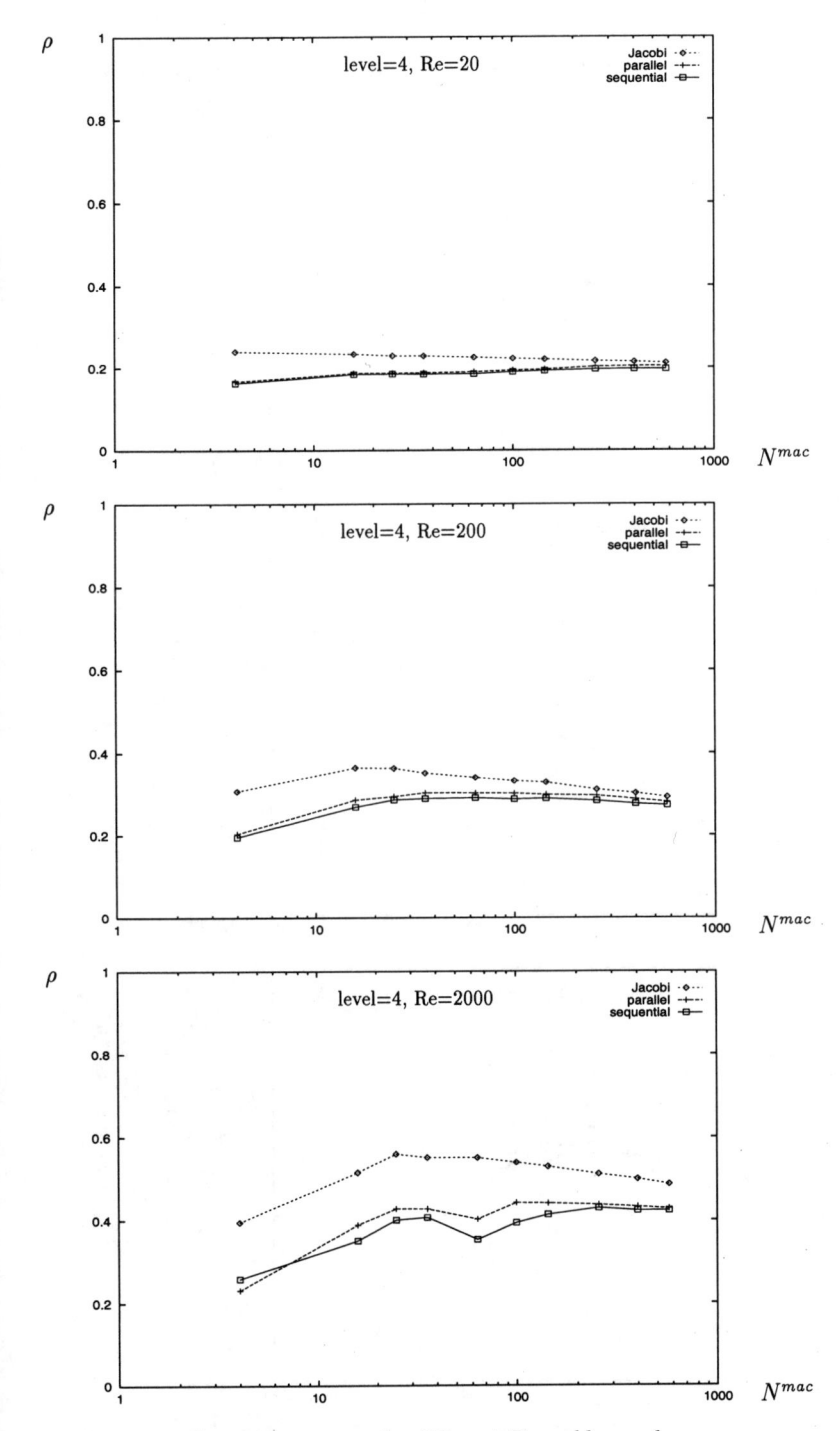

Figure 5: *Convergence rates for different Reynolds numbers*

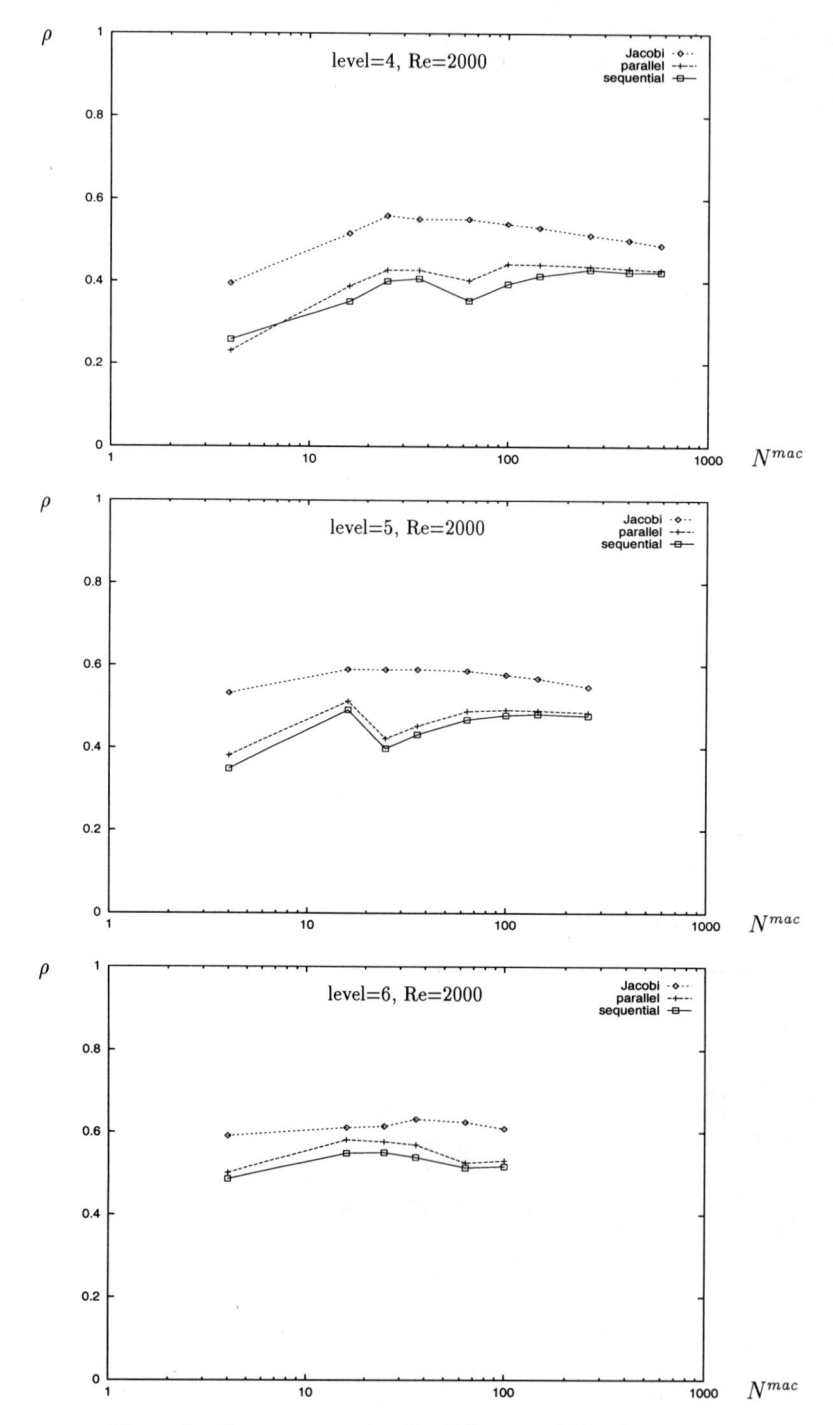

Figure 6: *Convergence rates for different grid levels*

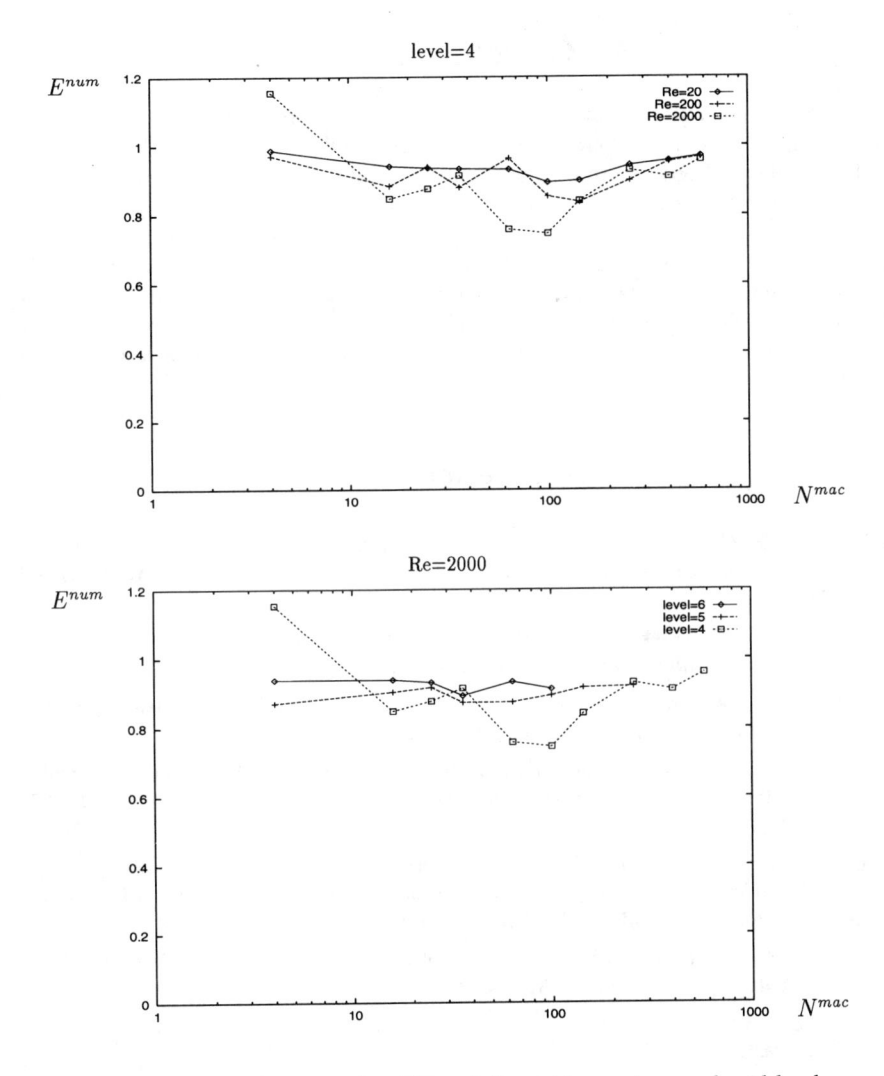

Figure 7: *Numerical efficiencies for different Reynolds numbers and grid levels*

which is a consequence of the small differences in the convergence rates of the parallel and sequential multigrid algorithm. The most important observation is that the numerical efficiency does not become small for an increasing number N^{mac} of subdomains which has been a main question of this paper.

7 Conclusions

The numerical tests indicate that for various Reynolds numbers, grid levels and macroelement decompositions of the domain the algorithmical difference between the sequential and the parallel multigrid method causes only a small difference in the multigrid convergence rates. As a consequence the parallel algorithm has a reasonable numerical efficiency even for a large number of subdomains and processors, respectively. That means, from the point of the numerical efficiency our parallel multigrid method could be a candidate for a suitable parallel Navier–Stokes solver on massive parallel systems and it has to be checked by further investigations whether it is possible to achieve also a reasonable efficiency of the parallel implementation of this solver for a large number of processors.

References

[1] M. Fortin, R. Glowinski, *Augmented Lagrangian Methods: Applications to the Numerical Solution of Boundary Value Problems.* North Holland, Amsterdam, New York, Oxford (1983).

[2] R. Rannacher, S. Turek, Simple Nonconforming Quadrilateral Stokes Element. *Numer. Methods Partial Differential Equations* **8,** 97-111 (1992).

[3] F. Schieweck, A parallel multigrid algorithm for solving the Navier–Stokes equations. *IMPACT Comput. Sci. Engrg.* **5,** 345-378 (1993).

[4] F. Schieweck, L. Tobiska, A nonconforming finite element method of upstream type applied to the stationary Navier–Stokes equation. *RAIRO Modél. Math. Anal. Numér.* **23,** 627–647 (1989).

[5] S. Turek, *Ein robustes und effizientes Mehrgitterverfahren zur Lösung der instationären, inkompressiblen, 2D Navier-Stokes-Gleichungen mit diskret divergenzfreien finiten Elementen.* Thesis, Preprint Nr. 642, SFB 123, Universität Heidelberg (1991).

[6] S. Vanka, Block-implicit multigrid calculation of two-dimensional recirculating flows. *Comput. Methods Appl. Mech. Engrg.* **59,** 29-48 (1986).

Parallel Iterative Solvers for Symmetric Boundary Element Domain Decomposition Methods

O. Steinbach
Mathematisches Institut A, Universität Stuttgart
Pfaffenwaldring 57, D–70569 Stuttgart, Fed. Rep. of Germany

Summary

The boundary integral equations concerning the potential problem can be written in a general form using the Calderon projector. This representation offers an explicit form of the symmetric Steklov–Poincaré operator, which maps the Dirichlet boundary data to the Neumann data. The resulting boundary integral variational formulation of the domain decomposition includes only the Steklov–Poincaré operators with respect to the local boundaries. Furthermore, this approach allows the coupling with the Finite Element Method, especially, if nonlinearities will appear.

Here we will describe two parallel iterative methods to solve the resulting discrete systems and we will discuss effective preconditioners. Numerical examples will show the efficiency of the announced algorithms.

1. Variational formulations

As a simple model problem let us consider the Dirichlet potential problem

$$\begin{aligned} -\operatorname{div} \alpha(x)\nabla u(x) &= 0 &&\text{for } x \in \Omega \subset \mathbb{R}^m, \\ u(x) &= g(x) &&\text{for } x \in \Gamma := \partial\Omega, \end{aligned} \quad (1.1)$$

where Ω is a bounded domain in $\mathbb{R}^m$ ($m = 2, 3$) with a piecewise Lipschitz continuous boundary Γ. Suppose

$$\alpha(x) = \alpha_i = \text{constant in } \Omega_i, \quad (1.2)$$

where the Ω_i are non–overlapping subdomains in a given domain decomposition

$$\overline{\Omega} = \bigcup_{i=1}^{p} \overline{\Omega}_i, \; \Gamma_i = \partial\Omega_i, \; \Omega_i \cap \Omega_j = \emptyset \text{ and } \Gamma_{ij} = \overline{\Omega}_i \cap \overline{\Omega}_j \text{ for } i \neq j. \quad (1.3)$$

Now we introduce the skeleton Γ_S and the global coupling boundary Γ_C by

$$\Gamma_S = \bigcup_{i=1}^{p} \Gamma_i, \; \Gamma_C = \Gamma_S \backslash \Gamma, \quad (1.4)$$

and a q–dimensional set ω of all cross points (coarse grid nodes).
The variational formulation of the model problem (1.1) reads:

Find $u \in H^1(\Omega)$ with $u_{|\Gamma} = g(x)$ such that

$$\int_\Omega \alpha(x) \nabla u(x) \nabla v(x) \, dx = 0 \tag{1.5}$$

for all test functions $v \in H_0^1(\Omega)$.

Before we apply the domain decomposition to (1.5), we should introduce boundary integral equations corresponding to the local potential problems. Using the fundamental solution of the Laplace equation,

$$E(x,y) = \begin{cases} -\frac{1}{2\pi} \log|x-y|, & m=2, \\ \frac{1}{4\pi} \frac{1}{|x-y|}, & m=3, \end{cases} \tag{1.6}$$

we define the local boundary integral operators, namely the simple layer potential

$$(V_i t_i)(x) = \int_{\Gamma_i} E(x,y) \, t_i(y) \, ds_y, \tag{1.7}$$

the double layer potential

$$(K_i u_{|\Gamma_i})(x) = \int_{\Gamma_i} \frac{\partial}{\partial n_y} E(x,y) \, u_{|\Gamma_i}(y) \, ds_y, \tag{1.8}$$

its adjoint

$$(K_i' t_i)(x) = \int_{\Gamma_i} \frac{\partial}{\partial n_x} E(x,y) \, t_i(y) \, ds_y, \tag{1.9}$$

and the hypersingular integral operator

$$(D_i u_{|\Gamma_i})(x) = -\frac{\partial}{\partial n_x} \int_{\Gamma_i} \frac{\partial}{\partial n_y} E(x,y) \, u_{|\Gamma_i}(y) \, ds_y. \tag{1.10}$$

Here, $(u_{|\Gamma_i}, t_i)$ are the Cauchy data with respect to the subdomain boundaries Γ_i for $i = 1, p$ and n_x describes the outward normal direction at the boundary point x. The local boundary integral equations can be written now by use of the Calderon projector [14].

$$\begin{pmatrix} u_{|\Gamma_i} \\ t_i \end{pmatrix} = \begin{pmatrix} \frac{1}{2}I - K_i & V_i \\ D_i & \frac{1}{2}I + K_i' \end{pmatrix} \begin{pmatrix} u_{|\Gamma_i} \\ t_i \end{pmatrix} = \mathbf{C_i} \begin{pmatrix} u_{|\Gamma_i} \\ t_i \end{pmatrix}. \tag{1.11}$$

¿From the invertibility of the simple layer potentials V_i, in the two–dimensional case we need therefore an additional condition, see e.g. [13], we get two explicit representations of the Steklov–Poincaré operators S_i:

$$\begin{aligned} t_i = S_i u_{|\Gamma_i} &= \left[V_i^{-1} \left(\tfrac{1}{2}I + K_i \right) \right] u_{|\Gamma_i} \\ &= \left[\left(\tfrac{1}{2}I + K_i' \right) V_i^{-1} \left(\tfrac{1}{2}I + K_i \right) + D_i \right] u_{|\Gamma_i} \end{aligned} \tag{1.12}$$

The properties of the boundary integral operators are well known [5], especially the Steklov–Poincaré operators are positive definite, bounded and selfadjoint.

Using the given domain decomposition (1.3) and Green's formula we may rewrite (1.5) into a variational formulation on the skeleton.

Find $u \in H^{1/2}(\Gamma_S)$ with $u_{|\Gamma} = g(x)$ such that

$$\sum_{i=1}^{p} \alpha_i \int_{\Gamma_i} S_i u_{|\Gamma_i} v_{|\Gamma_i} \, ds = 0 \qquad (1.13)$$

for all test functions $v \in H_0^{1/2}(\Gamma_S)$.

For the corresponding bilinear form we have the following result, from which follows the existence and uniqueness of the solution of (1.13) due to the Lax–Milgram theorem.

Theorem 1. *The bilinear form*

$$a_1(u, v) = \sum_{i=1}^{p} \alpha_i \int_{\Gamma_i} S_i u_{|\Gamma_i} v_{|\Gamma_i} \, ds \qquad (1.14)$$

is positiv definite and bounded with respect to the norm

$$\|u\|_{H^{1/2}(\Gamma_C)} = \left(\sum_{i=1}^{p} \|u_{|\Gamma_C \cap \Gamma_i}\|_{H^{1/2}(\Gamma_C \cap \Gamma_i)}^2 \right)^{1/2}, \qquad (1.15)$$

this means, that there hold inequalities

$$a_1(u, u) \geq \underline{c}_1 \|u\|_{H^{1/2}(\Gamma_C)}^2, \qquad (1.16)$$

$$a_1(u, v) \leq \overline{c}_1 \|u\|_{H^{1/2}(\Gamma_C)} \|v\|_{H^{1/2}(\Gamma_C)} \qquad (1.17)$$

for all functions $u, v \in H^{1/2}(\Gamma_C)$ and with positive constants $\underline{c}_1, \overline{c}_1$.

Finally we have to note, that it is not possible to discretize the bilinear form (1.14) directly, because both representations (1.12) of the local Steklov–Poincaré operators include a product of boundary integral operators, and moreover, the inverse of the simple layer potential.

2. Galerkin discretization of the symmetric formulation

The local Steklov–Poincaré operators in (1.14) correspond to Dirichlet problems with respect to the subdomains Ω_i. Such problems can be described by the first boundary integral equations in (1.11). Then we replace $S_i u_{|\Gamma_i}$ by t_i, and by using the second boundary integral equation in (1.11) we will get the well known symmetric formulation [4]:

Find $u \in H^{1/2}(\Gamma_S)$ with $u_{|\Gamma} = g$ and $t_i \in H^{-1/2}(\Gamma_i)$ for $i = 1, p$, such that

$$\sum_{i=1}^{p} \alpha_i \left\{ \langle D_i u_{|\Gamma_i}, v_{|\Gamma_i} \rangle_{\Gamma_i} + \tfrac{1}{2} \langle t_i, v_{|\Gamma_i} \rangle_{\Gamma_i} - \langle t_i, K_i v_{|\Gamma_i} \rangle_{\Gamma_i} \right\} = 0$$

$$\alpha_i \left\{ \langle V_i t_i, \tau_i \rangle_{\Gamma_i} - \tfrac{1}{2} \langle u_{|\Gamma_i}, \tau_i \rangle_{\Gamma_i} - \langle K_i u_{|\Gamma_i}, \tau_i \rangle_{\Gamma_i} \right\} = 0 \qquad (2.1)$$

for all test functions $v \in H_0^{1/2}(\Gamma_S)$, $\tau_i \in H^{-1/2}(\Gamma_i)$ and $i = 1, p$.

$\langle .,. \rangle_{\Gamma_i}$ denotes the usual L_2–scalar product over Γ_i. Existence and uniqueness is based on the Lax–Milgram theorem again, we have:

Theorem 2. *The bilinear form*

$$a_2(u,t;v,\tau) = \sum_{i=1}^{p} \alpha_i \left\{ \langle D_i u_{|\Gamma_i}, v_{|\Gamma_i} \rangle_{\Gamma_i} + \tfrac{1}{2} \langle t_i, v_{|\Gamma_i} \rangle_{\Gamma_i} + \langle t_i, K_i v_{|\Gamma_i} \rangle_{\Gamma_i} \right. \\ \left. + \langle V_i t_i, \tau_i \rangle_{\Gamma_i} - \tfrac{1}{2} \langle u_{|\Gamma_i}, \tau_i \rangle_{\Gamma_i} - \langle K_i u_{|\Gamma_i}, \tau_i \rangle_{\Gamma_i} \right\} \qquad (2.2)$$

is positiv definite and bounded with respect to the norm

$$\|(u,t)\|_{H^{1/2}(\Gamma_C) \times H^{-1/2}(\Gamma_1) \times \ldots \times H^{-1/2}(\Gamma_p)} = \\ \left(\sum_{i=1}^{p} \left\{ \|u_{|\Gamma_C \cap \Gamma_i}\|^2_{H^{1/2}(\Gamma_C \cap \Gamma_i)} + \|t_i\|^2_{H^{-1/2}(\Gamma_i)} \right\} \right)^{1/2}. \qquad (2.3)$$

For the discretization of (2.2) we need regular meshes $\Gamma_{h,i}$ of the local boundaries Γ_i. We request compatibility on the common parts $\Gamma_{h,ij}$, in the two–dimensional case these are the edges, in three dimensions the compatibility of the faces would imply the same for the edges.

Then the equivalent discrete system can be written in the block form

$$\begin{pmatrix} V_h & -\tfrac{1}{2} M_h - K_h \\ \tfrac{1}{2} M_h^\top + K_h^\top & D_h \end{pmatrix} \begin{pmatrix} \underline{t} \\ \underline{u} \end{pmatrix} = \begin{pmatrix} \underline{f}_t \\ \underline{f}_u \end{pmatrix}, \qquad (2.4)$$

where $\underline{t}$ denotes the vector of all local coefficients of the t_i and $\underline{u}$ the coupling values of the potential on the skeleton. The block entrees are defined by

$$\left. \begin{aligned} V_h &= \sum_{i=1}^{p} B_i^\top V_{h,i} B_i = \mathrm{diag}\,(V_{h,i})\,, \\ K_h &= \sum_{i=1}^{p} B_i^\top K_{h,i} A_i\,, \\ D_h &= \sum_{i=1}^{p} A_i^\top D_{h,i} A_i\,, \\ M_h &= \sum_{i=1}^{p} B_i^\top M_{h,i} B_i\,, \end{aligned} \right\} \qquad (2.5)$$

with the local stiffness matrices

$$\left. \begin{aligned} V_{h,i}[k,l] &= \langle V_i \varphi_k^\nu, \varphi_l^\nu \rangle_{\Gamma_i}\,, \\ K_{h,i}[k,l] &= \langle K_i \varphi_k^\mu, \varphi_l^\nu \rangle_{\Gamma_i}\,, \\ D_{h,i}[k,l] &= \langle D_i \varphi_k^\mu, \varphi_l^\mu \rangle_{\Gamma_i}\,, \\ M_{h,i}[k,l] &= \langle \varphi_k^\mu, \varphi_l^\nu \rangle_{\Gamma_i}\,, \end{aligned} \right\} \qquad (2.6)$$

which are derived from the boundary integral operators by the Galerkin method, ν and μ denote the degree of the ansatz and test functions. A_i, B_i are connectivity matrices, which describe the relations across the coupling boundaries. The system matrix in (2.4) is obviously block skew–symmetric, but positive definite. So we can use only a conjugate gradient type method to solve the system (2.4). To use a parallel preconditioned conjugate gradient method [10] we have to transform the matrix in (2.4). Two possibilities to do this will be described in the next sections.

3. Schur complement iteration

Note that the first block equation in (2.4) consists of local equations only, this means, that we can compute

$$\underline{t}_i = V_{h,i}^{-1}\left(\frac{1}{2}M_{h,i} + K_{h,i}\right)\underline{u}_{|\Gamma_i} + V_{h,i}^{-1}\underline{f}_{t|\Gamma_i} \qquad (3.1)$$

independently on the subdomains. Replacing the corresponding values in the second block equation, we get the Schur complement system

$$\left[\left(\frac{1}{2}M_h^\top + K_h^\top\right)V_h^{-1}\left(\frac{1}{2}M_h + K_h\right) + D_h\right]\underline{u} = \underline{f} \qquad (3.2)$$

with a symmetric and positive definite stiffness matrix S_h and the new right hand side

$$\underline{f} = \underline{f}_u - \left(\frac{1}{2}M_h^\top + K_h^\top\right)V_h^{-1}\underline{f}_t. \qquad (3.3)$$

To realize the matrix multiplication of S_h in an iteration process, we have to inverse the local simple layer potentials $V_{h,i}$. This action can be done in parallel, either by a factorization of the $V_{h,i}$ or by local preconditioned conjugate gradient methods. Such preconditioners $C_{V,i}$ have to be symmetric, positive definite and spectral equivalent to $V_{h,i}$, this means, that there hold inequalities

$$\underline{c}_{V,i}\left(C_{V,i}\underline{t}_i,\underline{t}_i\right) \leq \left(V_{h,i}\underline{t}_i,\underline{t}_i\right) \leq \overline{c}_{V,i}\left(C_{V,i}\underline{t}_i,\underline{t}_i\right) \qquad (3.4)$$

with positive constants $\underline{c}_{V,i}, \overline{c}_{V,i}$ for all vectors $\underline{t}_i$ according to the subdomain Ω_i and $i = 1, p$.

So we can use a parallel preconditioned conjugate gradient method to solve system (3.2), later on we will discuss different parallel preconditioners for S_h.

4. Bramble/Pasciak's transformation

Suppose, that we have a block diagonal matrix

$$C_V = \mathrm{diag}\,(C_{V,i})_{i=1,p}, \qquad (4.1)$$

with (3.4) then follows the spectral equivalence

$$\gamma_1\left(C_V\underline{t},\underline{t}\right) \leq \left(V_h\underline{t},\underline{t}\right) \leq \gamma_2\left(C_V\underline{t},\underline{t}\right) \qquad (4.2)$$

and we require

$$1 < \gamma_1 = \min_{i=1,p}\underline{c}_{V,i}. \qquad (4.3)$$

This condition can be fulfilled by scalar multiplications of the preconditioners $C_{V,i}$ after determining the smallest eigenvalue of the local preconditioned system matrix $C_{V,i}^{-1}V_{h,i}$

[20]. Applying the transformation proposed by Bramble/Pasciak in [3] to the system matrix in (2.4), the new matrix

$$A_h = \begin{pmatrix} C_V^{-1} V_h & -C_V^{-1}\left(\tfrac{1}{2}M_h + K_h\right) \\ \left(\tfrac{1}{2}M_h^T + K_h^T\right)\left(I - C_V^{-1}V_h\right) & \left(\tfrac{1}{2}M_h^T + K_h^T\right)C_V^{-1}\left(\tfrac{1}{2}M_h + K_h\right) + D_h \end{pmatrix} \quad (4.4)$$

is now self adjoint and positive definite with respect to the new inner product

$$\left[\begin{pmatrix} \underline{t} \\ \underline{u} \end{pmatrix}, \begin{pmatrix} \underline{\tau} \\ \underline{v} \end{pmatrix}\right] = ((V_h - C_V)\underline{t}, \underline{\tau}) + (\underline{u}, \underline{v}). \quad (4.5)$$

Moreover, there hold spectral equivalence inequalities

$$\lambda_1 \left[R\begin{pmatrix}\underline{t}\\\underline{u}\end{pmatrix}, \begin{pmatrix}\underline{t}\\\underline{u}\end{pmatrix}\right] \leq \left[A_h\begin{pmatrix}\underline{t}\\\underline{u}\end{pmatrix}, \begin{pmatrix}\underline{t}\\\underline{u}\end{pmatrix}\right] \leq \lambda_2 \left[R\begin{pmatrix}\underline{t}\\\underline{u}\end{pmatrix}, \begin{pmatrix}\underline{t}\\\underline{u}\end{pmatrix}\right] \quad (4.6)$$

with

$$R = \begin{pmatrix} I & 0 \\ 0 & \left(\tfrac{1}{2}M_h^T + K_h^T\right)V_h^{-1}\left(\tfrac{1}{2}M_h + K_h\right) + D_h \end{pmatrix} \quad (4.7)$$

and explicit known constants

$$\lambda_1 = \left(1 + \frac{\alpha}{2} + \sqrt{\alpha + \frac{\alpha^2}{4}}\right)^{-1}, \quad \lambda_2 = \frac{1 + \sqrt{\alpha}}{1 - \sqrt{\alpha}} \quad (4.8)$$

and α is given by

$$\alpha = 1 - \frac{1}{\gamma_2}. \quad (4.9)$$

The new right hand side for the transformed system is

$$\begin{pmatrix} \tilde{\underline{f}}_t \\ \tilde{\underline{f}}_u \end{pmatrix} = \begin{pmatrix} C_V^{-1}\underline{f}_t \\ \underline{f}_u - \left(\tfrac{1}{2}M_h^T + K_h^T\right)C_V^{-1}\underline{f}_t \end{pmatrix}. \quad (4.10)$$

Therefore we can use a parallel preconditioned conjugate gradient method with respect to the transformed matrix (4.4) in the special inner product (4.5) to solve the original system (2.4) [17]. Again we need a preconditioner for the discrete Schur complement S_h in (4.7).

5. Schur complement preconditioners

For preconditioning the Schur complement in (3.2), which was derived also as a part of the preconditioner for the transformed system (4.4), we decompose a function $u(x)$ with respect to a hierarchical basis, which is given by the domain decomposition (1.3),

$$u(x) = u_H(x) + \tilde{u}(x), \quad (5.1)$$

where we request

$$\tilde{u}(x) = 0 \quad \text{for all} \quad x_c \in \omega. \quad (5.2)$$

Obviously, this decomposition is not unique, so we can choose for the coarse grid function $u_H(x)$ the q–dimensional ansatz

$$u_H(x) = \sum_{k=1}^{q} u_k \varphi_k(x) \quad \text{with } u_k = u(x_{C,k}) \text{ and } k = 1, q, \tag{5.3}$$

with discrete harmonic basis functions $\varphi_k(x)$, which satisfy the following conditions:

- $\varphi_k(x_{C,l}) = \delta_{kl}$ for $k, l = 1, q$,
- $\varphi_k(x)$ piecewise linear on the skeleton,
- $\Delta \varphi_k(x) = 0$ in all subdomains Ω_l, $l = 1, p$.

(5.4)

In the two–dimensional case the bilinear form

$$b(u,v) = \sum_{i=1}^{p} \left\{ a_i(u_H, v_H) + \sum_{j=1}^{N_i} a_i(\tilde{u}_{|\Gamma_{ij}}, \tilde{v}_{|\Gamma_{ij}}) \right\} \tag{5.5}$$

is spectrally equivalent to the corresponding finite element domain decomposition form corresponding to (1.5) and to the the analogue boundary element form (1.14). N_i denotes the number of local cross nodes and we sum up over all edges Γ_{ij} of the subdomain boundary Γ_i. Moreover, the condition number is bounded by

$$C \left(1 + \ln (H/h)^2 \right), \tag{5.6}$$

where H describes the coarse mesh size and h the fine grid one [1],[2].

The realization of the preconditioner given in (5.5) includes first a transformation of the standard nodal basis into a hierarchical one, which corresponds to the function decomposition (5.1). Then we have to solve a coarse grid system given by

$$b_H(u,v) = \sum_{i=1}^{p} a_i(u_H, v_H) \tag{5.7}$$

and fine grid equations coming from

$$\tilde{b}(u,v) = \sum_{i=1}^{p} \sum_{j=1}^{N_i} a_i(\tilde{u}_{|\Gamma_{ij}}, \tilde{v}_{|\Gamma_{ij}}). \tag{5.8}$$

In the end we have to go back to the standard nodal basis. To solve the coarse grid system (5.7) we have to assembly the corresponding symmetric and positive definite stiffness matrix. This can be done by standard finite elements and would be very easy if the domain decomposition is uniform. But in general, this matrix can be generated by boundary elements again. According to (3.1) we have to solve local Dirichlet problems to realize the Steklov–Poincaré operators in (5.7) with respect to the hierarchical ansatz functions (5.4) and then we have to assembly over common boundaries and cross points, therefore global communication is necessary. This procedure to build up the global stiffness matrix is done once at the beginning and we can use the discretized boundary integral operators with respect to the standard nodal basis. The number of unknowns of the global system is less than q, so the action to inverse this system is not too expensive and can be done either by the simple Gauss elimination or an iterative scheme like the conjugate gradient method without preconditioning.

To realize the fine grid preconditioner (5.7) we have to note that this system is decoupled in the two–dimensional case. Then we find local preconditioners for the bilinear forms along the edges, e.g. the technique proposed by Dryja in [6]. A general technique is going back to [7],[8] and was used for boundary elements first in [14]. Then we have to solve Dirichlet–Neumann problems as well as pure Neumann problems with respect to the subdomains to realize the preconditioning of the fine grid nodes.

The examples given in the next section are based on a boundary element coarse grid system and the two dimensional edge preconditioning proposed by Dryja. As preconditioners $C_{V,i}$ for the local simple layer potentials $V_{h,i}$ we used circulant matrices, which are derived from the Galerkin discretization of the corresponding boundary integral operator on a circle [18].

6. Numerical results

As a simple example to show the efficiency of the described methods we consider problem (1.1) with a constant coefficient function $\alpha(x) \equiv 1$ and domain decompositions of the unit square into 4 and 16 subdomains. Method 1 is the preconditioned parallel conjugate gradient iteration with respect to the symmetric Schur complement system (3.2). Method 2 is the corresponding pcg iteration with respect to the transformed system matrix (4.4) and to the special chosen inner product (4.5). Both methods are iterative schemes which stop at a predefined bound with respect to a residual norm. For a relative error reduction of $\varepsilon = 10^{-6}$ we got the following results.

Table 1. Iteration numbers.

Subdomains	4		16	
Nodes	Method 1	Method 2	Method 1	Method 2
16	7	13	14	18
32	9	15	15	20
64	10	15	16	20
128	10	16	16	21
256	10	16	17	22

In Table 1 the first column describes the number of nodes with respect to the subdomains, this means, that 256 nodes on 16 subdomain boundaries correspond to 1024 boundary nodes of the original Dirichlet boundary Γ, moreover, an implementation on a *Intel Paragon* with up to 64 processors respective subdomains and a local discretization of 1024 nodes leads to 8192 boundary nodes and this corresponds to around 4 million finite element nodes. Such problems are often unsolvable on convential computers like workstations. On the other hand, one needs powerful processors with large memory, so workstation clusters would be an alternative to classical parallel computers.

A measure of the quality of a parallel algorithm and its implementation on a parallel system is the efficiency, which compares the work to solve a problem on one processor with the work to solve the equivalent problem on p processors, especially one subdomain per processor. The corresponding definition is

$$\text{Efficiency} = \frac{T_1}{p \times T_p}. \qquad (6.1)$$

In Table 2 the first column denotes the number of nodes on the original Dirichlet boundary. So the local discretization for 512 global nodes and 16 subdomains includes only 128 boundary nodes. Here we see the influence of the coarse grid system, we have small local work, but the work according the coarse grid system will grow with the number of subdomains.

Table 2. Efficiency.

Subdomains	4		16	
Nodes	Method 1	Method 2	Method 1	Method 2
64	0.62	0.65	0.35	0.38
128	0.72	0.78	0.51	0.57
256	0.80	0.84	0.63	0.73
512	0.89	0.89	0.73	0.81

All results presented here were computed on a *Parsytec MC3 DE* system with 28 processors T805.

References

[1] I. Babuska, A. Craig, J. Mandel and J. Pitkäranta, *Efficient preconditioning for the p-version finite element method in two dimensions*, SIAM J. Numer. Anal. **28:3** (1991), 624–661.

[2] J. H. Bramble, J. E. Pasciak and A. H. Schatz, *The construction of preconditioners for elliptic problems by substructuring* I, Math. Comput. **47** (1986), 103–134.

[3] J. H. Bramble and J. E. Pasciak, *A preconditioning technique for indefinite systems resulting from mixed approximations of elliptic problems*, Math. Comput. **50** (1988), 1–17.

[4] M. Costabel, *Symmetric methods for the coupling of finite elements and boundary elements*, Boundary Elements IX (C. A. Brebbia, G. Kuhn and W. L. Wendland eds.), Springer, Berlin, 1987, pp. 411–420.

[5] M. Costabel, *Boundary integral operators on Lipschitz domains: Elementary results*, SIAM J. Math. Anal. **19** (1988), 613–626.

[6] M. Dryja, *A capacitance matrix method for Dirichlet problem on polygon region*, Numer. Math. **39** (1982), 51–64.

[7] M. Dryja and O. B. Widlund, *Schwarz methods of Neumann–Neumann type for three-dimensional elliptic finite element problems*, Comm. Pure Appl. Math. (submitted).

[8] R. Glowinski and M. F. Wheeler, *Domain decomposition and mixed finite element methods for elliptic problems*, First International Symposium on Domain Decomposition Methods for Partial Differential Equations (R. Glowinski et.al. eds.), SIAM, Philadelphia, 1988, pp. 144–172.

[9] G. Haase, *Die nichtüberlappende Gebietszerlegungsmethode zur Parallelisierung und Vorkonditionierung iterativer Verfahren.*, PhD–Thesis, TU Chemnitz, 1993.

[10] G. Haase, U. Langer and A. Meyer, *Parallelisierung und Vorkonditionierung des CG–Verfahrens durch Gebietszerlegung.*, GAMM–Seminar "Numerische Algorithmen auf Transputer–Systemen", Heidelberg, Teubner, 1991.

[11] G. C. Hsiao, B. Khoromskij and W. L. Wendland, *Boundary integral operators and domain decomposition*, in preparation.

[12] G. C. Hsiao, O. Steinbach and W. L. Wendland, *Boundary element domain decomposition with Glowinski–Wheeler preconditioning*, in preparation.

[13] G. C. Hsiao and W. L. Wendland, *A finite element method for some integral equations of the first kind*, J. Math. Anal. Appl. **58** (1977), 449–481.

[14] G. C. Hsiao and W. L. Wendland, *Domain decomposition in boundary element methods*, Fourth International Symposium on Domain Decomposition Methods for Partial Differential Equations (R. Glowinski et.al., eds.), SIAM, Philadelphia, 1991, pp. 41–49.

[15] G. C. Hsiao and W. L. Wendland, *Domain decomposition via boundary element methods*, Numerical Methods in Engineering and Applied Sciences Part I (H. Alder et. al., eds.), CIMNE, Barcelona, 1992, pp. 198–207.

[16] B. N. Khoromskij and W. L. Wendland, *Spectrally equivalent preconditioners for boundary equations in substructuring techniques*, East–West J. Numer. Math. **1** (1992), 1–25.

[17] U. Langer, *Parallel iterative solution of symmetric coupled FE/BE-equations via domain decomposition*, SIAM J. Sci. Stat. Comput. (submitted).

[18] S. Rjasanow, *Effective iterative solution methods for boundary element equations*, Boundary Elements XIII (C. A. Brebbia and G. S. Gipson, eds.), CMP, Southampton, 1991, pp. 889–899.

[19] A. H. Schatz, V. Thomée and W. L. Wendland, *Mathematical Theory of Finite and Boundary Element Methods*, Birkhäuser, Basel, 1990.

[20] A. A. Samarskii and E. S. Nikolaev, *Numerical Methods for Grid Equations*, Birkhäuser, Basel, 1989.

[21] B. F. Smith, *Domain Decomposition Algorithms for the Partial Differential Equations of Linear Elasticity*, PhD–Thesis, Courant Institute, New York, 1991.

[22] O. Steinbach, *Gebietsdekompositionsmethoden in der BEM.*, Diplomarbeit, TU Chemnitz. Preprint 17, DFG–Schwerpunkt "Randelementmethoden", 1992.

[23] H. Yserentant, *On the multi–level splitting of finite element spaces*, Numer. Math. **49** (1986), 379–412.

The Equivalence of A Posteriori Error Estimators

R. VERFÜRTH

Fakultät für Mathematik, Ruhr-Universität Bochum, D-44780 Bochum, Germany

Summary

For finite element discretzations of partial differential equations, various a posteriori error estimators have been proposed in the literature. They can be classified as follows: (1) estimators based on evaluating the residual of the discrete solution with respect to the strong form of the pde, (2) estimators based on the solution of auxiliary local problems of the same type as the original pde and involving either natural or essential boundary conditions, (3) estimators based on the evelution of the residual of the discrete solution with respect to another higher-order discretization, (4) estimators based on a higher order recovery of the gradient. We prove that all these estimators are equivalent in the sense that their ratios can be bounded from above and from below by constants which only depend on the coefficients of the differential operator, on the element geometry, and on the polynomial degree of the finite element approximation.

1. Introduction

The efficiency of a numerical method for the solution of partial differential equations strongly depends on the choice of an "optimal" discretization, the use of a fast and efficient algorithm for the solution of the discrete problem, and a simple, but reliable method for judging the quality of the obtained numerical solution. These three objectives are often interdependent. The first and last one are related to the problem of a posteriori error estimation, i.e. of extracting from the given data of the problem and the computed numerical solution reliable bounds on the error of the numerical solution. Of course, the computation of the a posteriori error estimates should be much less costly than the solution of the original discrete problem.

Within the framework of finite element methods various strategies of a posteriori error estimation have been devised during the last 15-20 years. They can roughly be classified as follows:

(1) *residual estimates:* Estimate the error of the computed numerical solution by a suitable norm of its residual with respect to the strong form of the differential equation (cf. e.g. [1]).

(2) *solution of local problems:* Solve locally discrete problems similar to, but simpler than the original problem and use appropriate norms of the local solutions for error estimation (cf. e.g. [1, 3]).

(3) *hierarchical estimates:* Solve global auxiliary problems which involve the residual of the computed approximate solution and a higher order finite element space which either involves higher order approximations or is based on an auxiliary refined mesh (cf. e.g. [2, 4]).

(4) *averaging methods:* Use some local averaging technique for error estimation (cf. e.g. [11]).

For a simple model problem - the Poisson equation in two dimesnsions - we will prove that all these methods are equivalent in the sense that, up to multiplicative constants, they yield the same upper and lower bounds on the error of the numerical solution. The restriction to this model problem is made in order not to overload the presentation. Most of our results carry over to more complex problems such as e.g. the elasticity or the Stokes equations (cf. [10]). In this context it should be noted that, in order to be efficient, an a posteriori error estimation should yield upper and lower bounds on the error. Clearly, upper bounds are sufficient to ensure that the numerical solution achieves a prescribed tolerance. Lower bounds, however, are essential to guarantee that the error is not overestimated and that its local distribution is correctly resolved. Often, only upper bounds are established in the literature.

Various methods are used for constructing a posteriori error estimators and for proving that they yield upper and/or lower bounds on the error. These methods often depend on a particular class of problems and discretizations. A close inspection, however, reveals that they all rely on the following general principles:
(1) The stability and continuity of the infinite dimensional variational problem imply that a suitable norm of the error is bounded from above and from below by a corresponding dual norm of the residual of the computed approximate solution.
(2) The consistency of the discretization and sharp interpolation error estimates with respect to natural norms yield upper bounds on the dual norm of the residual.
(3) Local cut-off functions are used to prove that the error estimators also yield lower bounds on the error.

It is the aim of this paper to elucidate these common principles. In this context we are satisfied with proving that the upper and lower bounds differ by a multiplicative constant which is independent of the mesh-size. We neither intend to derive optimal estimates for this constant nor to prove efficiency of the error estimators, i.e. that the ratio of the true and the estimated error asymptotically tends to 1.

2. Preliminaries

As a model problem we consider the Poisson equation with homogeneous Dirichlet boundary conditions

$$-\Delta u = f \text{ in } \Omega, \quad u = 0 \text{ on } \Gamma \qquad (2.1)$$

in a connected, bounded, polygonal domain $\Omega \subset \mathbb{R}^2$ with boundary Γ.

For any open subset ω of Ω with Lipschitz boundary γ we denote by $L^2(\omega)$, $H^1(\omega)$, and $L^2(\gamma)$ the standard Lebesgue- and Sobolev-spaces equipped with the norms $\|.\|_{0,\omega} := \|.\|_{L^2(\omega)}$, $\|.\|_{1,\omega} := \|.\|_{H^1(\omega)}$, and $\|.\|_{0,\gamma} := \|.\|_{L^2(\gamma)}$. The inner products of $L^2(\omega)$ and $L^2(\gamma)$ are denoted by $(.,.)_\omega$ and $(.,.)_\gamma$.

Set

$$V := H_0^1(\Omega) := \{u \in H^1(\Omega) : u = 0 \text{ on } \Gamma\}.$$

The standard weak formulation of problem (2.1) then is: Find $u \in V$ such that

$$(\nabla u, \nabla v)_\Omega = (f, v)_\Omega \quad \forall v \in V. \qquad (2.2)$$

It is well-known that problem (2.2) admits a unique solution.

Denote by $\mathcal{T}_h$ a family of admissible and shape regular triangulations of Ω, i.e., any two triangles share at most a common edge or a common vertex and the minimal angle of all triangles is bounded away from zero. Denote by $P_{m,d}$, $m \geq 0, d \geq 1$, the space of polynomials of degree $\geq m$ in d variables and set

$$V_h := \{u \in V : u|_T \in P_{k,2} \, \forall T \in \mathcal{T}_h\}.$$

Here, $k \geq 1$ is fixed. We then consider the following finite element discretization of problem (2.2): Find $u_h \in V_h$ such that

$$(\nabla u_h, \nabla v_h)_\Omega = (f_h, v_h)_\Omega \quad \forall v_h \in V_h. \tag{2.3}$$

Here, f_h denotes the L^2-projection of f onto V_h. One easily checks that problem (2.3) admits a unique solution.

Next, we introduce some notations which will be used in the derivation and comparison of the error estimators. For $T \in \mathcal{T}_h$ we denote by $\mathcal{E}(T)$ and $\mathcal{N}(T)$ the set of its edges and vertices. Set

$$\mathcal{E}_h := \bigcup_{T \in \mathcal{T}_h} \mathcal{E}(T) \quad, \quad \mathcal{N}_h := \bigcup_{T \in \mathcal{T}_h} \mathcal{N}(T).$$

We split $\mathcal{E}_h$ and $\mathcal{N}_h$ in the form

$$\mathcal{E}_h = \mathcal{E}_{h,\Omega} \cup \mathcal{E}_{h,\Gamma} \quad, \quad \mathcal{N}_h := \mathcal{N}_{h,\Omega} \cup \mathcal{N}_{h,\Gamma}$$

with

$$\mathcal{E}_{h,\Gamma} := \{e \in \mathcal{E}_h : e \subset \Gamma\} \quad, \quad \mathcal{N}_{h,\Gamma} := \{x \in \mathcal{N}_h : x \in \Gamma\}.$$

For $T \in \mathcal{T}_h$ and $e \in \mathcal{E}_h$ we denote by h_T and h_e their diameter and their length. The minimal angle condition implies that the ratios $h_T/h_e, T \in \mathcal{T}_h, e \in \mathcal{E}(T)$, and $h_T/h_{T'}, T, T' \in \mathcal{T}_h, \mathcal{N}(T') \neq \emptyset$, are bounded from below and from above. For $T \in \mathcal{T}_h, e \in \mathcal{E}_h$, and $x \in \mathcal{N}_h$ set

$$\omega_T := \bigcup_{\mathcal{E}(T) \cap \mathcal{E}(T') \neq \emptyset} T', \, \omega_e := \bigcup_{e \in \mathcal{E}(T')} T', \, \omega_x := \bigcup_{x \in \mathcal{N}(T')} T',$$

$$\tilde{\omega}_T := \bigcup_{\mathcal{N}(T) \cap \mathcal{N}(T') \neq \emptyset} T', \, \tilde{\omega}_e := \bigcup_{e \cap T' \neq \emptyset} T'.$$

Denote by ψ_T and ψ_e cut-off functions which are uniquely defined by

$$\begin{aligned}\text{supp}\,\psi_T \subset T, \, \psi_T \in P_{3,2}, \, \psi_T \geq 0, \, \max_{x \in T} \psi_T(x) = 1, \\ \text{supp}\,\psi_e \subset \omega_e, \, \psi_{e|T'} \in P_{2,2} \, \forall T' \subset \omega_e, \, \psi_e \geq 0, \, \max_{x \in e} \psi_e(x) = 1.\end{aligned} \tag{2.4}$$

For any $e \in \mathcal{E}_{h,\Omega}$ we define a prolongation operator $P : C(e) \to C(\omega_e)$ as follows: Denote by T_1, T_2 the two triangles adjacent to e and let $\sigma \in C(e)$. Then the restriction of $P\sigma$ to $T_i, i = 1, 2$, is constant along lines parallel to the edge of T_i which has the left end-point of e as its right end-point. Here, the orientation of edges is induced by the exterior normal of the corresponding triangles.

The following lemma collects the main properties of the cut-off functions ψ_T and ψ_e and of the prolongation operator P. A proof may be found in [9; Lemma 4.1] and [10; Lemma 5.1].

2.1 Lemma. Let $m \in \mathbb{N}$. There exist constants $c_1, ..., c_8$ which only depend on m and on the minimal angle in the triangulation such that the following inequalities hold for all $T \in \mathcal{T}_h$, $e \in \mathcal{E}(T)$, $v \in P_{m,2}$, $\sigma \in P_{m,1}$:

$$\|\psi_T v\|_{0,T} \leq \|v\|_{0,T}, \quad \|\psi_T^{1/2} v\|_{0,T} \geq c_1 \|v\|_{0,T},$$
$$\|\psi_e \sigma\|_{0,e} \leq \|\sigma\|_{0,e}, \quad \|\psi_e^{1/2} \sigma\|_{0,e} \geq c_2 \|\sigma\|_{0,e},$$
$$c_3 h_T \|\nabla(\psi_T v)\|_{0,T} \leq \|\psi_T v\|_{0,T} \leq c_4 h_T \|\nabla(\psi_T v)\|_{0,T},$$
$$c_5 h_T \|\nabla(\psi_e P\sigma)\|_{0,T} \leq \|\psi_e P\sigma\|_{0,T} \leq c_6 h_T \|\nabla(\psi_e P\sigma)\|_{0,e},$$
$$c_7 h_e^{1/2} \|\sigma\|_{0,e} \leq \|\psi_e^{1/2} P\sigma\|_{0,T} \leq c_8 h_e^{1/2} \|\sigma\|_{0,e}.$$

Given $e \in \mathcal{E}_{h,\Omega}$ and $\varphi \in L^2(\omega_e)$ with $\varphi|_T \in C(T), \forall T \subset \omega_e$, we denote by $[\varphi]_e$ the jump of φ across e in an arbitrary, but fixed direction. With the solution u_h of problem (2.3) we finally associate edge and element residuals which are defined by

$$R_e(u_h) := \begin{cases} -[\partial_n u_h]_e & \forall e \in \mathcal{E}_{h,\Omega}, \\ 0 & \forall e \in \mathcal{E}_{h,\Gamma}, \end{cases} \quad (2.5)$$

and

$$R_T(u_h) := f_h + \Delta u_h \quad \forall T \in \mathcal{T}_h. \quad (2.6)$$

In comparing the various error estimators we will frequently use the following result which is proven in [6] and [5; Ex. 3.2.3].

2.2 Lemma. There is an interpolation operator $I_h : V \to V_h$ which satisfies for all $v \in V$ and all $T \in \mathcal{T}_h, e \in \mathcal{E}_h$ the following estimates

$$\|v - I_h v\|_{0,T} \leq c_9 h_T \|v\|_{1,\tilde\omega_T},$$
$$\|v - I_h v\|_{0,e} \leq c_{10} h_e^{1/2} \|v\|_{1,\tilde\omega_e}.$$

3. A residual Error Estimator

The residual error estimator of this section evaluates the residual of the solution u_h of Problem (2.3) with respect to the strong form of the differential equation (2.1). It was first proposed and analyzed by Babuška and Rheinboldt [1].
Let $u \in V$ and $u_h \in V_h$ be the unique solutions of problems (2.2) and (2.3). The stability of problem (2.2) implies that

$$\|u - u_h\|_{1,\Omega} \leq \sup_{v \in V} \frac{(\nabla(u - u_h), \nabla v)_\Omega}{\|v\|_{1,\Omega}} = \sup_{v \in V} \frac{(f, v)_\Omega - (\nabla u_h, \nabla v)_\Omega}{\|v\|_{1,\Omega}}. \quad (3.1)$$

Its continuity on the other hand yields for all $v \in V$

$$(f, v)_\Omega - (\nabla u_h, \nabla v)_\Omega = (\nabla(u - u_h), \nabla v)_\Omega \leq \|u - u_h\|_{1,\mathrm{supp} v} \|v\|_{1,\Omega} \quad (3.2)$$

and in particular

$$\sup_{v \in V} \frac{(f, v)_\Omega - (\nabla u_h, \nabla v)_\Omega}{\|v\|_{1,\Omega}} \leq \|u - u_h\|_{1,\Omega}.$$

The consistency of problem (2.3) implies that

$$(f, v_h)_\Omega - (\nabla u_h, \nabla v_h)_\Omega = 0 \tag{3.3}$$

holds for all $v_h \in V_h$. Finally, elementwise integration by part gives for all $v \in V$

$$\begin{aligned}
(f,v)_\Omega & - (\nabla u_h, \nabla v)_\Omega \\
&= \sum_{T \in \mathcal{T}_h} (f + \Delta u_h, v)_T - \sum_{e \in \mathcal{E}_{h,\Omega}} ([\partial_n u_h]_e, v)_e \\
&= \sum_{T \in \mathcal{T}_h} (R_T(u_h), v)_T + \sum_{e \in \mathcal{E}_h} (R_e(u_h), v)_e + \sum_{T \in \mathcal{T}_h} (f - f_h, v)_T.
\end{aligned} \tag{3.4}$$

From estimates (3.1), (3.3), and (3.4), equations (2.5) and (2.6), Lemma 2.2, and the Cauchy-Schwarz inequality we obtain an upper bound for the error

$$\|u - u_h\|_{1,\Omega} \le c \{ \sum_{T \in \mathcal{T}_h} h_T^2 \|R_T(u_h)\|_{0,T}^2 + \sum_{e \in \mathcal{E}_h} h_e \|R_e(u_h)\|_{0,e}^2 \\
+ \sum_{T \in \mathcal{T}_h} h_T^2 \|f - f_h\|_{0,T}^2 \}^{1/2}. \tag{3.5}$$

Now, consider an arbitrary triangle $T \in \mathcal{T}_h$. Set $w_T := \psi_T R_T(u_h)$. Lemma 2.1, equation (3.4), and inequality (3.2) then imply that

$$\begin{aligned}
c_1^2 \|R_T(u_h)\|_{0,T}^2 &\le (R_T(u_h), w_T)_T \\
&= (f, w_T)_\Omega - (\nabla u_h, \nabla w_T)_\Omega - (f - f_h, w_T)_\Omega \\
&\le [c_3^{-1} h_T^{-1} \|u - u_h\|_{1,T} + \|f - f_h\|_{0,T}] \|R_T(u_h)\|_{0,T}.
\end{aligned} \tag{3.6}$$

Next, consider an arbitrary edge $e \in \mathcal{E}(T)$. Set $w_e := \psi_e P(R_e(u_h))$. Lemma 2.1, equation (3.4), and estimate (3.2) yield

$$\begin{aligned}
c_2^2 \|R_e(u_h)\|_{0,e}^2 &\le (R_e(u_h), w_e) \\
&= (f, w_e)_\Omega - (\nabla u_h, \nabla w_e)_\Omega - (f - f_h, w_e)_\Omega \\
&\le c_8 h_e^{1/2} [c_5^{-1} h_e^{-1} \|u - u_h\|_{1,\omega_e} + \|R_T(u_h)\|_{0,\omega_e} + \|f - f_h\|_{0,\omega_e}] \\
&\quad \|R_e(u_h)\|_{0,e}.
\end{aligned} \tag{3.7}$$

Estimates (3.5) - (3.7) give the following result.

3.1 Proposition. *Define the residual error estimator $\eta_{R,T}, T \in \mathcal{T}_h$, by*

$$\eta_{R,T} := \{ h_T^2 \|R_T(u_h)\|_{0,T}^2 + \frac{1}{2} \sum_{e \in \mathcal{E}(T)} h_e \|R_e(u_h)\|_{0,e}^2 \}^{\frac{1}{2}}.$$

Then the following upper and lower bounds on the error hold with constants $\underline{c}$ and $\overline{c}$ which only depend on the polynomial degree k of V_h and on the minimal angle in $\mathcal{T}_h$:

$$\|u - u_h\|_{1,\Omega} \le \overline{c} \{ \sum_{T \in \mathcal{T}_h} [\eta_{R,T}^2 + h_T^2 \|f - f_h\|_{0,T}^2] \}^{1/2},$$

$$\eta_{R,T} \le \underline{c} \{ \|u - u_h\|_{1,\omega_T}^2 + \sum_{T' \subset \omega_T} h_{T'}^2 \|f - f_h\|_{0,T'}^2 \}^{1/2}.$$

3.2 Remark. The function f_h may be replaced by any other finite dimensional approximation of f such as, e.g., its L^2-projection onto piecewise constant functions. □

4. Error Estimators Based on the Solution of Local Problems

The first estimator of this section is originally due to Babuška and Rheinboldt [1] and is based on the solution of local Dirichlet-problems. For any $x \in \mathcal{N}_{h,\Omega}$ set

$$V_x := \text{span}\{\psi_T v, \psi_e P\sigma : x \in \mathcal{N}(T), v \in P_{\max\{k-2,0\},2}, x \in e \in \mathcal{E}_{h,\Omega}, \sigma \in P_{k-1,1}\}. \quad (4.1)$$

Denote by $v_x \in V_x$ the unique solution of

$$(\nabla v_x, \nabla w)_{\omega_x} = (f_h, w)_{\omega_x} - (\nabla u_h, \nabla w)_{\omega_x} \quad \forall w \in V_x \quad (4.2)$$

and set

$$\eta_{D,x} := \|\nabla v_x\|_{0,\omega_x}. \quad (4.3)$$

Problem (4.2) may be viewed as a local higher order variant of problem (2.3). The function $u_h + v_x$ may also be interpreted as a discrete approximation to the solution φ of the following local Dirichlet-problem

$$-\Delta\varphi = f \text{ in } \omega_x, \ \varphi = u_h \text{ on } \partial\omega_x \cap \Omega, \ \varphi = 0 \text{ on } \partial\omega_x \cap \Gamma.$$

The second estimator of this section is originally due to Bank and Weiser [3] and is based on the solution of local Neumann-problems. For any $T \in \mathcal{T}_h$ set

$$V_T := \text{span}\{\psi_T v, \psi_e P\sigma : v \in P_{\max\{k-2,0\}}; \ e \in \mathcal{E}(T) \cap \mathcal{E}_{h,\Omega}, \sigma \in P_{k-1,1}\}. \quad (4.4)$$

Denote by $v_T \in V_T$ the unique solution of

$$(\nabla v_T, \nabla w)_T = (R_T(u_h), w)_T + (R_e(u_h), w)_{\partial T} \quad \forall w \in V_T \quad (4.5)$$

and set

$$\eta_{N,T} := \|\nabla v_T\|_{0,T}. \quad (4.6)$$

Note that problem (4.5) admits a unique solution since the functions in V_T vanish at the vertices of T. The function v_T may be interpreted as a discrete approximation to the solution φ of the following Neumann-problem

$$-\Delta\varphi = R_T(u_h) \text{ in } T, \ \partial_n\varphi = R_E(u_h) \text{ on } \partial T \cap \Omega, \ \varphi = 0 \text{ on } \partial T \cap \Gamma.$$

The estimators $\eta_{D,x}, \eta_{N,T}$, and $\eta_{R,T}$ are essentially equivalent:

4.1 Proposition. *There are constants $c_{D,1}, c_{D,2}, c_{N,1}$, and $c_{N,2}$ which only depend on the polynomial degree k of V_h and on the minimal angle on $\mathcal{T}_h$ such that the following estimates hold for all $x \in \mathcal{N}_{h,\Omega}$ and all $T \in \mathcal{T}_h$*

$$\eta_{D,x} \leq c_{D,1}\{\sum_{x \in \mathcal{N}(T)} \eta_{R,T}^2\}^{1/2}, \quad \eta_{R,T} \leq c_{D,2}\{\sum_{x \in \mathcal{N}(T) \cap \mathcal{N}_{h,\Omega}} \eta_{D,x}^2\}^{1/2},$$

$$\eta_{N,T} \leq c_{N,1}\eta_{R,T}, \quad \eta_{R,T} \leq c_{N,2}\{\sum_{\mathcal{E}(T) \cap \mathcal{E}(T') \neq \emptyset} \eta_{N,T'}^2\}^{1/2}.$$

Proof. Since the functions in V_x vanish at all vertices the norms $\{\sum_{x \in \mathcal{N}(T)} h_T^{-2} \|\cdot\|_{0,T}^2\}^{1/2}$ and $\|\cdot\|_{1,\omega_x}$ are equivalent on V_x with constants which only depend on k and on the minimal angle. Moreover, elementwise integration by parts yields

$$(f_h, w)_{\omega_x} - (\nabla u_h, \nabla w)_{\omega_x} = \sum_{x \in \mathcal{N}_T} (R_T(u_h), w)_T + \sum_{x \in e \in \mathcal{E}_h} (R_e(u_h), w)_e \quad \forall w \in V_x.$$

Inserting v_x as a test-function in equation (4.2) and taking into a account these two observations yields the first estimate. The second one follows from inequalities (3.6) and (3.7) and the fact that the functions w_T and w_e of section 3 are contained in V_x. Inserting v_T as a test-function in equation (4.5) and observing that $\|\cdot\|_{1,T}$ and $h_T^{-1}\|\cdot\|_{0,T}$ are equivalent norms on V_T establishes the third estimate. The fourth estimate finally follows with the same arguments as the second one. □

5. Hierarchical Basis Error Estimators

Hierarchical basis error estimators as proposed e.g. in [2, 4] bound the error $u - u_h$ by solving global auxiliary problems which involve the residual of u_h and a higher order finite element space $W_h \subset V$. Here, "higher order" means that the functions in W_h either have a higher polynomial degree than those in V_h or that they correspond to a refinement of the triangulation $\mathcal{T}_h$. This notion will be made clearer by the examples which are given at the end of this section.

To make things more precise, we assume that the space W_h satisfies the following conditions:

(1) The norms $\|\cdot\|_{1,\Omega}$ and $\{\sum_{T \in \mathcal{T}_h} h_T^{-2} \|\cdot\|_{0,T}^2\}^{1/2}$ are equivalent on W_h.
(2) W_h contains the functions $w_T, T \in \mathcal{T}_h$, $w_e, e \in \mathcal{E}_{h,\Omega}$, of section 3.

Condition (1) is satisfied if, e.g., the functions in W_h vanish at the vertices of the triangulation. It assures that the H^1-scalar product can efficiently be preconditioned by a smesh-dependent L^2-scalar product. As we will see later, condition (2) can be modified.

Let $\overline{v}_h \in W_h$ be the unique solution of

$$(\nabla \overline{v}_h, \nabla w_h)_\Omega = (f_h, w_h)_\Omega - (\nabla u_h, \nabla w_h)_\Omega \quad \forall w_h \in W_h \tag{5.1}$$

and set

$$\eta_H := \|\nabla \overline{v}_h\|_{0,\Omega} \quad . \quad \eta_{H,T} := \|\nabla \overline{v}_h\|_{0,T}. \tag{5.2}$$

In contrast to problems (4.2) and (4.5) problem (5.1) is a global one. Thanks to condition (1), however, it can rather efficiently be solved by a preconditioned conjugate-gradient algorithm. The function $u_h + \overline{v}_h$ may be interpreted as a higher order approximation to the solution u of (2.2) which is obtained by replacing in problem (2.3) V_h by $V_h \oplus W_h$. The following proposition shows that $\eta_{H,T}$ yields upper and lower bounds on the error which are similar to those given by $\eta_{R,T}$. In combination with Proposition 3.1, it shows that $\eta_{H,T}$ and $\eta_{R,T}$ are equivalent.

5.1 Proposition. *There are two constants $c_{H,1}$ and $c_{H,2}$ which only depend on the constants in the norm-equivalence of condition (1), on the polynomial degree k, and on* the minimal angle such that

$$\eta_H \leq c_{H,1}\{\|u - u_h\|_{1,\Omega}^2 + \sum_{T \in \mathcal{T}_h} h_T^2 \|f - f_h\|_{0,T}^2\}^{1/2},$$

$$\eta_{R,T} \leq c_{H,2}\{\sum_{\mathcal{E}(T') \cap \mathcal{E}(T) \neq \emptyset} \eta_{H,T'}^2\}^{1/2}.$$

Proof. The first estimate immediately follows from inequality (3.2), equations (5.1) and (5.2), and condition (1).
Observe that the right-hand side of (5.1) admits a representation analogous to (3.4):

$$(f_h, w_h)_\Omega - (\nabla u_h, \nabla w_h)_\Omega = \sum_{T \in \mathcal{T}_h} (R_T(u_h), w_h)_T + \sum_{e \in \mathcal{E}_h} (R_e(u_h), w_h)_e \quad \forall w_h \in W_h.$$

Together with Lemma 2.1 and condition (2) this yields for all $T \in \mathcal{T}_h$ and $e \in \mathcal{E}_{h,\Omega}$

$$\begin{aligned} c_1^2 \|R_T(u_h)\|_{0,T}^2 &\leq (R_T(u_h), w_T)_T = (\nabla \bar{v}_h, \nabla w_T)_\Omega \\ &\leq c_3^{-1} h_T^{-1} \eta_{H,T} \|R_T(u_h)\|_{0,T}, \\ c_2^2 \|R_e(u_h)\|_{0,e}^2 &\leq (R_e(u_h), w_e)_e \\ &= (\nabla \bar{v}_h, \nabla w_e)_\Omega - (R_T(u_h), w_e)_\Omega \\ &\leq c_8 h_e^{1/2} \{\sum_{e \in \mathcal{E}(T')} c_5^{-2} h_{T'}^{-2} \eta_{H,T'}^2 + \|R_{T'}(u_h)\|_{0,T'}^2\}^{1/2} \|R_e(u_h)\|_{0,e}. \end{aligned}$$

This proves the second estimate. □

When using higher order finite elements, i.e. $k > 1$, condition (2) might be prohibitive. It can be replaced by the following saturation assumption:
(2') There is a constant $\beta < 1$ such that

$$\|\nabla(u - u_h - \bar{v}_h)\|_{0,\Omega} \leq \beta \|\nabla(u - u_h)\|_{0,\Omega}.$$

Condition (2') and the triangle inequality immediately yield

$$\|\nabla(u - u_h)\|_{0,\Omega} \leq \eta_H + \beta \|\nabla(u - u_h)\|_{0,\Omega}$$

and thus

$$\|u - u_h\|_{1,\Omega} \leq \frac{1}{1 - \beta} \eta_H.$$

Combined with the first estimate of Proposition 5.1, which only used condition (1), this shows that η_H still yields upper and lower bounds on the error. In conjunction with Proposition 3.1 it also establishes the equivalence of η_H and $\eta_R := \{\sum_{T \in \mathcal{T}_h} \eta_{R,T}^2\}^{1/2}$. Note, that one can prove that condition (2') is indeed equivalent to the fact that η_H yields an upper bound on the error $u - u_h$ (cf. [4]).

We end this section with some examples of spaces W_h.

5.2 Example: (1) $W_h := \text{span}\{\psi_T v, \psi_e P\sigma : T \in \mathcal{T}_h, v \in P_{\max\{k-2,0\},2}; e \in \mathcal{E}_{h,\Omega}, \sigma \in P_{k-1,1}\}$ satisfies conditions (1) and (2).

(2) $W_h := \{\varphi \in V : \varphi|_T \in P_{k+1,2} \, \forall T \in \mathcal{T}_h, \varphi(x) = 0 \, \forall x \in \mathcal{N}_h\}$ satisfies conditions (1) and (2').

(3) Let $\mathcal{T}_{h/2}$ be the triangulation which is obtained by a regular refinement of $\mathcal{T}_h$, i.e., each $T \in \mathcal{T}_h$ is cut into four congruent triangles by joining the midpoints of its edges. Let $\mathcal{N}_{h/2}$ be the set of all vertices in $\mathcal{T}_{h/2}$. Then $W_h := \{\varphi \in V : \varphi|_T \in P_{k,2} \, \forall T \in \mathcal{T}_h, \varphi(x) = 0 \, \forall x \in \mathcal{N}_{h/2} \setminus \mathcal{N}_h\}$ satisfies conditions (1) and (2'). □

6. An Error Estimator Based on a Higher Order Recovery of the Gradient

Throughout this section we assume that $k = 1$, i.e., V_h consists of linear finite elements. The estimator of this section was originally proposed by Zienkiewicz and Zhu [11]. The analysis of this section follows the lines of [8]. Similar estimators may be found in [7]. Let W_h be the space of the piecewise linear vectorfields and set $X_h := W_h \cap C(\Omega, \mathbb{R}^2)$. Note, that $\nabla V_h \subset W_h$. Define a mesh-dependent scalar product $(.,.)_h$ on W_h by

$$(\underline{v}, \underline{w})_h := \sum_{T \in \mathcal{T}_h} \frac{|T|}{3} \{ \sum_{x \in \mathcal{N}(T)} \underline{v}|_T(x) \cdot \underline{w}|_T(x) \}. \tag{6.1}$$

Here, $|T|$ is the area of T. One easily checks that

$$(\underline{v}, \underline{w})_h = (\underline{v}, \underline{w})_\Omega \tag{6.2}$$

if both arguments are elements of W_h and at least one of them is piecewise constant. Moreover,

$$\frac{1}{4} \|\underline{v}\|_{0,\Omega}^2 \leq (\underline{v}, \underline{v})_h \leq \|\underline{v}\|_{0,\Omega}^2 \quad \forall \underline{v} \in W_h \tag{6.3}$$

and

$$(\underline{v}, \underline{w})_h = \frac{1}{3} \sum_{x \in \mathcal{N}_h} |\omega_x| \underline{v}(x) \cdot \underline{w}(x) \quad \forall \underline{v}, \underline{w} \in X_h, \tag{6.4}$$

where $|\omega_x|$ is the area of ω_x.

Let $Gu_h \in X_h$ be the $(.,.)_h$-projection of ∇u_h onto X_h, i.e.

$$(Gu_h, \underline{v}_h)_h = (\nabla u_h, \underline{v}_h)_h \quad \forall \underline{v}_h \in X_h. \tag{6.5}$$

Equations (6.2) and (6.4) imply that

$$Gu_h(x) = \sum_{x \in \mathcal{N}(T)} \frac{|T|}{|\omega_x|} \nabla u_h|_T \quad \forall x \in \mathcal{N}_h. \tag{6.6}$$

Thus, Gu_h may easily be computed by a local averaging of ∇u_h. We finally set

$$\eta_{Z,T} := \|Gu_h - \nabla u_h\|_{0,T} \quad , \quad \eta_Z := \{\sum_{T \in \mathcal{T}_h} \eta_{Z,T}^2\}^{1/2}. \tag{6.7}$$

In order to prove that η_Z is an error estimator, we will compare it with a modification $\tilde{\eta}_{R,T}$ of the residual error estimator of section 3 which is obtained by omitting the element residuals:

$$\tilde{\eta}_{R,T} := \{\frac{1}{2} \sum_{e \in \mathcal{E}(T) \cap \mathcal{E}_{h,\Omega}} h_e \|R_e(u_h)\|_{0,e}^2\}^{1/2} \quad , \quad \tilde{\eta}_R := \{\sum_{T \in \mathcal{T}_h} \tilde{\eta}_{R,T}^2\}^{1/2}. \tag{6.8}$$

6.1 Proposition. *There are constants $c_{Z,1}, ..., c_{Z,4}$ which only depend on the smallest angle such that*

$$\|u - u_h\|_{1,\Omega} \leq c_{Z,1}\{\tilde{\eta}_R^2 + \sum_{T \in \mathcal{T}_h} h_T^2 \|f\|_{0,T}^2\}^{1/2},$$

$$\tilde{\eta}_{R,T} \leq c_{Z,2}\{\|u - u_h\|_{1,\omega_T}^2 + \sum_{\mathcal{E}(T') \cap \mathcal{E}(T) \neq \emptyset} h_{T'}^2 \|f\|_{0,T'}^2\}^{1/2},$$

$$c_{Z,3}\tilde{\eta}_R \leq \eta_Z \leq c_{Z,4}\tilde{\eta}_R.$$

Proof. Observe that $\triangle u_h$ vanishes elementwise since u_h is piecewise linear. The first estimate then follows from the first equality in (3.4), inequality (3.1), and Lemma 2.2. The second estimate follows from the proof of inequality (3.7) simply by disregarding the approximation f_h of f.

In order to prove the third estimate, we recall that the boundary residuals $R_e(u_h)$ are piecewise constant and rewrite $\tilde{\eta}_R$:

$$\tilde{\eta}_R^2 = \sum_{T \in \mathcal{T}_h} \frac{1}{2} \sum_{e \in \mathcal{E}(T)} h_e \|R_e(u_h)\|_{0,e}^2 = \sum_{e \in \mathcal{E}_h} h_e^2 |R_e(u_h)|^2.$$

Due to the minimal angle condition, the right-hand side of the above equation is bounded from above and from below by multiples of $\sum_{x \in \mathcal{N}_h} |\omega_x| \tilde{\eta}_{R,x}^2$ where

$$\tilde{\eta}_{R,x} := \{\sum_{x \in e \in \mathcal{E}_h} |R_e(u_h)|^2\}^{1/2}.$$

Thanks to inequality (6.3) on the other hand, η_Z is bounded from above and from below by multiples of $(Gu_h - \nabla u_h, Gu_h - \nabla u_h)_h$. But equations (6.2), (6.4), (6.5), and (6.6) imply that

$$(Gu_h - \nabla u_h, Gu_h - \nabla u_h)_h = (\nabla u_h, \nabla u_h)_h - (Gu_h, Gu_h)_h$$

$$= \sum_{x \in \mathcal{N}_h} \{\sum_{x \in \mathcal{N}(T)} \frac{|T|}{3}|\nabla u_{h|T}|^2 - \frac{|\omega_x|}{3}|\sum_{x \in \mathcal{N}(T)} \frac{|T|}{|\omega_x|}\nabla u_{h|T}|^2\}$$

$$= \frac{1}{3} \sum_{x \in \mathcal{N}_h} |\omega_x| \eta_{Z,x}^2$$

where

$$\eta_{Z,x} := \{\sum_{x \in \mathcal{N}(T)} \frac{|T|}{|\omega_x|}|\nabla u_{h|T}|^2 - |\sum_{x \in \mathcal{N}(T)} \frac{|T|}{|\omega_x|}\nabla u_{h|T}|^2\}^{1/2}.$$

Therefore, it is sufficient to prove that

$$\underline{c}\tilde{\eta}_{R,x}^2 \leq \eta_{Z,x} \leq \bar{c}\tilde{\eta}_{R,x}^2 \qquad (6.9)$$

holds for all $x \in \mathcal{N}_h$.

We only consider an interior node x. Boundary nodes are treated in the same way. Let N by the number of triangles in ω_x. Enumerate the triangles in a counter-clockwise ordering starting with 0 and set $T_N := T_0$. Let

$$v_i := \partial_x u|_{T_i}, \ w_i := \partial_y u|_{T_i}, \ \mu_i := \frac{|T_i|}{|\omega_x|}, \ 0 \leq i \leq N,$$

$$v := (v_1, ..., v_N)^T, \ w := (w_1, ..., w_N)^T.$$

Since the tangential components of ∇u_h are continuous across edges, we have

$$\tilde{\eta}_{R,x}^2 = \sum_{i=1}^{N} \{(v_i - v_{i-1})^2 + (w_i - w_{i-1})^2\} = v^t B v + w^t B w,$$

$$\tilde{\eta}_{Z,x}^2 = \sum_{i=1}^{N} \mu_i v_i^2 - (\sum_{i=1}^{N} \mu_i v_i)^2 + \sum_{i=1}^{N} \mu_i w_i^2 - (\sum_{i=1}^{N} \mu_i w_i)^2 = v^t A v + w^t A w.$$

with symmetric, positive semi-definite $N \times N$ matrices A and B. Since $\sum_{i=1}^{N} \mu_i = 1$, we have $e := (1, ..., 1)^t \in \ker(A) \cap \ker(B)$. For any $z \in \mathbb{R}^N$ set $\tilde{z} := (z_1 - z_N, ..., z_{N-1})^t \in \mathbb{R}^{N-1}$ and denote by $\tilde{A}, \tilde{B}$ the matrices which are obtained by deleting the last row and the last column of A, B. We then have for all $z \in \mathbb{R}^N$

$$z^t A z = (z - z_N e)^t A (z - z_N e) = \tilde{z}^t \tilde{A} \tilde{z},$$
$$z^t B z = (z - z_N e)^t B (z - z_N e) = \tilde{z}^t \tilde{B} \tilde{z}.$$

One easily checks that $\tilde{B}$ is the matrix corresponding to the central, second order difference quotient and that its eigenvalues are given by $\lambda_k(\tilde{B}) = 2\sin^2(k\pi/2N), 1 \leq k \leq N-1$. Gerschegorin's theorem on the other hand implies that the spectrum of $\tilde{A}$ is contained in $[\min \mu_i^2, 2]$. This establishes inequality (6.9). $\square$

7. References

[1] I. Babuška, W. C. Rheinboldt : Error estimates for adaptive finite element computations. SIAM J. Numer. Anal. 15, 736-754 (1978)

[2] R.E. Bank, R.K. Smith: A posteriori error estimates based on hierarchical bases. SIAM J. Numer. Anal. 30, 921-935 (1993)

[3] R. E. Bank, A. Weiser : Some a posteriori error estimators for elliptic partial differential equations. Math. Comput. 44, 283-301 (1985)

[4] F. Bornemann, B. Erdmann, R. Kornhuber. A posteriori error estimates for elliptic problems in two and three space dimensions. Konrad-Zuse Zentrum für Informationstechnik Berlin, Preprint SC 93-29 (1993)

[5] P. G. Ciarlet : The finite element method for elliptic problems. North Holland, Amsterdam 1978

[6] P. Clément : Approximation by finite element functions using local regularization. RAIRO Anal. Numér. 9, 77-84 (1975)

[7] R. Duran, M. A. Muschietti, R. Rodriguez : On the asymptotic exactness of error estimators for linear triangular elements. Numer. Math. 59, 107-127(1991)

[8] R. Rodriguez : Some remarks on Zienkiewicz-Zhu estimator. Int. J. Numer. Meth. in PDE (to appear)

[9] R. Verfürth : A Posteriori Error Estimation and Adaptive Mesh-Refinement Techniques. J. Comput. Appl. Math. (to appear)

[10] R. Verfürth : A Posteriori Error Estimates for Non-Linear Problems. Finite Element Discretizations of Elliptic Equations. Math. Comput. (to appear)

[11] O. C. Zienkiewicz, J. Z. Zhu : A simple error estimator and adaptive procedure for practical engineering analysis. Int. J. Numer. Meth. Engrg. 24, 337-357 (1987)

List of Participants:

Ansorge, R. Universität Hamburg, Institut für Angewandte Mathematik, Bundesstraße 55, 20146 Hamburg

Axelsson, Owe Faculty of Mathematics and Informatics, Katholieke Universiteit Nijmegen, NL-6525 ED Nijmegen

Bänsch, Eberhard Universität Freiburg, Inst. f. Angewandte Mathematik, Hermann-Herder-Str. 10, 79104 Freiburg

Bank, Randolph E. Dept of Mathematics, University of California at San Diego, La Jolla, CA 92093, USA

Barlag, C. Institut für Strömungsmechanik und Elektron. Rechnen im Bauwesen, Appelstr. 9a, 30167 Hannover

Bastian, Peter ICA III, Universität Stuttgart, Pfaffenwaldring 27, 70569 Stuttgart

Becker, Roland Institut f. Angew. Mathematik, Im Neuenheimer Feld 294, 69120 Heidelberg

Birken, Klaus Universität Stuttgart, Computer Center, Allmandring 30, 70550 Stuttgart

Blömer, Carda GKW Ingenieurgesellschaft mbH, Am Birkenbusch 9a, 44803 Bochum

Bollhöfer, Matthias TU Chemnitz-Zwickau, FB Mathematik, Postf. 964, 09009 Chemnitz

Brackenridge, K. O.U.C.L., Wolfson Building, Parks RD, Oxford OX1 3QD, U.K.

Brandt, Volker Mathematisches Institut, Universität Bonn, 53115 Bonn

Burmeister, Jens Institut für Praktische Mathematik und Informatik, Universität Kiel, Olshausenstr. 40-60, 24098 Kiel

Buthmann, Ute Institut für Praktische Mathematik und Informatik, Universität Kiel, Olshausenstr. 40-60, 24098 Kiel

Coclici, Cristian Mathemat. Inst. 17, Universität Stuttgart, Pfaffenwaldring 57, 70569 Stuttgart

Dahmen, W. Institut für Geometrie und Praktische Mathematik, Templergraben 55, RWTH Aachen, 52056 Aachen

Düsterhöft, Carsten Fachbereich Physik, Universität Kassel, 34013 Kassel

Emmrich, Etienne Institut f. Analysis und Numerik, Universität Magdeburg, Postfach 41 20, 39016 Magdeburg

Engl, Gabriele Mathematisches Institut, Techn. Universität München, Arcisstr. 21, 80290 München

Faermann, Birgit Institut für Praktische Mathematik und Informatik, Universität Kiel, Olshausenstr. 40-60, 24098 Kiel

Fellehner, Stefan Wattenbergstr. 20 21075 Hamburg

Fischer, Bernd Institut für Angewandte Mathematik, Universität Hamburg, Bundesstraße 55, 20146 Hamburg

Geiben, Monika Inst. f. Angew. Mathematik, Uni. Freiburg, Hermann-Herder-Str 10, 79104 Freiburg

Griebel, Michael Institut für Informatik der Technischen Universität München, Arcisstr. 21, Postfach 20 24 20, 80290 München 2

Grönner, Jürgen Fb - 12, Universität Essen, Schützenbahn 70, 45127 Essen

Groh, Ulrich Techn. Universität Chemnitz-Zwickau, Fachbereich Mathematik, Postf. 964, 09009 Chemnitz

Haack, Chr. Technische Universität Hamburg-Harburg, Arbeitsbereich Meerestechnik II, Eißendorfer Str. 42, 21073 Hamburg

Haag, Reinhard Inst. f. Wiss. Rechnen, Universität Heidelberg, 69120 Heidelberg

Hackbusch, Wolfgang Institut für Praktische Mathematik und Informatik, Universität Kiel, Olshausenstr. 40-60, 24098 Kiel

Hänel, D. Institut f. Verbrennung und Gasdynamik, Universität Duisburg, Lotharstr. 1, 47048 Duisburg

Hahn, H. Inst. f. Theoretische Physik, TU Braunschweig, Postfach 33 29, 38023 Braunschweig

Hahne, Manfred Inst. f. Angew. Mathematik, Universität Hannover, Postfach 6009, 30060 Hannover

Hansen, Knut Uni Kiel, Inst. f. Experimentalphysik, 24098 Kiel

Heinrichs, W. Ehreshover Str. 10, 50735 Köln

Heisig, Michael Inst. f. Wiss. Rechnen, Universität Heidelberg, 69120 Heidelberg

Helf, Clemens Rechenzentrum der Univ. Stuttgart, Abt. Numerik f. Hochleistungsrechner, Allmandring 30, 70550 Stuttgart

Hemforth, Frank Mathematik XIII, Ruhr-Univiversität, 44801 Bochum

Hemker, P.W. Mathematisch Centrum, P.O. Box 4079, NL-1009 AB Amsterdam, Niederlande

Hinkelmann, Reinhard Inst. f. Strömungsmechanik, Uni. Hannover, 30167 Hannover

Hofmann, Wolfgang Inst. f. Wiss. Rechnen, Uni. Heidelberg, 69120 Heidelberg

Johannsen, Klaus Inst. f. wiss. Rechnen, Uni. Heidelberg, 69120 Heidelberg

Junkherr, Jörg Institut für Praktische Mathematik und Informatik, Universität Kiel, Olshausenstr. 40-60, 24098 Kiel

Kantiem, Kerstin Uni. of Warschau, Dept. of Mathematic, UL. Banacha 2, 02097 Warschau

Karahan, Sedat Technische Fakültät der Universität Kiel, Kaiserstr. 2, 24143 Kiel

Karpurkin, A. Universität Magdeburg, Postfach 4120, 39016 Magdeburg

Katzer, Edgar Institut für Analysis, Universität Magdeburg, Postfach 4120, 39016 Magdeburg

Kersken, Hans-Peter Alfred-Wegener-Institut, Rechenzentrum, Postfach 120161, 27515 Bremerhaven

Kilian, Susanne Institut f. Angewandte Mathematik, INF 293, 69120 Heidelberg

Klawonn, Axel Courant Institute, 251 Mercer St. New York, NY 100 11, USA

Klein, Olaf FB 10 Bauwesen Ingenieurmathematik, Universität-GHS-Essen, Universitätsstr. 2, D-45117 Essen

Klinger, Thomas Uni Kiel, Institut f. Experimentalphysik, 24098 Kiel

König, Christoph Ruhr-Universität Bochum, KIB Ag IV, 44780 Bochum

v. Kopylow, Andrea FB 18 - Physik, Universität Kassel, Heinrich-Plett-Str. 40, 34132 Kassel

Kornhuber, Ralf Konrad Zuse Zentrum Berlin, Heilbronner Str. 10, 10711 Berlin

Krause, Elke Alfred-Wegener-Institut, Rechenzentrum, Postfach 120161, 27515 Bremerhaven

Kröner, Dietmar Institut für Angewandte Mathematik Albert-Ludwigs-Universität Hermann-Herder-Str. 10 79104 Freiburg

Küster, Uwe Rechenzentrum, der Universität Stuttgart, Allemandring 3a, W-7000 Stuttgart 80

Kunoth, Angela Institut für Informatik, Freie Universität Berlin, Arnimalle 2-6, 14195 Berlin

Langer, Andreas Institut für Mathematik, Universität Dortmund, 44221 Dortmund

Linde, Jürgen Institut für Mathematik, Universität Dortmund, 44221 Dortmund

Lube, Gert Institut für Numerische und Angewandte Mathematik, Georg-August-Universität Göttingen, Lotzestraße 16-18, 37083 Göttingen

Luksch, Peter Institut für Informatik, TU München, Postfach 20 24 20, 80290 München

Maar, Bernd Uni. Heidelberg, Inst. f. Wiss. Rechnen, 69120 Heidelberg

Mahrenholtz, O. Arbeitsbereich Meerestechnik II, TU Hamburg-Harburg, Eißendorfer Str. 38, 21073 Hamburg

Martensen, E. Mathematisches Inst. II, Universität Karlsruhe (TH), Englerstr. 2, 76128 Karlsruhe

Mehrmann, Volker Techn. Universität Chemnitz-Zwickau, Fachbereich Mathematik, Postfach 964, 09009 Chemnitz

Meinel, Stefan Techn. Universität Chemnitz-Zwickau, Fachbereich Mathematik, Postfach 964, 09009 Chemnitz

Mewis, Peter Institut f. Strömungsmechanik, Universität Hannover, 30060 Hannover

Miller, John J.H. Department of Mathematics, University of Dublin, Trinity College, Dublin 2, IRELAND

Mohring, Jan Wasserlochstücke 14, 67661 Kaiserslautern

Mordhorst, Uwe Rechenzentrum der Universität Kiel, Olshausenstr. 40, 24098 Kiel

Nabben, Reinhard Universität Bielefeld, Fakultät f. Mathematik, Postf. 10 01 31, 33501 Bielefeld

Neuß, Nicolas Institut für Wissenschaftliches Rechnen, Im Neuenheimer Feld 294, 69120 Heidelberg

Nikolakis, D. Bachstr. 123, 22083 Hamburg

Oosterlee, Kees Ges. f. Mathem. u. Datenverarb., GMD / I1.T, Postfach 13 16, 53731 St. Augustin

Rannacher, Rolf Institut für Angewandte Mathematik, Universität Heidelberg, Im Neuenheimer Feld 293/294, 69120 Heidelberg

Ratke, R. Institut für Strömungsmechanik und Elektr. Rechnen im Bauwesen, Universität Hannover, Appelstr. 9 A, 30167 Hannover

Raw, Michael J. ASC Ltd, 554 Parkside Dr., Unit 4, Waterloo, Ontario, Canada, N2L 5Z4

Reichert, Henrik Institut für Computeranwendungen (ICA 3), Universität Stuttgart, Pfaffenwaldring 27, 70550 Stuttgart

Reusken, Arnold TU Eindhoven, Dept. Math. and Computing Science, Postbus 513, NL - 5600 MB Eindhoven

Rumpf, Martin Institut f. Angewandte Mathematik, Hermann-Herder-Str. 10, 79104 Freiburg

Sarzin, Regina Ruhr-Universität Bochum, Mathematik XIII, 44801 Bochum

Scheuerer, Georg ASC Advanced Scientific Computing GmbH, Am Gangsteig 26, 83607 Holzkirchen

Schieweck, Friedhelm Inst. f. Analysis/Numerik, Universität Magdeburg, PSF 4120, 39016 Magdeburg

Schlegel, Volker Techn. Universität Hamburg-Harburg, Arbeitsbereich Meerestechnik II, Eißendorfer Str. 42, 21073 Hamburg

Schlüter, H.-J. Universität - GH Duisburg, FB 7 / Schiffstechnik, Lotharstraße, 47057 Duisburg

Schmidt, Alfred Inst. f. Angewandte Mathematik, Universität Freiburg, Hermann-Herder-Strasse 10, 79104 Freiburg

Schreiber, Peter Institut für Angewandte Mathematik, Universität Heidelberg, Im Neuenheimer Feld 293, 69120 Heidelberg

Schreiner, W. Fachber. Mathematik, Uni. Kaiserslautern, 67653 Kaiserslautern

Siebert, Kunibert G. Institut für Angewandte Mathematik, Hermann-Herder-Str. 16, 79104 Freiburg

Siehl, Barbara Institut f. Angewandte Mathematik, Universität Bonn, Wegelerstr. 6, 53115 Bonn

Steinbach, O. Universität Stuttgart, Mathematisches Institut A, Pfaffenwaldring 57, 70569 Stuttgart

Stephan, Ernst P. Inst. f. Angewandte Mathematik, Universität Hannover, Postf. 6009, 30060 Hannover

Stevenson, R.P. Department of Mathematics and Computer Science, Eindhoven University of Technology, P.O.Box 513, NL-5600 MB Eindhoven

Stoyan, Gisbert ELTE Universität Budapest, Bogdanfy u.10/B, H-1117 Budapest, Ungarn

Tiwari, Sudarshan Fachber. Mathematik, Uni. Kaiserlslautern, 67653 Kaiserslautern

Turek, Stefan Institut f. Angew. Mathematik, Universität Heidelberg, Im Neuenheimer Feld 294, 69120 Heidelberg

Urban, Karsten Inst. f. Geometrie und Prakt. Mathem., RWTH Aachen, Templergraben 55, 52056 Aachen

Wagner, Christian Institut f. Wissenschaftl. Rechnen, Uni. Heidelberg, 69120 Heidelberg

Wittum, Gabriel Institut für Computeranwendungen (ICA 3), Universität Stuttgart, Pfaffenwaldring 27, 70550 Stuttgart

Zimmer, Stefan Institut für Informatik, Techn. Universität München, 80290 München

Zulehner, Walter Institut für Mathematik, Johannes Kepler Universität Linz, A-4040 Linz, Österreich

How to contact the authors by email:

Bank, Randolph E. reb@sdna1.ucsd.edu

Bey, Jürgen juergen@na.mathematik.uni-tuebingen.de

Geiben, Monika mgeiben@mathematik.uni.freiburg.de

Hemker, P.W. pieth@cwi.nl

Katzer, Edgar edgar.katzer@mathematik.uni-magdeburg.d400.de

Meyer, Arnd meyer@imech.tu-chemnitz.de

Miller, J.J.H. jackie@maths.tcd.ie

Rannacher, Rolf rannacher@agaia.iwr.uni-heidelberg.de

Raw, Michael mjraw@asc.on.ca

Reusken, Arnold wsanar@win.tue.nl

Schieweck, Friedhelm
    friedhelm.schieweck@mathematik.uni-magdeburg.d400.de

Stephan, Ernst P. stephan@server.ifam.uni-hannover.de

Steinbach, O. steinbach@mathematik.uni-stuttgart.de

Turek, Stefan ture@gaia.iwr.uni-heidelberg.de

Wittum, Gabriel wittum@ica.uni-stuttgart.de

Addresses of the Editors of the Series "Notes on Numerical Fluid Mechanics"

Prof. Dr. Ernst Heinrich Hirschel (General Editor)
Herzog-Heinrich-Weg 6
D-85604 Zorneding
Federal Republic of Germany

Prof. Dr. Kozo Fujii
High-Speed Aerodynamics Div.
The ISAS
Yoshinodai 3-1-1, Sagamihara
Kanagawa 229
Japan

Prof. Dr. Bram van Leer
Department of Aerospace Engineering
The University of Michigan
3025 FXB Building
1320 Beal Avenue
Ann Arbor, Michigan 48109-2118
USA

Prof. Dr. Keith William Morton
Oxford University Computing Laboratory
Numerical Analysis Group
8-11 Keble Road
Oxford OX1 3QD
Great Britain

Prof. Dr. Maurizio Pandolfi
Dipartimento di Ingegneria Aeronautica e Spaziale
Politecnico di Torino
Corso Duca Degli Abruzzi, 24
I-10129 Torino
Italy

Prof. Dr. Arthur Rizzi
Royal Institute of Technology
Aeronautical Engineering
Dept. of Vehicle Engineering
S-10044 Stockholm
Sweden

Dr. Bernard Roux
Institut de Mécanique des Fluides
Laboratoire Associé au C.R.N.S. LA 03
1, Rue Honnorat
F-13003 Marseille
France

Brief Instruction for Authors

Manuscripts should have well over 100 pages. As they will be reproduced photomechanically they should be produced with utmost care according to the guidelines, which will be supplied on request. In print, the size will be reduced linearly to approximately 75 per cent. Figures and diagrams should be lettered accordingly so as to produce letters not smaller than 2 mm in print. The same is valid for handwritten formulae. Manuscripts (in English) or proposals should be sent to the general editor, Prof. Dr. E. H. Hirschel, Herzog-Heinrich-Weg 6, D-85604 Zorneding.

Notes on Numerical Fluid Mechanics (NNFM)   Volume 49

Series Editors: Ernst Heinrich Hirschel, München (General Editor)
Kozo Fujii, Tokyo
Bram van Leer, Ann Arbor
Keith William Morton, Oxford
Maurizio Pandolfi, Torino
Arthur Rizzi, Stockholm
Bernard Roux, Marseille

Volume 44   Multiblock Grid Generation – Results of the EC/BRITE-EURAM Project EUROMESH, 1990–1992 (N. P. Weatherill / M. J. Marchant / D. A. King, Eds.)
Volume 45   Numerical Methods for Advection – Diffusion Problems (C. B. Vreugdenhil / B. Koren, Eds.)
Volume 46   Adaptive Methods – Algorithms, Theory and Applications. Proceedings of the Ninth GAMM-Seminar, Kiel, January 22–24, 1993 (W. Hackbusch / G. Wittum, Eds.)
Volume 47   Numerical Methods for the Navier-Stokes Equations (F.-K. Hebeker, R. Rannacher, G. Wittum, Eds.)
Volume 48   Numerical Simulation in Science and Engineering. Proceedings of the FORTWIHR Symposium on High Performance Scientific Computing, München, June 17–18, 1993 (M. Griebel / Ch. Zenger, Eds.)
Volume 49   Fast Solvers for Flow Problems. Proceedings of the Tenth GAMM-Seminar, Kiel, January 14–16, 1994 (W. Hackbusch / G. Wittum, Eds.)
Volume 50   Computational Fluid Dynamics on Parallel Systems. Proceedings of a CNRS-DFG Symposium in Stuttgart, December 9 and 10, 1993 (S. Wagner, Ed.)